Nobel Symposium NS 160 – Chemistry and Physics of Heavy and Superheavy Elements

EPJ Web of Conferences Volume 131 (2016)

Kristianstad, Sweden
29 May - 3 June 2016

Editors:

D. Rudolph

ISBN: 978-1-5108-3368-5

TABLE OF CONTENTS

Scientific Programme

Sunday May 29th

ARRIVAL

11:00-17:00	**REGISTRATION AT GAMLA BISKOPSHUSET, LUND**
15:00-17:00	**RECEPTION AT GAMLA BISKOPSHUSET, LUND**
15:15	GÖRAN BEXELL, FORMER VICE-CHANCELLOR, LUND UNIVERSITY *THE 350-YEAR ANNIVERSARY OF LUND UNIVERSITY (15 MIN)*
17:15	**BUS TRANSFER TO BÄCKASKOG CASTLE**
19:30	**DINNER**

Monday May 30th

9.00-10.35 INTRODUCTION

Discussion Leader: D. Rudolph

D. Rudolph, Lund University. *Welcome*

P. Pyykkö, University of Helsinki. *Does the Periodic Table of Elements Look Alright? (35+10 min)*

M. Leino, University of Jyväskylä. *Production and Properties - Towards the Island of Stability (35+10 min)*

10.35-11.00 COFFEE

11.00-12.25 NUCLEAR STRUCTURE

Discussion Leader: S. Åberg

P.-H. Heenen, Brussels. *Beyond Mean-field Correlations and the Description of Superheavy Nuclei (25+5 min)*

R.M. Clark, LBNL Berkeley. *Decay Spectroscopy of Heavy and Superheavy Nuclei at LBNL (25+5 min)*

U. Forsberg, Lund University. *Recoil-α(-α)-fission Events Observed in the Reaction ^{48}Ca+^{243}Am (20+5 min)*

12.25 LUNCH

15:00-16.30 NUCLEAR STRUCTURE

Discussion Leader: A. Sobiczewski

J. Dobaczewski, York. *Density Functionals for Heavy and Superheavy Nuclei (25+5 min)*

R.-D. Herzberg, Liverpool. *In-Beam Spectroscopy of Heavy and Superheavy Nuclei (25+5 min)*

F.P. Heßberger, GSI Darmstadt. *Nuclear Structure of the Transactinides – Investigated by Decay Spectroscopy (25+5 min)*

16.30-17.00 COFFEE

17:00-18.25 FISSION BARRIERS

Discussion Leader: W. Nazarewicz

A. Lopez-Martens, CSNSM Orsay. *Stability and Synthesis of Superheavy Nuclei: Winning the Battle Against Fission (25+5 min)*

P. Möller, Los Alamos. *The Limits of the Nuclear Chart as Set by Fission and Alpha Decay (25+5 min)*

J. Khuyagbaatar, GSI Darmstadt. *Fission in the Landscape of the Heaviest Nuclei: Some Recent Examples (20+5 min)*

19.00 DINNER

Tuesday May 31st

9.00-10.15 PRODUCTION AND REACTIONS

Discussion Leader: J.V. Kratz

M. Schädel, GSI Darmstadt . *Prospects of Heavy and Superheavy Element Production via Inelastic Nucleus-nucleus Collisions - from ^{238}U + ^{238}U to ^{18}O + ^{254}Es (35+10 min)*

A.K. Nasirov, JINR Dubna. *Entrance Channel Effects in Superheavy Element Production (25+5 min)*

10.15-10.45 COFFEE

10.45-12.15 PRODUCTION AND REACTIONS

Discussion Leader: H. Sakai

W. Loveland, Oregon State. *Characterizing the Mechanism(s) of Heavy Element Synthesis Reactions (25+5 min)*

D. Hinde, ANU Canberra. *Quasifission in Heavy and Superheavy Element Formation Reactions (25+5 min)*

M. Kowal, NCNR Warsaw. *Fusion-fission Probabilities, Cross Sections and Structure Notes of Superheavy Nuclei (25+5 min)*

12.15 LUNCH

15.00-16.35 ATOMIC STRUCTURE

Discussion Leader: M. Itkis

E. Eliav, Tel Aviv, Israel. *Atomic Theory – Models and Methods (35+10 min)*

T. Sato, JAEA Tokai. *Ionization Potentials of the Heaviest Actinide (20+5 min)*

M. Laatiaoui, Helmholtz Institute Mainz. *Laser Spectroscopy of Nobelium (20+5 min)*

16.35-17.00 COFFEE

17:00-18.40 INTERFACES

Discussion Leader: C. Fahlander

M. Block, GSI Darmstadt. *High-precision Mass Measurements and Advanced Ion-manipulation Techniques (35+10 min)*

L.G. Sarmiento, Lund. *Nuclear Spectroscopy with Geant4: The Superheavy Challenge (20+5)*

K. Rykaczewski, ORNL Oak Ridge. *The ORNL Actinide Materials and a New Detection System for Superheavy Nuclei (25+5 min)*

19.00 DINNER

Wednesday June 1st (“public session”, press participation possible)

9.00-10.15 DISCOVERY ASPECTS

Discussion Leader: L. Öhrström

S. Hofmann, GSI Darmstadt. *The Discovery of Elements 107-112 (35+10 min)*

K.E. Gregorich, LBNL Berkeley. *How Reliable Are Isotopic Assignments of Alpha-decay Chains? (25+5 min)*

10.15-10.40 COFFEE

10.40-12:05 DISCOVERY ASPECTS

Discussion Leader: L.-I. Elding

J. Reedijk, Leiden & IUPAC. *Rules and Regulations for Validation, Assignation and Name giving of Newly Discovered Chemical Elements (15+10 min)*

C. Jarlskog, Lund & IUPAP. *Procedure Around Naming New Elements (10+5 min)*

V. Utyonkov, FLNR Dubna. *The Discovery of Elements 113-118 (35+10 min)*

12.10-13:00 (LOCAL) PRESS CONFERENCE

12.30-14.00 LUNCH

14:30-18.00 EXCURSIONS

19.00 DINNER

Thursday June 2nd

9.00-10.30 CHEMISTRY

Discussion Leader: P. Armbruster

A. Türler, University of Bern & PSI Villigen. *Advances in Chemical Investigations of Heaviest Elements (35+10 min)*

V. Pershina, GSI Darmstadt. *Theoretical Chemistry of Superheavy Elements: Support for Experiment (35+10 min)*

10.30-11.00 COFFEE

11.00-12.30 CHEMISTRY

Discussion Leader: M.A. Stoyer

A. Yakushev, GSI Darmstadt. *Gas-phase Chemistry of Element 114, Flerovium (25+5 min)*

P. Schwerdtfeger U Massey. *Toward an Accurate Description of Chemical and Physical Properties of the Superheavy Elements (25+5 min)*

R. Eichler, PSI Villigen. *Complex Chemistry with Metal Complexes (25+5 min)*

12.30 LUNCH

15.00-16:25 CHEMISTRY

Discussion Leader: J.P. Omtvedt

H. Haba, RIKEN . *Superheavy Element Chemistry Behind GARIS at RIKEN (25+5 min)*

Y. Nagame, JAEA Tokai. *Chemical Properties of Rf and Db in the Aqueous Phase (25+5 min)*

J. Even, University of Groningen. *Chemistry Aided Nuclear Physics Experiments (20+5 min)*

16:25-17.00 COFFEE

17.00-18:20 PREREQUISITES

Discussion Leader: I. Ragnarsson

S. Dmitriev, FLNR Dubna. *Status and Perspectives of the Dubna Superheavy Element Factory (25+5 min)*

B.G. Carlsson, Lund University. *How to Calculate Alpha-Decay Rates in the Future? (20+5 min)*

J.M. Gates, LBNL Berkeley. *Prospects of A and Z Identification Experiments at LBNL (20+5 min)*

19.00 NOBEL SYMPOSIUM DINNER

Friday June 3rd

9.00-10.30 STATUS AND PERSPECTIVES

Discussion Leader: M. Leino

K. Morita, RIKEN Tokyo. *Discovery of Element 113 and Future Research Direction at RIKEN (35+10 min)*

Ch.E. Düllmann, University of Mainz. *Search for Elements Beyond Z=118 and Future SHE Research Opportunities at GSI (35+10)*

10.30-11.00 COFFEE

11.00-12.30 FUTURE OF HEAVY ELEMENT RESEARCH

Discussion Leader: P. Pyykkö

W. Nazarewicz, Michigan State University. *Key Questions to Understand and Reach the Island of Stability (35+10 min)*

DISCUSSION

12.30 LUNCH

END OF SYMPOSIUM

Participants

ORGANIZING COMMITTEE

Prof. Lars Ivar Elding, Lund University, Sweden
Prof. Claes Fahlander, Lund University, Sweden
Prof. Dirk Rudolph, Lund University, Sweden
Prof. Sven Åberg, Lund University, Sweden
Katarina Lindqvist, Lund University, Sweden
Yulia Lindholm, Lund University, Sweden

SUPPORT STAFF

Dr. Pavel Golubev, Lund University, Sweden
Nataša Lalović, Lund University, Sweden
Christian Lorenz, Lund University, Sweden
Anton Roth, Lund University, Sweden
Daniel Ward, Lund University, Sweden

INVITED SCIENTISTS

Prof. Peter Armbruster, GSI Darmstadt, Germany
Prof. Michael Block, GSI Darmstadt, Germany
Dr. Gillis Carlsson, Lund University, Sweden
Dr. Roderick M. Clark, LBNL Berkeley, USA
Prof. Sergey N. Dmitriev, FLNR Dubna, Russia
Prof. Jacek Dobaczewski, University of York, United Kingdom
Prof. Christoph E. Düllmann, University of Mainz, Germany
Oscar Edlund, the Nobel Foundation, Sweden
Dr. Robert Eichler, PSI Villigen, Switzerland
Prof. Ephraim Eliav, Tel Aviv University, Israel
Dr. Julia Even, University of Groningen, Netherlands
Dr. Ulrika Forsberg, Lund University, Sweden
Dr. Jacklyn M. Gates, LBNL Berkeley, USA
Dr. Kenneth E. Gregorich, LBNL Berkeley, USA
Dr. Hiromitsu Haba, RIKEN Tokyo, Japan
Prof. Paul-Henri Heenen, University of Brussels, Belgium
Prof. Rolf-Dietmar Herzberg, University of Liverpool, United Kingdom

Dr. Fritz Peter Heßberger, GSI Darmstadt, Germany
Prof. David Hinde, ANU Canberra, Australia
Prof. Sigurd Hofmann, GSI Darmstadt, Germany
Prof. Mikhail Itkis, FLNR Dubna, Russia
Prof. Cecilia Jarlskog, Lund University, Sweden
Dr. Jadambaa Khuyagbaatar, GSI Darmstadt, Germany
Dr. Michał Kowal, NCBJ Warsaw, Poland
Prof. Jens-Volker Kratz, University of Mainz, Germany
Dr. Mustapha Laatiaoui, HIM Mainz, Germany
Prof. Matti Leino, University of Jyväskylä, Finland
Dr. Araceli Lopez-Martens, CSNSM Orsay, France
Prof. Walter D. Loveland, Oregon State University, USA
Dr. Peter Möller, Los Alamos, USA
Prof. Kosuke Morita, RIKEN Tokyo, Japan
Prof. Yuichiro Nagame, JAEA Tokai, Japan
Dr. Avazbek K. Nasirov, FLNR Dubna, Russia
Prof. Witold Nazarewicz, Michigan State University, USA
Prof. Lars Öhrström, Chalmers Gothenburg, Sweden
Prof. Jon Petter Omtvedt, University of Oslo, Norway
Dr. Valeria Pershina, GSI Darmstadt, Germany
Prof. Pekka Pyykkö, University of Helsinki, Finland
Prof. Ingemar Ragnarsson, Lund University, Sweden
Prof. Jan Reedijk, University Leiden, Netherlands
Dr. Krzysztof Rykaczewski, ORNL Oak Ridge, USA
Prof. Hideyuki Sakai, RIKEN Tokyo, Japan
Dr. Luis G. Sarmiento, Lund University, Sweden
Dr. Tetsuya K. Sato, JAEA Tokai, Japan
Dr. Matthias Schädel, GSI Darmstadt, Germany
Prof. Peter Schwerdtfeger, Massey University, New Zealand
Prof. Adam Sobiczewski, NCBJ Warsaw, Poland
Dr. Mark A. Stoyer, LLNL Livermore, USA
Prof. Andreas Türler, University of Bern, Switzerland
Dr. Vladimir Utyonkov, FLNR Dubna, Russia
Dr. Alexander Yakushev, GSI Darmstadt, Germany

International Advisory Committee

Prof. Sergey N. Dmitriev
Flerov Laboratory of Nuclear Reactions, Dubna, Russia

Dr. Fritz Peter Heßberger
GSI Helmholtzzentrum, Darmstadt, Germany

Prof. Cecilia Jarlskog
Lund University, Lund, Sweden

Prof. Björn Jonson
Chalmers University of Technology, Gothenburg, Sweden

Prof. Matti Leino
University of Jyväskylä, Jyväskylä, Finland

Prof. Sven Lidin
Lund University, Lund, Sweden

Prof. Walter D. Loveland
Oregon State University, Corvallis, USA

Prof. Witold Nazarewicz
Michigan State University, USA; Warsaw University, Poland

Prof. Pekka Pyykkö
University of Helsinki, Helsinki, Finland

Prof. Hideyuki Sakai
RIKEN & University of Tokyo, Tokyo, Japan

Dr. Matthias Schädel
GSI Helmholtzzentrum, Darmstadt, Germany

Dr. Dawn A. Shaughnessy
Lawrence Livermore National Laboratory, Livermore, USA

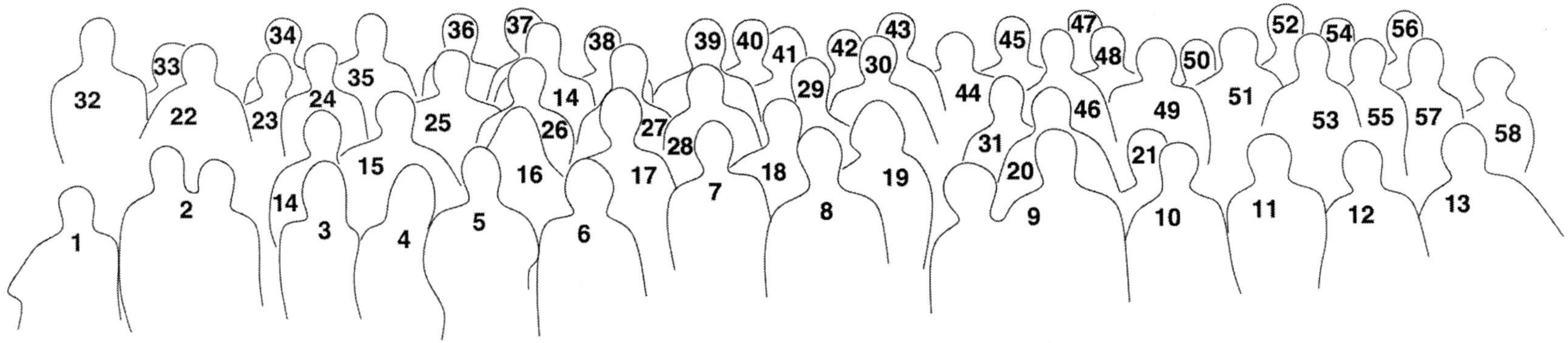

1 Ephraim Eliav
2 Carola* & Kenneth E. Gregorich
3 Nataša Lalović
4 Araceli Lopez–Martens
5 Luis G. Sarmiento
6 Valeria Pershina
7 Peter Möller
8 Tetsuya K. Sato
9 Monique Bernas* & Peter Armbruster
10 Pekka Pyykkö
11 Sigurd Hofmann
12 Mikhail Itkis
13 Sergey N. Dmitriev
14 Nancy* & Mark A. Stoyer
15 Sven Åberg
16 Ulrika Forsberg
17 Yuichiro Nagame
18 Dirk Rudolph
19 Julia Even
20 Kosuke Morita
21 Roderick M. Clark
22 David Hinde
23 Hideyuki Sakai
24 Jacklyn M. Gates
25 Pavel Golubev
26 Paul–Henri Heenen
27 Michael Block
28 Jens–Volker Kratz
29 Christoph E. Düllmann
30 Lars Ivar Elding
31 Claes Fahlander
32 Michał Kowal
33 Christian Lorenz
34 Anton Roth
35 Robert Eichler
36 Matti Leino
37 Andreas Türler
38 Krzysztof Rykaczewski
39 Jadambaa Khuyagbaatar
40 Jan Reedijk
41 Avazbek K. Nasirov
42 Lars Öhrström
43 Fritz Peter Heßberger
44 Walter D. Loveland
45 Rolf–Dietmar Herzberg
46 Mustapha Laatiaoui
47 Jon Petter Omtvedt
48 Witold Nazarewicz
49 Peter Schwerdtfeger
50 Vladimir Utyonkov
51 Jacek Dobaczewski
52 Daniel Ward
53 Matthias Schädel
54 Ingemar Ragnarsson
55 Adam Sobiczewski
56 Gillis Carlsson
57 Hiromitsu Haba
58 Yukari Onishi*

missing in the picture:
Oscar Edlund
Cecilia Jarlskog
Alexander Yakushev
Jochen Nägele*
Katriina Leino *
Yulia Lindholm
Katarina Lindqvist

* accompanying person

edp sciences

EPJ Web of Conferences

Statement of Peer review

In submitting conference proceedings to EPJ Web of Conferences, I certify to the Publisher that I adhere to the **Policy on Publishing Integrity** of the journal in order to safeguard good scientific practice in publishing.

1. All articles have been subjected to peer review administered by the proceedings editors.
2. Reviews have been conducted by expert referees, who have been requested to provide unbiased and constructive comments aimed, whenever possible, at improving the work.
3. Proceedings editors have taken all reasonable steps to ensure the quality of the materials they publish and their decision to accept or reject a paper for publication has been based only on the merits of the work and the relevance to the journal.

Title, date and place of the conference

Nobel Symposium NS 160
Chemistry and Physics of Heavy and Superheavy Elements
May 29 - June 3, 2016
Bäckaskog Castle, Scania, Sweden

Proceedings editor(s):

Prof. Dirk Rudolph
Department of Physics
Lund University
Sweden

Date and editor's signature

Editorial

Science often develops most vigorously through challenging studies of extreme phenomena. Superheavy elements fall into such a category.

What is the heaviest element that can exist in Nature?

Driven by the continued search for an anticipated "island of stability" of superheavy atomic nuclei and the understanding of their underlying nuclear (in)stability and atomic structure hence chemical properties, the past decades have seen a tremendous progress in experimental ingenuity and theoretical methodology to study and characterize superheavy elements. Therefore, we are very grateful that the Nobel Foundation [1] approved and, jointly with the Knut and Alice Wallenberg Foundation [2], provided the financial resources to organize and conduct the Nobel Symposium NS160, entitled *Chemistry and Physics of Heavy and Superheavy Elements*. These symposia "are devoted to areas of science where breakthroughs are occurring or deal with other topics of primary cultural or social significance" [1]. About three symposia are held each year, roughly every fourth symposium promotes a topic in physics as primary research area, and from about every third symposium a contemporary Nobel Price is being awarded.

In fact, forty-two years ago in 1974, the first Nobel Symposium NS27 in the field of superheavy elements was organized by late Sven Gösta Nilsson [4]. One Nobel Laureate, Glenn T. Seaborg, attended the symposium, together with Aage Bohr and Ben Mottelson, who received the price one year later in 1975. Bridging from past to present, four at the time young researchers brought four decades of theoretical (Peter Möller, Ingemar Ragnarsson, Adam Sobiczewski) and experimental (Peter Armbruster) superheavy element experience to the Nobel Symposium NS160. Furthermore, pioneering work at heavy-ion accelerators has led to the discovery of thirteen new elements in the meantime, starting with element 106, seaborgium, to element 118, oganesson.

Upon arrival in Lund, Sweden, participants gathered at "Biskopshuset" for a presentation of Lund University by the former vice chancellor, Göran Bexell. The reception was hosted by the Science Faculty and the Technical Faculty of Lund University. Thereafter, the Nobel Symposium NS160 continued at beautiful and sunny Bäckaskog Castle near Kristianstad, Scania, Sweden, between May 29th and June 3rd, 2016 [3]. Fifty-one invited scientists provided forty-one presentations, which were sometimes bold, eventually brave, possibly of intriguing simplicity, and at best included all such aspects providing a lively atmosphere. Thirty-six contributions are summarized in this volume.

Following the session on discovery aspects in the morning of Wednesday, June 1st, a brief press conference was held for public outreach. There, any interference was carefully avoided with the announcement of naming propositions for elements 113, 115, 117, and 118, which happened one week later by officials of the International Union for Pure and Applied Chemistry (IUPAC).

There might be the perception that nuclear and atomic spectroscopy, one-atom-at-a-time chemistry experiments, or nuclear reaction studies, simply follow in the wake of new-element discoveries. In turn, many contributions show that with the present end of the

^{48}Ca-beam era, increased theoretical understanding based on more profound and comprehensive experimental data on all of the above is pivotal of any targeted hunt for even heavier elements and/or for reaching the "island of stability". New computational tools and capabilities accompany these theoretical and experimental efforts. Ever increasing sensitivity is being called for, based on continued high-technology R&D on heavy-ion accelerator components, target production, separation techniques, and sophisticated radiation detectors – not to speak of revised or new identification schemes for the predicted shorter-lived (go North for the glory of new elements!) and longer-lived (go East to find the island!) species.

Finally, and to possibly add a grain of salt, on the last transparency of the last presentation of the symposium, Witold Nazarewicz mentioned the importance and incentives of basic science. Indeed, "WOW"-effects define the major driving force of young scientists to reach breakthrough of one or another kind. Undoubtedly, superheavy-element research will continue to allow for numerous "WOW"-effects. Nevertheless, long-term scientific credibility and reputation form another viable basis for sustainability, resources, opportunities, and hence success – opposed to eventually rushed short-term glorification, especially when sought for solely on behalf of the scientists signed responsible for the discoveries.

Anyhow, let's go for the "WOW"! The many young scientists having attended the Nobel Symposium NS160 should be encouraged and able to reach the "island of stability" in the course of the coming decades, i.e. they may celebrate their achievements at the next Nobel Symposium devoted to superheavy element research in the middle of the century.

Dirk Rudolph, Lars-Ivar Elding, Claes Fahlander and Sven Åberg

References

[1] `https://www.nobelprize.org/nobel_organizations/nobelfoundation/symposia/`.

[2] `https://www.wallenberg.com/Kaw/en`.

[3] `http://www.nuclear.lu.se/ns160/`.

[4] Proceedings of the Nobel Symposium NS27, *Super-Heavy Elements – Theoretical Predictions and Experimental Generation*, eds. Sven Gösta Nilsson and Nils Robert Nilsson, Physica Scripta Volume **10A** (1974).

EPJ Web of Conferences **131**, 01001 (2016) DOI: 10.1051/epjconf/201613101001
Nobel Symposium NS160 – Chemistry and Physics of Heavy and Superheavy Elements

Is the Periodic Table all right ("PT OK")?

Pekka Pyykkö[a]

Department of Chemistry, University of Helsinki, POB 55 (A.I. Virtasen aukio 1), 00014 Helsinki, Finland

Abstract. The history of the Periodic Table and its predecessors spans almost 200 years. The present IUPAC PT for $Z = 1-118$ is still adequate. The remarkable measurement for the Lr atom does not change the chemistry. The extensions up to $Z = 172$ are discussed and compared. New data for ions are presented. The "Madelung rule" is found to be surprisingly good even in that range.

1. Introduction

The standard Periodic Table (PT) for the first 118 elements is shown in Fig. 1. Note that we have followed the IUPAC in putting the 15 elements La-Lu in a lanthanide row, and the 15 elements Ac-Lr in an actinide row, corresponding to a hole in Group 3 for Periods 6 and 7, respectively. This is entirely logical if we count the f-elements from f^0 to f^{14}.

A fine point is that for instance Th is counted as an f-element, despite of having a $7s^26d^2$ atomic ground state. In molecules or solids it may have a certain amount of $5f$ character. In the same sense E118 (Og), [Li, Be, Mg, Zn], [Cs, Ca, Sr, Ba] and Th could be called *pre-s*, *pre-p*, *pre-d* and *pre-f* elements, respectively [2]. The example for E118 was the existence of the atomic anion [3], Og^-, using the preliminary proposed name of Oganesson, Og.

As discussed in the lecture by Sato, the atomic ground state of the Lr atom is $(7p_{1/2})^1$, see Ref. [4], not $6d^1$, like the $5d^1$ ground state for Lu. Note that their experimental first ionization potential was $4.96(^{0.08}_{0.07})$ eV, compared with their own theoretical value of 4.963(15) eV. The $6d$ level is only calculated to lie only 0.18 eV higher. However, the three first ionization potentials in Fig. 2 and the nature of the covalent bonds to Lu and Lr in a number of molecules are very similar [5]. Therefore changes in the PT are not proposed.

A brief history of the PT is given below in Sect. 3.

2. How to continue the PT beyond $Z = 118$?

2.1 The long-row proposal of Fricke *et al.* [6]

In 1971 Fricke *et al.* [6] reported Dirac-Slater (DS) calculations on neutral atoms up to $Z = 172$. This method could be called a primitive density functional approach, using the so-called Slater exchange. The limit at 172 is imposed by the diving of the lowest, $1s$ state in the lower, positron-like continuum. The nuclei were simply assumed to be there and assigned realistic Fermi charge distributions.

They found considerable occupational overlap of the $8s$, $8p$, $7d$, $6f$, and $5g$ shells, and no big gaps between any of them. This motivated putting the elements 119–164 on one

[a] e-mail: Pekka.Pyykko@helsinki.fi.

Periodic Table 1-118, April 2016

Period	1	2	3	4	5	6	7	8	9	10	11	12	13	14	15	16	17	18	Orbitals
1	1 H																	2 He	1s
2	3 Li	4 Be											5 B	6 C	7 N	8 O	9 F	10 Ne	2s2p
3	11 Na	12 Mg											13 Al	14 Si	15 P	16 S	17 Cl	18 Ar	3s3p
4	19 K	20 Ca	21 Sc	22 Ti	23 V	24 Cr	25 Mn	26 Fe	27 Co	28 Ni	29 Cu	30 Zn	31 Ga	32 Ge	33 As	34 Se	35 Br	36 Kr	4s3d4p
5	37 Rb	38 Sr	39 Y	40 Zr	41 Nb	42 Mo	43 Tc	44 Ru	45 Rh	46 Pd	47 Ag	48 Cd	49 In	50 Sn	51 Sb	52 Te	53 I	54 Xe	5s4d5p
6	55 Cs	56 Ba	57-71	72 Hf	73 Ta	74 W	75 Re	76 Os	77 Ir	78 Pt	79 Au	80 Hg	81 Tl	82 Pb	83 Bi	84 Po	85 At	86 Rn	6s5d6p
7	87 Fr	88 Ra	89-103	104 Rf	105 Db	106 Sg	107 Bh	108 Hs	109 Mt	110 Ds	111 Rg	112 Cn	113	114 Fl	115	116 Lv	117	118	7s6d7p

6	57 La	58 Ce	59 Pr	60 Nd	61 Pm	62 Sm	63 Eu	64 Gd	65 Tb	66 Dy	67 Ho	68 Er	69 Tm	70 Yb	71 Lu	4f
7	89 Ac	90 Th	91 Pa	92 U	93 Np	94 Pu	95 Am	96 Cm	97 Bk	98 Cf	99 Es	100 Fm	101 Md	102 No	103 Lr	5f

Figure 1. The standard Periodic Table for the 118 first elements. Adapted from Pyykkö [1].

very long row, followed by a shorter row for elements 165–172. The new feature was the large spin-orbit splitting of the $8p$ and $9p$ shells and the large relativistic stabilization of the $9s$ shell.

2.2 The short-rows proposal of Pyykkö [1]

It would offer at least a typographical advantage to have shorter rows. Moreover, the effective charges of these heavy atoms in compounds can often be expected to be positive, possibly strongly positive. For simplicity, one can study the electronic behaviour of monoatomic cations. These calculations were carried out by the Extended Average Level Multiconfiguration Dirac-Fock (EAL MCDF) method using the code of Desclaux and Indelicato [7]. Again, a realistic nuclear model was assumed, assuming that the nuclei are there.

All available calculations (see [1]) predict for E119 and E120 an $8s^1$ and $8s^2$ electron configuration, respectively. In the extreme case, assume that there is only one valence electron and ask, what does it want to do? If one raises the nuclear charge, the single valence electron remains an $8s$ for $Z = 120-122$ but then prefers $6f$ for $Z = 123-124$ and finally turns to $5g$ for $Z \geq 125$, see Fig. 3. Similar results lead to a formal assignment of all the elements $121-138$ to a $5g$ series, often in a *pre-5g* sense [1]. Recall the 'actinide' Th.

A second example motivating this assignment of a $5g$ row is the calculation of Indelicato *et al.* [8], who find an $8s^2(8p_{1/2})^2 5g^{18}$ closed-shell electron configuration for the isoelectronic series to the E140 atom of $Z = 143-148$. Thirdly, neutral-atom calculations start to occupy the first $5g$ states around E125 [6, 9–11] or E126 [12].

We summarize the short-row 172-element PT in Fig. 4.

2.3 The Madelung mnemonic rule

It is almost disappointing that the simple ($n + l$, n) "Madelung rule" still seems to hold. Shells are filled in the order of: 1) Increasing $N = n + l$ and, 2) For same N, largest n.

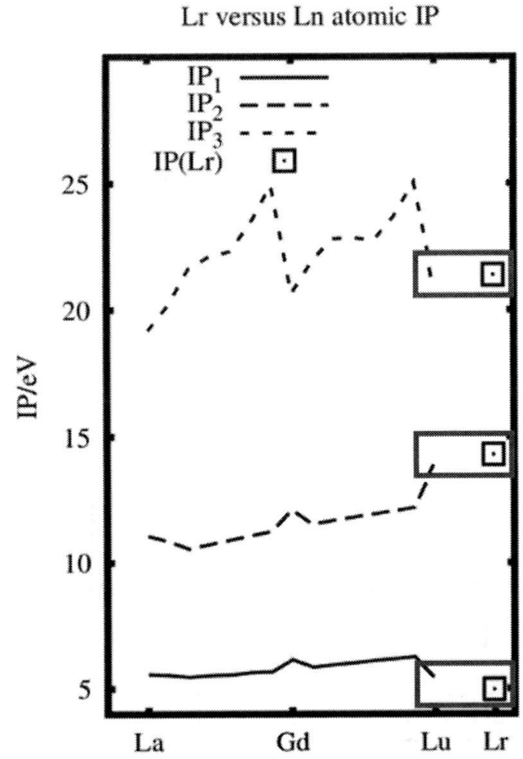

Figure 2. The ionization potentials IP_{1-3} of the lanthanides La-Lu and of Lr. Note the similarity of the values for Lu and Lr. From Ref. [5].

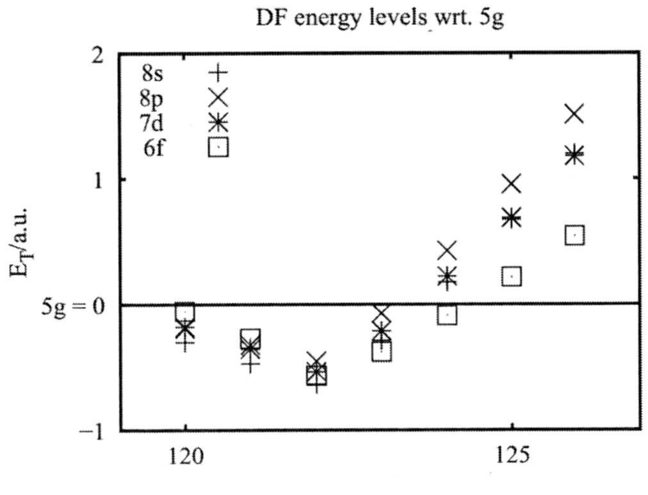

Figure 3. The lowest occupied states of a [118e]+1e alkali-like atom for various nuclear charges. The $5g$ energy is defined as zero. Previously unpublished.

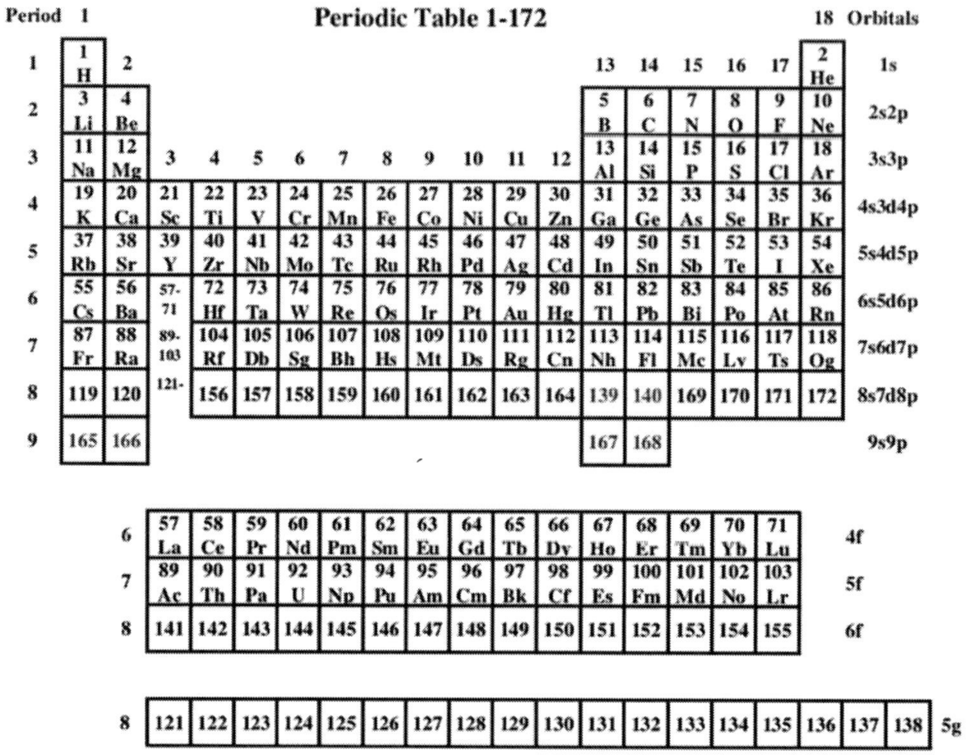

Figure 4. The PT proposed by Pyykkö. Modified from [1] by adding the year-2016 preliminary name proposals, and by emphasizing the non-monotonous cases by colour.

The "Madelung rule" up to Z = 172

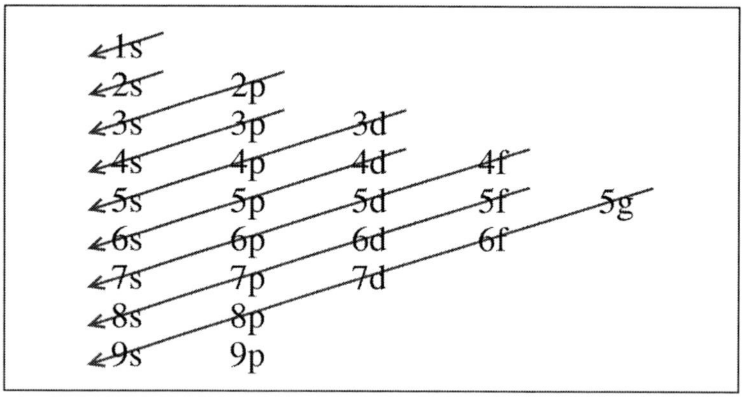

Figure 5. The "Madelung rule" applied in the present case. Previously unpublished.

The occupation order in the 172-element PT of Fig. 4 closely follows the "Madelung" occupation order in Fig. 5. For the 8th and 9th Period in Fig. 4 this amounts to

$$8s < 5g < 6f < 7d < 8p << 9s. \quad (1)$$

This is only broken up by the large spin-orbit splitting for the $8p$ and $9p$ shells, and the large relativistic stabilization of the $9s$ shell.

Actually the "Madelung" occupation order is also close to that obtained from the old quantum theory by Niels Bohr. In his Nobel lecture [13] he talks about the *number of quanta*, meaning in modern atomic theory the number of radial half-waves plus the number of angular nodes.

For more literature on the Madelung rule, see Ostrovsky [14].

3. History of the Periodic Table

For literature on the history of the PT we recommend the book of Scerri [15] and the paper of Wang and Schwarz [16]. The earliest work was based on atomic weights. The first fragments were the "triads" of Döbereiner: (Ca, Sr, Ba) in 1817 and (Li, Na, K), (S, Se, Te), (Cl, Br, I) in 1829. Gmelin in 1843 had a table of 55 elements, having oxygen in the correct group, and discussing chemical properties. The 1860 Karlsruhe conference (with a 150-year celebration in Karlsruhe in 2010) was instrumental in summarizing tha data. Following the *vis tellurique* of de Chancourtois in 1862 and the *octaves* of Newlands in 1863 and 1865, Meyer (1864) had a table of 28 elements with gaps. Mendeleev in 1869 had a table explicitly predicting missing elements with atomic weights of 45, 68 and 70. These were discovered shortly afterwards and were named Sc, Ga and Ge, respectively. The group of noble gases was reported by Ramsay in 1900. Finally, Seaborg introduced the actinide row in 1945. So far the 6*d* elements have been chemically boringly similar to their 5*d* analogues, but this may change towards the end of the series [17].

4. Conclusion

The Periodic Table for the first 118 elements seems well settled, if one chooses 15-element lanthanide and actinide rows, corresponding to f^0-to-f^{14} trivalent ions. Then one has to note that Th is counted as an actinide, although the neutral atom still has no 5*f* electrons. In compounds Th may have some 5*f* character.

Beyond the element 118 (preliminarily named Oganesson), the next two are expected to be *s*-elements, an alkali metal and an alkaline earth. For the subsequent elements we here consider at least two choices. In the present, "short rows" suggestion [1], the elements 121–138 are counted as 5*g* elements. Then follow the $8p_{1/2}$ elements 139–140, the 6*f* elements 141–155 and the 7*d* elements 156–164. In the "long rows" suggestion of Fricke *et al.* [6] overlaps between these shell occupations are allowed. Fricke considered neutral atoms while the present author included ions.

Actually the Periodic Table is about chemistry, and one should consider compounds. Little information is yet available beyond element 118.

References

[1] P. Pyykkö, Phys. Chem. Chem. Phys. **13**, 161 (2011).
[2] P. Pyykkö, Chem. Rev. **112**, 371 (2012).
[3] E. Eliav, U. Kaldor, Y. Ishikawa, P. Pyykkö, Phys. Rev. Lett. **77**, 5350 (1996).
[4] T.K. Sato *et al.*, Nature **520**, 209 (2015).
[5] W.H. Xu and P. Pyykkö, Phys. Chem. Chem. Phys. **18**, 17351 (2016).
[6] B. Fricke, W. Greiner, J.T. Waber, Theor. Chim. Acta **21**, 235 (1971).
[7] J.P. Desclaux and P. Indelicato, *The relativistic atomic program MCDFGME V 2005.10* (Published at `http://dirac.spectro.jussieu.fr/mcdf/` on August 17, 2005).
[8] P. Indelicato, J. Bieroń, P. Jönsson, Theor. Chem. Acc. **129**, 495 (2011).

[9] J.B. Mann, J. Chem. Phys. **51**, 841 (1969).
[10] J.T. Waber, D.T. Cromer, D. Liberman, J. Chem. Phys. **51**, 664 (1969).
[11] V.I. Nefedov, M.B. Trzhaskovskaya, V.G. Yarzhemskii, Doklady Phys. Chem. **408**, 149 (2006), original in Russian, **408**, 488 (2006).
[12] K. Umemoto and S. Saito, J. Phys. Soc. Japan **65**, 3175 (1996).
[13] N. Bohr, Nature, Supplement **No. 2801**, 29 (1923).
[14] V.N. Ostrovsky, Ann. N. Y. Acad. Sci. **988**, 182 (2003).
[15] E.R. Scerri, *The Periodic Table. Its Story and Its Significance* (Oxford U. P., Oxford, 2007).
[16] S.G. Wang and W.H.E. Schwarz, Angew. Chem. Int. Ed. **48**, 3404 (2009).
[17] A. Türler and V. Pershina, Chem. Rev. **113**, 1237 (2013).

EPJ Web of Conferences **131**, 01002 (2016) DOI: 10.1051/epjconf/201613101002
Nobel Symposium NS160 – Chemistry and Physics of Heavy and Superheavy Elements

Production and properties towards the island of stability

Matti Leino

Department of Physics, University of Jyväskylä, PO Box 35, 40014 University of Jyväskylä, Finland

Abstract. The structure of the nuclei of the heaviest elements is discussed with emphasis on single-particle properties as determined by decay and in-beam spectroscopy. The basic features of production of these nuclei using fusion evaporation reactions will also be discussed.

1. Introduction

In this short review, some examples of nuclear structure physics and experimental methods relevant for the study of the heaviest elements will be presented. The emphasis is on single-particle properties.

Several excellent review articles on the topics of this article have appeared recently. A number of them can be found in a special volume of Nuclear Physics A (Vol. 944). Articles which deal with the topics discussed here include [1–5]. Other recent review articles include, for instance, Refs. [6–9].

2. Nuclear structure

Nuclear structure physics deals with details of the interactions between the protons and the neutrons constituting the nucleus of an atom. Nuclei of the heaviest elements are of particular interest since their very existence is based on quantum mechanical effects creating a barrier against fission. Also, in the heaviest nuclei, very high single particle angular momenta appear, sometimes in combination with low angular momenta, leading to interesting nuclear interactions. Both single-particle and collective phenomena are of interest. Here, emphasis will be on single-particle properties when discussing structure details. Collective phenomena, especially those involved in nuclear fission, will also be touched upon. A particularly interesting but also challenging question is the influence of nuclear structure on reaction cross sections. This problem is, however, beyond the scope of this article.

2.1 Single-particle spectrum

The single-particle spectrum of a nucleus determines some of its most important properties. It constitutes a critical test of a theory. As a function of energy, there appear regions of larger density of single-particle levels, so called shells, separated by energy gaps. This is the shell structure of a nucleus. The closure of a shell indicates higher than average nuclear binding and stability. One of the key objectives of the study of the heaviest nuclei, or superheavy nuclei (here also, SHN) is the shell structure and its influence on nuclear half-lives and production cross sections.

When the structure of the nucleus is studied as a function of nuclear deformation, it is observed that energy gaps appear also at nonsphericity, albeit at different proton and neutron numbers [10]. Such phenomena persist also at high deformation, possibly leading to the creation of a fission barrier also for nuclei which according to the classical liquid drop theory would fission essentially instantaneously, i.e. those with proton number Z higher than 104 or so [1, 8]. Such shell effects or shell corrections to the ground-state mass constitute the main part of the barrier against fission. Shell-correction energies in the region of SHN can typically reach 6–8 MeV.

The heaviest nucleus in nature with closed proton and neutron shells is ^{208}Pb, a particularly tightly bound so called doubly magic nucleus. The question which has been driving the study of SHN since the 1960's is whether such nuclei exist beyond ^{208}Pb and if so, where precisely do they appear on the chart of nuclei. This problem is still incompletely understood although presently we know that there are stabilizing nuclear shells at neutron number $N = 152$ and 162 in the region of rather strongly deformed nuclei [11]. The key problem is whether spherical SHN exist. Neutron number $N = 184$ is the most promising candidate based on calculations and there are strong indications that experiments have already taken us to the shores of the so called island of stability with the synthesis of elements up to proton number $Z = 118$ [12].

2.2 Results from the nobelium region

Many nuclear structure details have been learned from the study of nuclei in the so called nobelium (No) region. This is the region around ^{254}No which nuclide can be synthesized in the reaction ^{48}Ca + ^{208}Pb (two doubly magic nuclei) with the rather high production cross section of about 2 μb. Several other nuclei in this region also have a sufficiently high yield for spectroscopic studies in ^{48}Ca- or ^{48,50}Ti-induced reactions. The present cross section limit for so called in-beam studies is around 10 nb [4]. In-beam studies give information about nuclear dynamics, in particular from rotational properties. Apart from selected results, such experiments will not be discussed further here (see, however, e.g. the review articles [4, 13]).

The nuclei in the No region are themselves interesting from the nuclear structure point of view. They may also give information on single-particle levels of nuclei in the spherical SHN region: Spherical single-particle orbitals split into components according to the projection K of the angular momentum onto the nuclear symmetry axis. Deformation brings these levels down into the more accessible No region. Furthermore, a natural requirement on a theory which is hoped to realistically describe spherical SHN is that it also describes nuclei in the No region sufficiently well. One example of a related question is the location of the spherical proton gap. According to some calculations, if the spin-orbit interaction between the $2f_{5/2}$ and $2f_{7/2}$ spin-orbit partners is strong, the gap is expected at $Z = 114$ while for a weaker interaction it is expected at $Z = 120$ [14].

Of particular interest are the so called K isomers [15]. Such relatively long-lived states are due to the presence of neutron or proton orbitals with large K near the Fermi surface. Then there opens up the possibility that – for example due to pair breaking – there is at low excitation energy a state with high spin which has no decay route down except by breaking the K conservation rule. This leads to strong hindrance for the decay. K isomers provide information e.g. on position of single-particle states and on the pairing interaction. As an example, the lowest two-quasi particle excitations are expected at an excitation energy of 2Δ where Δ is the pair-gap parameter [4]. K-isomer data in the No region support the existence of deformed shell gaps at $Z = 100$ and $N = 152$ [16].

A basic question related to excited nuclear states is whether they are generated by the excitation of protons or neutrons, or both. This question may be difficult to answer. Branching ratios of transitions depopulating the state in question are sometimes helpful. An important example is the 3^+ state in ^{254}No at 988 keV excitation [14]. Experimental branching ratios clearly identify it as a two-proton excitation involving the Nilsson orbitals $1/2^-[521]$ and $7/2^-[514]$. The $1/2^-[521]$ proton orbital stems from the spherical $2f_{5/2}$ orbital and informs us of its energy [14].

2.3 Decay studies

The most important decay types in the SHN region are α decay and spontaneous fission although the possibility of electron capture decay which may be difficult to observe should in some cases be kept in mind. Alpha decay and spontaneous fission will be briefly discussed in the following. From the point of view of half-life calculations, α decay is by far the easier one of these two.

2.3.1 *Alpha decay*

The half-life T_α in α decay depends mainly on the decay energy Q_α. A good approximation is given by the Geiger-Nuttall (or Viola-Seaborg) type of formula [17]

$$log_{10}T_\alpha = a/\sqrt{Q_\alpha} + b. \tag{1}$$

Here, the values of the constants a and b are linear functions of the proton number of the decaying nucleus.

Equation (1) works very well for ground state to ground state decays in even-even nuclei. For odd-mass and odd-odd nuclei one can talk about a hindrance for transitions between different single-particle states and define the hindrance factor $HF = T_{\text{exp}}/T_{\text{calc}}$ where T_{exp} and T_{calc} are the experimental and calculated half-life, respectively. Alpha decay preferably populates that single-particle state in the daughter nucleus which has the same single-particle configuration as the parent state, i.e. $HF \approx 1$. If, for example, there is a parity change and the nucleon spin changes direction (spin flip), typically $HF > 1000$. The barrier penetration probability depends strongly on the energy and in the region of SHN one MeV of excitation energy in the daughter nucleus typically reduces the decay constant by two to three orders of magnitude.

Systematic studies of α-decay energies reaching over long isotope chains can give evidence of shell or sub-shell closures [17]. Combined data from both cold and hot fusion reactions (see Sect. 3) reaching from $Z = 106$ to $Z = 118$ show clear evidence of a shell at $N = 162$ and also show a strong decrease of α-particle energy when N increases beyond 165, an indication of approaching a closed shell, possibly at $N = 184$ [18]. These observations go hand in hand with the increasing half-lives as $N = 184$ is approached. On the other hand, there seems to be no effect at $N = 172$, given by some calculations as the location of the shell closure (for a review, see e.g. Ref. [9]).

2.3.2 *Spontaneous fission*

The current understanding is that spontaneous fission will ultimately limit the number of chemical elements [1]. This is because of the long-range repulsive Coulomb interaction between the protons in the nucleus. In the region of SHN, there is a hindrance against fission caused by single-particle quantum effects which build up a fission barrier. Half-lives depend on the height and width of the barrier, on the existence of an outer barrier, and on the collective inertia of the nucleus [19–22]. Spontaneous fission is very much a dynamical problem, and calculation of the fission half-life is difficult. This is not helped by the fact that fission barriers are not observables that can be directly measured [20]. Once again, odd-mass and odd-odd nuclei are more difficult to treat than even-even nuclei. Very roughly, an odd nucleon increases the fission half-life by a few orders of magnitude.

Just like in α decay, systematic determination of spontaneous fission half-lives can give us information on shell and sub-shell closures. For the elements Fm and No, there is a clear maximum in the spontaneous fission half-life at $N = 152$. At Rf there is a significant change in the behaviour of the half-life as a function of N, understood by the diminishing of the second barrier below the ground state [1]. The increase of the half-lives with N for Rf, Sg and Hs when N approaches 162 is understood by the sub-shell nature of this neutron number. Likewise, the increase of the half-life by roughly two orders of magnitude when going from ^{282}Cn to ^{284}Cn and from ^{286}Fl to ^{288}Fl shows the influence of the predicted closed $N = 184$ shell. The stability seems to increase more rapidly for spontaneous fission than for α decay when $N = 184$ is approached.

2.4 Mass measurements

Mass measurements yield the nuclear binding energy without any knowledge about the nuclear state in question. Traditionally the masses of the heaviest nuclides have been obtained via α-decay studies which give the energy released in the process. If the decay chain is long, the accumulated uncertainty can be large. If the nuclides in the chain are not of even-even type, the essential part of the decay scheme must be known which often is a challenge.

The state-of-the-art are Penning Trap Mass Spectrometry studies performed at GSI using SHIPTRAP. Mass excesses of ^{252,253,254,255}No have been measured with uncertainties of 13–16 keV [5]. Detection rates of on the order of one particle per minute are sufficient.

2.5 *X*-ray identification

The products of hot fusion reactions lack connection to previously known nuclei with firmly established proton and neutron numbers [1]. One possible way of providing such unambiguous information is to observe X rays of the daughter nuclide in coincidence with the α decay of the parent [23]. First candidates for such a Z determination were recently observed in decay chains of element 115 produced in the ^{48}Ca + ^{243}Am reaction [24]. Although photons with energies compatible with the calculated energies of Bh K-X-rays were observed, statistics were not sufficient for completely excluding e.g. γ rays as the source of these photons (see also Ref. [25]).

3. Production

The only successful method for producing SHN is heavy-ion induced fusion evaporation [6]. Two varieties have been used during the past few decades: In what is called cold fusion, lead and bismuth nuclei are bombarded with the most neutron rich naturally occurring isotopes of Ti, Cr, Fe, Ni and Zn to make isotopes of Rf to element 113. The produced isotopes are rather neutron deficient which means their neutron separation energies, B_n, are high and fission barriers, B_f, are low. Both features diminish the production cross section. Also, the half-lives are short because of the long distance to the beta stability line. The idea for cold fusion came from Oganessian [26]. In hot fusion reactions, man-made isotopes such as ^{244}Pu, ^{248}Cm and ^{249}Cf are bombarded with the rare isotope ^{48}Ca [1]. The terms cold and hot refer to the excitation energy of the compound nucleus. Because of the high binding energy of

^{208}Pb and ^{209}Bi, the excitation energy of the compound system in cold fusion is low, typically 15 MeV, and the compound system gets rid of its excitation energy by the evaporation of only one neutron. In hot fusion, the excitation energy is typically 35–40 MeV and evaporation of three to five neutrons is normally observed.

There is a significant difference in the Coulomb energy which is proportional to the product $Z_1 \cdot Z_2$ between cold and hot fusion reactions: If we compare the reactions ^{76}Ge + ^{208}Pb and ^{48}Ca + ^{244}Pu to make element 114 we find a difference of eight in the neutron number of the compound nucleus. Because of the Coulomb hindrance to fusion, this means a factor of five to six orders of magnitude in compound nucleus formation [1]. The cold fusion reaction cross section goes down by a factor of eight orders of magnitude when going from No to element 113 [6]. Consequently, spherical SHN seem to be impossible to reach using cold fusion reactions. The main drawback using hot fusion reactions are the fission losses in the long evaporation chain. In each evaporation step, fission competes and typically wins by a factor of 10 or so.

Calculation of the probability, or cross section, for producing evaporation residues, results of neutron evaporation, is very difficult. It is also difficult to compare results of calculations with experimental results, especially for the process leading to the formation of the compound system. This is due to the difficulty of identifying products from competing processes such as fission and so called quasi fission. Also, because of the very small yields, the statistical uncertainty of the evaporation residue cross sections is high.

There are several methods for calculating evaporation residue cross sections [2, 27–29]. In the following, use is made of especially the review article of Zagrebaev and Greiner [2].

The process leading to the successful formation of the evaporation residue is normally divided into three stages. This division is justified by the different time scales of the phases. The three stages are

- The nuclei overcome the Coulomb barrier (termed contact or capture).
- The system evolves into a (spherical) compound nucleus (fusion).
- The compound nucleus cools down by neutron emission (survival).

The sub-barrier energy region is important during the contact stage. Channel coupling of the relative motion of the two nuclei with internal degrees of freedom (vibration, rotation) can enhance the cross section by orders of magnitude. This stage is relatively well understood, and accuracy of predictions is typically within a factor of two to three. Even this is, of course, undesirable from the point of view of planning experiments [30].

Compound nucleus formation is the least well understood reaction stage, mainly because of the large probability of quasi-fission [29]. There are significant differences in the theoretical treatment of the transformation from two nuclei to the compound nucleus as to how the nuclei lose their individuality [31].

Calculation of the survival stage is based on the statistical model. Typically, fission, neutron evaporation and γ emission are taken into account. Fission is the dominating factor, and calculation of the fission width and in particular the fission barrier is the most uncertain part of the model. The value of the shell-damping parameter which takes into account the weakening of the shell effect with increasing excitation energy is especially important for SHN. The decisive factor in survival is the ratio of the neutron evaporation width to the fission width $\Gamma_n/\Gamma_f = \exp(B_f - B_n)/T$ where T is the nuclear temperature.

It is worth noting that when the calculations of all three stages are combined, different theories may yield similar final evaporation residue cross sections but there may be significant differences in the probabilities for the individual stages [30].

As discussed in detail in Ref. [2], some rather general observations can be made when comparing cold and hot fusion reactions:

- Contact cross sections are roughly a factor of ten higher for hot fusion.
- The hot fusion probability is usually much higher than the cold fusion probability. Moreover, the cold fusion probability decreases much faster with increasing compound system proton number.
- The survival probability for the three-neutron evaporation channel (hot fusion) is much smaller than the survival probability of the one-neutron evaporation channel (cold fusion).

To put this on a quantitative level, here are some experimental evaporation residue cross sections and predictions:

- The lowest measured cross section is ~20 fb for the production of element 113 in cold fusion in the reaction ^{70}Zn + ^{209}Bi [32].
- The cross section for the production of element 118 in the hot fusion reaction ^{48}Ca + ^{249}Cf is ~0.5 pb [12].
- There are no positive results on element 119 or 120 [1, 33, 34]. The cross section limits for element 120 are on the order of 0.1–1 pb.
- One of the most promising candidates for the production of element 120 is the reaction ^{50}Ti + ^{249}Cf. Two representative cross-section predictions are 40 fb [35] and 6 fb [36].

4. Conclusion

The amount of information which has been gathered during the past ten to fifteen years both in the No region and on the heaviest elements is extensive. Rather detailed level schemes have been built for certain nuclei in the No region, and intriguing structure details especially from α-γ coincidence measurements have emerged for even heavier nuclei. The influence of the $Z = 108$ and $N = 152, 162$ shells has been firmly established and there are clear indications of a closed neutron shell at higher N, possibly $N = 184$. Firm identification of hot fusion products by X-ray methods seems to be a possibility in the near future in selected cases.

Useful discussions with M. Bender, P.T. Greenlees, R.-D. Herzberg, F.P. Heßberger, T.L. Khoo and A. Sobiczewski are gratefully acknowledged.

References

[1] Yu.Ts. Oganessian and V.K. Utyonkov, Nucl. Phys. **A944**, 62 (2015).
[2] V.I. Zagrebaev and W. Greiner, Nucl. Phys. **A944**, 257 (2015).
[3] M. Asai, F.P. Heßberger, A. Lopez-Martens, Nucl. Phys. **A944**, 308 (2015).
[4] Ch. Theisen, P.T. Greenlees, T.-L. Khoo, P. Chowdhury, T. Ishii, Nucl. Phys. **A944**, 333 (2015).
[5] M. Block, Nucl. Phys. **A944**, 471 (2015).
[6] Yu.Ts. Oganessian and V.K. Utyonkov, Rep. Prog. Phys. **78**, 036301 (2015).
[7] A. Sobiczewski, Radiochim. Acta **99**, 395 (2011).
[8] S. Hofmann, J. Phys. G **42**, 114001 (2015).
[9] M. Bender and P.-H. Heenen, J. Phys. Conf. Series **420**, 012002 (2013).

[10] R.R. Chasman, I. Ahmad, A.M. Friedman, J. Erskine, Rev. Mod. Phys. **49**, 833 (1977).
[11] G. Münzenberg, Nucl. Phys. **A944**, 5 (2015).
[12] Yu.Ts. Oganessian *et al.*, Phys. Rev. Lett. **109**, 162501 (2012).
[13] R.-D. Herzberg and P. T. Greenlees, Prog. Part. Nucl. Phys. **61**, 674 (2008).
[14] R.-D. Herzberg *et al.*, Nature **442**, 896 (2006).
[15] G.D. Dracoulis, Phys. Scr. **T152**, 014015 (2013).
[16] P.T. Greenlees *et al.*, Phys. Rev. C **78**, 021303(R) (2008).
[17] A.N. Andreyev *et al.*, Phys. Rev. Lett. **110**, 242502 (2013).
[18] V.K. Utyonkov *et al.*, Phys. Rev. C **92**, 034609 (2015).
[19] M. Kowal, P. Jachimowicz, A. Sobiczewski, Phys. Rev. C **82**, 014303 (2010).
[20] A. Staszczak, A. Baran, W. Nazarewicz, Phys. Rev. C **87**, 024320 (2013).
[21] B.S. Bhandari and Y.B. Bendardaf, Phys. Rev. C **45**, 2803 (1992).
[22] R. Smolańczuk, J. Skalski, A. Sobiczewski, Phys. Rev. C **52**, 1871 (1995).
[23] C.E. Bemis *et al.*, Phys. Rev. Lett. **31**, 647 (1973).
[24] D. Rudolph *et al.*, Phys. Rev. Lett. **111**, 112502 (2013).
[25] J.M. Gates *et al.*, Phys. Rev. C **92**, 021301(R) (2015).
[26] Yu.Ts. Oganessian, Lect. Notes Phys. **33**, 221 (1975).
[27] G.G. Adamian, N.V. Antonenko, W. Scheid, V.V. Volkov, Nucl. Phys. **A627**, 361 (1997).
[28] W.J. Świątecki, K. Siwek-Wilczyńska, J. Wilczyński, Phys. Rev. C **71**, 014602 (2005).
[29] G. Giardina *et al.*, J. Phys. Conf. Series **282**, 012006 (2011).
[30] W. Loveland, Eur. Phys. J. A **51**, 120 (2015).
[31] V.I. Zagrebaev, Y. Aritomo, M.G. Itkis, Yu.Ts. Oganessian, M. Ohta, Phys. Rev. C **65**, 014607 (2001).
[32] K. Morita *et al.*, J. Phys. Soc. Jpn **81**, 103201 (2012).
[33] Yu.Ts. Oganessian *et al.*, Phys. Rev. C **79**, 024603 (2009).
[34] S. Hofmann *et al.*, Eur. Phys. J. A **52**, 180 (2016).
[35] V.I. Zagrebaev and W. Greiner, Phys. Rev. C **78**, 034610 (2008).
[36] K. Siwek-Wilczyńska, T. Cap, M. Kowal, A. Sobiczewski, J. Wilczyński, Phys. Rev. C **86**, 014611 (2012).

Beyond-mean-field correlations and the description of superheavy elements

Paul-Henri Heenen[1,a], *Benjamin* Bally[2], *Michael* Bender[3], and *Wouter* Ryssens[1]

[1] PNTPM, CP229, Université Libre de Bruxelles, 1050 Bruxelles, Belgium
[2] ESNT, CEA Saclay, IRFU/Service de Physique Nucléaire, 91191 Gif-sur-Yvette, France
[3] IPNL, Université Lyon 1, CNRS/IN2P3, 69622 Villeurbanne, France

Abstract. The role of beyond-mean-field correlations from the restoration of broken symmetries and configuration mixing for the description of superheavy elements is illustrated.

1. Introduction

Self-Consistent Mean-Field (SCMF) models are one of the standard approaches to describe a wide range of phenomena in nuclear structure and dynamics [1]. They are the only available microscopic method that at present can be applied in a systematic manner to all bound atomic nuclei irrespective of their mass and their proton-to-neutron ratio. For superheavy elements (SHE) this method is even the only microscopic model presently available for large-scale studies [2–4].

In one way or the other, the product state that provides the solution of the SCMF equations always breaks some symmetries of the nuclear Hamiltonian, for example through deformation of the density distribution or the presence of BCS-type pairing correlations. This feature is at the heart of the predictive and explanatory power of the nuclear SCMF method, as it can be exploited to grasp correlations associated with collective degrees of freedom in an intuitive manner at moderate computational cost. The physical many-body states, however, are eigenstates of angular momentum, parity, proton and neutron numbers, which in general might all be broken by SCMF states. For some applications, however, it is important to restore the states' quantum numbers within what is usually called a "beyond-mean-field method". Exact projection on angular momentum, parity, and particle number then ensures that calculated energy spectra and transition moments for electromagnetic and other transitions can be unambiguously compared with experimental data. Similarly, for many nuclei the mixing of states with different intrinsic deformation is necessary for their correct description, for example to describe fluctuations in shape degrees of freedom or the wide-spread phenomenon of shape coexistence. Such mixing can be most easily achieved within the Generator Coordinate Method (GCM), which can be naturally combined with projection into a single framework [5, 6].

The beyond-mean-field method that we use has already been applied to many regions of the nuclear chart [7–10]. A complete calculation proceeds in three steps. The first one consists in carrying out a set of mean-field calculations with suitably varied constraints on the shape of the nucleus. The resulting intrinsic wave functions are then projected on particle numbers and

[a] Corresponding author: phheenen@ulb.ac.be

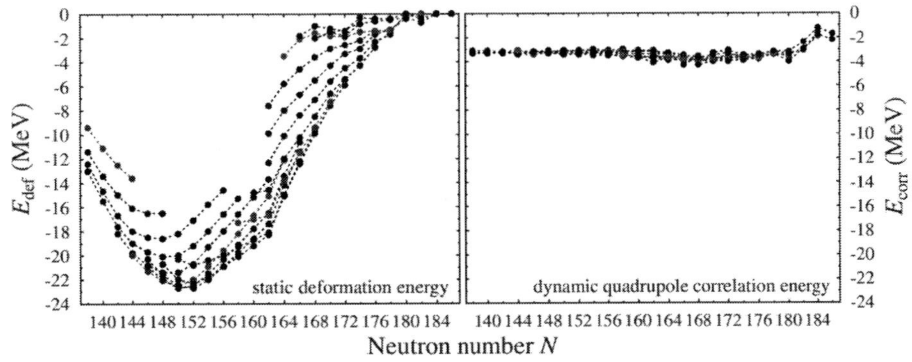

Figure 1. Static and dynamic deformation energies (see text) of heavy and superheavy nuclei around the valley of stability, plotted as a function of their neutron number. Lines connect nuclides within an isotopic chain. Those marked in red correspond to $Z = 90$, 100, 110, and 120, see Fig. 2.

angular momentum. Finally, the symmetry restored states are mixed. The final result of such a calculation is a spectrum of states in the laboratory frame of reference. In what follows, we will give a few illustrative results for properties of very heavy and superheavy nuclei obtained at the various stages of such method.

2. Systematic study of even-even SHE

A systematic beyond-mean-field calculation of even-even superheavy nuclei has been carried out under the same conditions as in the earlier study [7] that addressed the properties of lighter even-even nuclei up to $A \approx 250$. The SLy4 parameterisation of the Skyrme EDF is used throughout. The mean-field configurations are limited to axial shapes with conserved parity, which, after projection on good particle number and angular momentum $J = 0$, are then mixed within the GCM.

The resulting static and the dynamic deformation energies are plotted in Fig. 1. The static deformation energy is defined as the difference between the binding energy obtained for the spherical mean-field configuration and the lowest energy found when allowing for deformed shapes. The dynamic deformation energy is the additional gain obtained for the $J^{\pi} = 0^{+}$ ground state after projection and configuration mixing, see also Ref. [7]. The static deformation energy can reach up to 22.7 MeV and varies quickly from one isotope to the other. By contrast, the variation of the dynamic correlation energy is less structured. As in earlier studies [4, 11–13], spherical mean-field ground states are only found in the direct vicinity of the neutron shell closure at $N = 184$, while nuclei with a proton shell closure, which for SLy4 is at $Z = 126$, are found to be deformed, with deformation energies of a few MeV. As observed for lighter nuclei in Ref. [7], the dynamic correlation energy plays a crucial role only for transitional nuclei close to spherical shell closures, where it is of the same order or even larger than the static deformation energy. In such case, it might change the topography of energy surfaces and visibly modify binding energy differences [7].

For superheavy nuclei, the most important binding energy difference is the Q_{α} value, which often is the first observable that becomes available for newly discovered nuclides. Figure 2 presents results for Q_{α} values obtained at four different levels of modelling. The upper left panel shows Q_{α} values obtained from a liquid drop model (LDM) whose parameters correspond to the nuclear matter properties of the EDF used for the microscopic calculations [14]. No structure can be seen; the very smooth global trends are entirely determined by the scaling of Coulomb and symmetry energy with proton and neutron number.

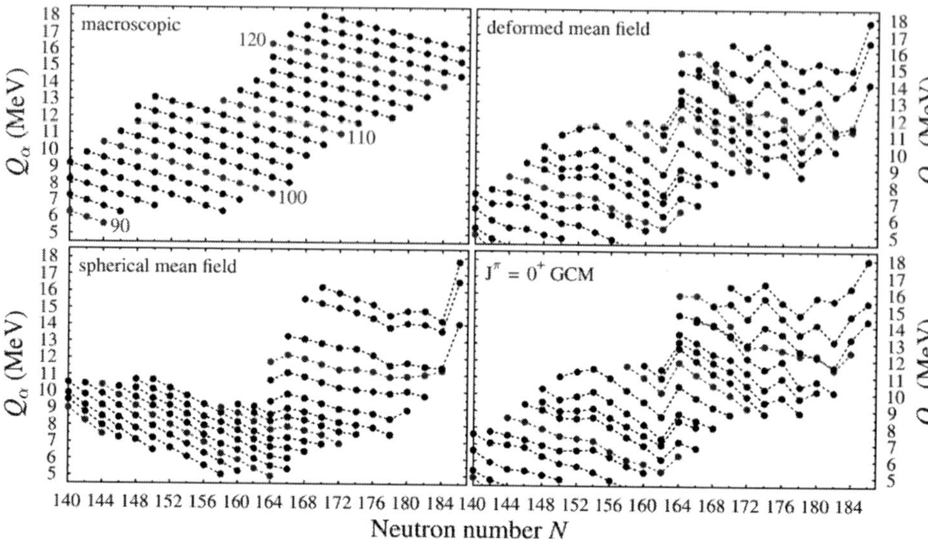

Figure 2. Q_α values obtained at four different levels of modelling (see text). Lines connect nuclides within isotopic chains. Those drawn in red correspond to $Z = 90$, 100, 110, and 120 as indicated in the upper left panel.

The effect of spherical shell structure can be seen by comparing with the lower left panel that displays the Q_α values obtained when imposing spherical symmetry on the mean-field states. When plotted as in Fig. 2, large jumps within an isotopic chain indicate a neutron shell effect, whereas large gaps between the isotopic chains indicate a proton shell effect. Not all of those visible are related to actual shell closures, however. As pointed out in Ref. [15], spherical configurations of superheavy nuclei have the unusual feature that variations in the degeneracy of spherical shells around the Fermi energy, which result from the sequence of high-j and low-j orbits in the spectrum, are sufficient to provide significant extra binding without the appearance of visible shell closures. And indeed, the spreading of the curves representing spherical Q_α values for $Z > 114$ cannot be correlated to the opening of gaps in the single-particle spectrum, see Fig. 3. Instead, it results from the small degeneracy of the low-j orbits around the Fermi energy. Comparing this panel with the upper right one shows the effect of static quadrupole deformation. As most nuclei are deformed, the density of single-particle levels is much more uniform than for spherical shapes. As a result, the Q_α values from deformed mean-field calculations are on the average much closer to the LDM values than to those from spherical mean-field calculations. In the region of nuclei with large deformation energy, the remaining jumps and gaps between the lines indicate deformed neutron ($N = 150$, 162) and proton ($Z = 108$) shell closures, not all of which agree with experiment [3, 12, 16]. For the heavier transitional nuclei the direct connection between shell structure and Q_α remains less direct. As expected from Fig. 1, the dynamic quadrupole correlations obtained from constructing the $J = 0$ projected GCM ground states only have a minor impact on Q_α values, see the lower right panel. Still, compared to the deformed mean-field results, for the heavy transitional nuclei all curves are slightly shifted in a non-uniform manner, most obviously for the $Z = 120$ chain drawn in red. How these calculated values compare with experiment has already been discussed in Ref. [4].

The interplay of single-particle shell structure, ground-state deformation and dynamic quadrupole collectivity in these calculations is illustrated in Fig. 3 for $^{290}116$ and $^{294}118$, the two heaviest even-even nuclei identified so far [17]. When calculated with the SLy4

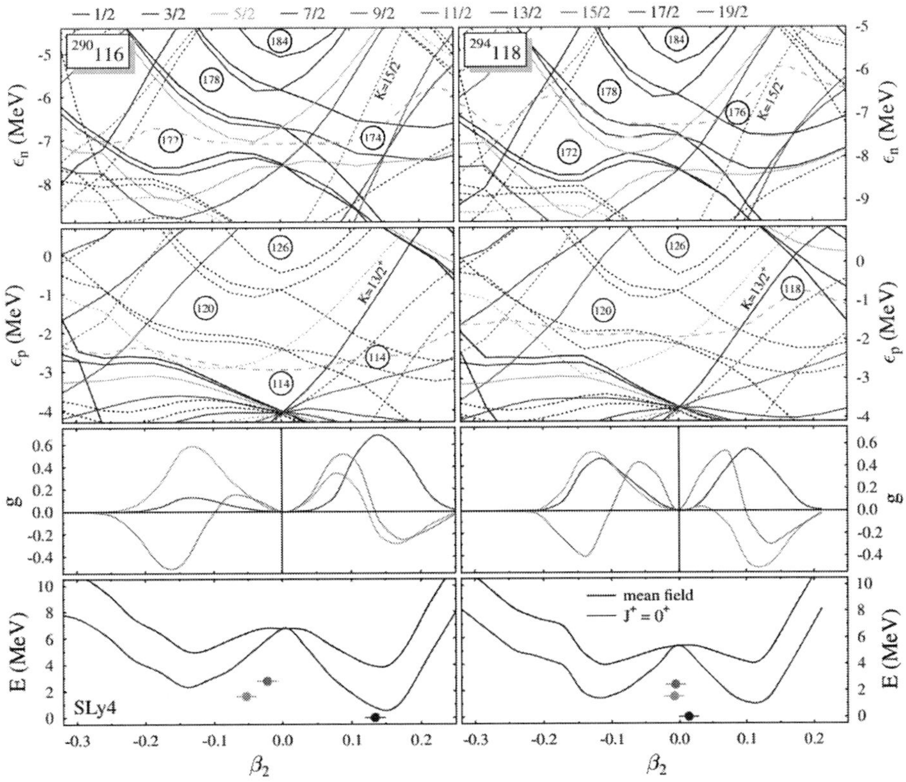

Figure 3. From top to bottom: Nilsson diagrams of neutron (ϵ_n) and proton (ϵ_p) single-particle energies, collective GCM wave functions g of the three lowest 0^+ states, and the mean field (black) and angular momentum $J=0$ projected deformation energy curves obtained for $^{290}116$ (left panels) and $^{294}118$ (right panels) as a function of the intrinsic axial quadrupole deformation β_2. For the definition of β_2 and $g(\beta_2)$ see Ref. [7]. Positive (negative) β_2 values indicate prolate (oblate) shapes. The colour of the levels in the Nilsson diagrams indicates the expectation value of angular-momentum along the nucleus' symmetry axis. Solid (dotted) lines correspond to levels of positive (negative) parity. The respective Fermi energies are indicated by dashed orange lines. The numbers put into various gaps in the Nilsson diagrams indicate the number of particles that can be put into the levels below. The bottom panel also show the energies of the GCM states drawn at the average deformation of the intrinsic states they are constructed from using the expressions from Ref. [7]. The corresponding wave functions $g(\beta_2)$ in the panel above are plotted in the same colour.

parameterisation of the Skyrme EDF, both nuclei have soft deformation energy surfaces with slightly deformed minima. Calculations of the full β-γ plane that will be reported elsewhere [19] indicate that the prolate and oblate minima visible for $^{294}118$ remain separated by a small barrier in γ direction, while for $^{290}116$ the oblate structure is actually a saddle point. The energy gain from projection on angular momentum $J=0$ increases with deformation, which for $^{294}118$ inverts the order of the two minima.

The collective GCM wave functions of the lowest 0^+ states are significantly different in both nuclei and illustrate the distinction between deformation softness ($^{294}118$) and shape coexistence ($^{290}116$). The wave functions of $^{294}118$ are almost equally spread over both minima. By contrast, the ground state of $^{290}116$ is mainly situated on the prolate side, whereas its first excited 0^+ is mainly oblate. The origin of this difference can be identified in the Nilsson diagram. As has been argued for a long time, the phenomenon of shape coexistence relates to configurations with a different number of occupied intruder orbitals [18], i.e. single-particle states with a parity opposite to the one of the last major shell. The Nilsson diagrams

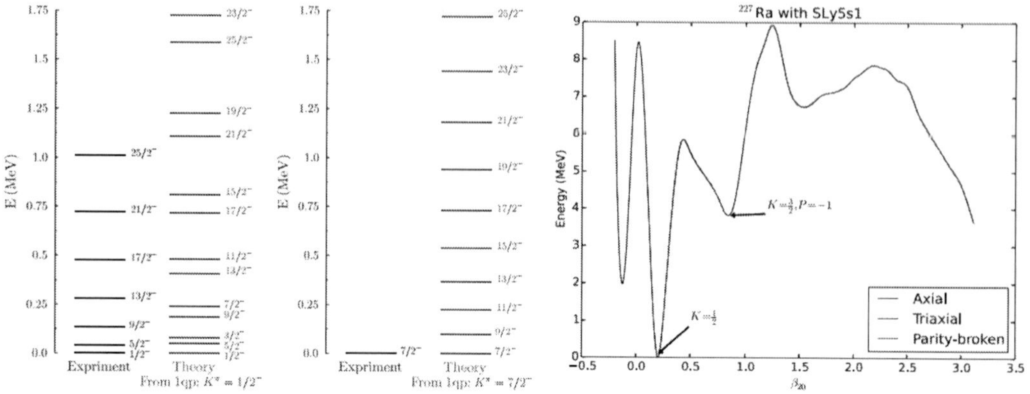

Figure 4. Left: energies obtained after projection on angular momentum and particle numbers of two different one-quasiparticle states of ^{251}Md. Right: fission barrier for ^{227}Ra constructed from the lowest one-quasiparticle configuration found at each deformation. Colours indicate the symmetries taken by the solutions found.

of both nuclei are very similar, although there are many small differences in the relative position of the single-particle levels. The main difference relevant for our discussion is the change in position of the Fermi energies that indicate where the smooth distributions of occupation numbers as determined by pairing correlations fall below 1/2. Its shift has two important consequences for configurations around the prolate minimum. Firstly, the up-sloping neutron $K^\pi = 15/2^-$ level originating from the spherical $1j_{15/2^-}$ intruder sub-shell is still below the Fermi energy for ^{294}118, but far above for ^{290}116. Secondly, the $K^\pi = 13/2^+$ proton level from the spherical $i_{13/2^+}$ intruder sub-shell is next to the Fermi energy for ^{294}118, but much higher up for ^{290}116. There are no other intruder levels crossing the Fermi energy in the range of deformations over which the lowest 0^+ states are spread. As a consequence, about the same number of proton and neutron intruder levels has large occupancy on the prolate and the oblate sides for ^{294}118, whereas for prolate configurations of ^{290}116 there are roughly two neutron and proton intruder orbitals less that are occupied for oblate ones, which leads to two distinct configurations that are only weakly mixed in ^{290}116. Such difference might have a visible impact on the α transition probability; however, no significant deviation from the empirical relation between Q_α values and lifetimes is found for the observed transition between both nuclei [17]. A more detailed analysis of these results will be given elsewhere [19].

3. Prospective calculations for odd-mass nuclei

The application of beyond-mean-field methods to odd nuclei is still in its infancy and requires new developments in particular to build effective interactions. The need to block one-quasiparticle states makes their description already much more complicated at the level of SCMF calculations.

As a first application of the projection techniques for odd nuclei that we have recently developed [10] to very heavy nuclei, the left panel of Fig. 4 shows the spectra obtained from the projection of two different blocked one-quasiparticle HFB configurations obtained for ^{251}Md with the effective interaction SLyMR0 [20]. There is a clear difference between the spectra obtained from the projection of these two near-axial states with $K^\pi \approx 7/2^-$ and $K^\pi \approx 1/2^-$, respectively. While the former is very regular, a very large signature splitting is found for the latter. Only levels in the favoured band built on the $K^\pi \approx 1/2^-$ level have been observed so far [21]; compared to the data, the projected energy spectrum is sligthly too spread out.

Besides α decay, fission is the overall most important decay mode of superheavy nuclei. While there are many calculations of the fission barriers of heavy even-even nuclei, there are almost none for odd ones. How to follow their fission path is not a trivial problem. Indeed, the one-quasiparticle configuration that provides the lowest energy may change its properties along the fission path, making it much more complicated to ensure that the path is continuous. Non-axial deformations can connect axial configurations with different K-value, while octupole deformations mix states of opposite parity. The fission barrier of ^{227}Ra as obtained from exploratory blocked HFB calculations [22] with the recent SLy5s1 parameterization of the Skyrme EDF [23] is plotted in the right panel of Fig. 4. At each deformation the energy of the lowest configuration found is given. For the ground state of this nucleus, parity is broken and the lowest quasiparticle corresponds to $K = 1/2$, while at large deformations, the second minimum corresponds to $K^{\pi} = 3/2^{-}$.

Part of the computations were performed using HPC resources from of the MCIA (Mésocentre de Calcul Intensif Aquitain) of the Université de Bordeaux and of the Université de Pau et des Pays de l'Adour. This research has been supported in part by the PAI-P7-12 of the Belgian Office for Scientific Policy.

References

[1] M. Bender, P.-H. Heenen, P.-G. Reinhard, Rev. Mod. Phys. **75**, 121 (2003).
[2] M. Bender and P.-H. Heenen, J. Phys. Conf. Series **420**, 012002 (2013).
[3] J. Dobaczewski, A.V. Afanasjev, M. Bender, L.M. Robledo, Y. Shi, Nucl. Phys. **A944**, 388 (2015).
[4] P.-H. Heenen, J. Skalski, A. Staszczak, D. Vretenar, Nucl. Phys. **A944**, 415 (2015).
[5] M. Bender, Eur. Phys. J. ST **156**, 217 (2008).
[6] J. L. Egido, Phys. Scripta **91**, 073003 (2016).
[7] M. Bender, G.F. Bertsch, P.-H. Heenen, Phys. Rev. C **73**, 034322 (2006).
[8] M. Bender and P.-H. Heenen, Phys. Rev. C **78**, 024309 (2008).
[9] J.M. Yao, M. Bender, P.-H. Heenen, Phys. Rev. C **87**, 034322 (2013).
[10] B. Bally, B. Avez, M. Bender, P.-H. Heenen, Phys. Rev. Lett. **113**, 162501 (2014).
[11] S. Ćwiok, J. Dobaczewski, P.-H. Heenen, P. Magierski, W. Nazarewicz, Nucl. Phys. **A611**, 211 (1996).
[12] T. Bürvenich *et al.*, Eur. Phys. J. **A3**, 139 (1998).
[13] S. Ćwiok, P.-H. Heenen, W. Nazarewicz, Nature, **433**, 705 (2005).
[14] P.-G. Reinhard, M. Bender, W. Nazarewicz, T. Vertse, Phys. Rev. C **73**, 014309 (2006).
[15] M. Bender, W. Nazarewicz, P.-G. Reinhard, Phys. Lett. **B515**, 42 (2001).
[16] M. Asai, F.P. Heßberger, A. Lopez-Martens, Nucl. Phys. **A944**, 308 (2015).
[17] Yu.Ts. Oganessian and V.K. Utyonkov, Nucl. Phys. **A944**, 62 (2015).
[18] K. Heyde and J.L. Wood, Rev. Mod. Phys. **83**, 1467 (2011).
[19] M. Bender and P.-H. Heenen, in preparation.
[20] J. Sadoudi, M. Bender, K. Bennaceur, D. Davesne, R. Jodon, T. Duguet, Phys. Scripta **T154**, 014013 (2013).
[21] A. Chatillon *et al.*, Phys. Rev. Lett. **98**, 132503 (2007).
[22] W. Ryssens, Thesis, Université Libre de Bruxelles (2016).
[23] R. Jodon, M. Bender, K. Bennaceur, J. Meyer, preprint `arXiv:1606.01410`.

EPJ Web of Conferences **131**, 02002 (2016) DOI: 10.1051/epjconf/201613102002
Nobel Symposium NS160 – Chemistry and Physics of Heavy and Superheavy Elements

Decay studies of heavy and superheavy nuclei

Roderick M. Clark[a]
Nuclear Science Division, Lawrence Berkeley National Laboratory, Berkeley, CA 94720, USA

Abstract. In this contribution I discuss recent results on the decay of excited states, via electromagnetic transitions (γ-ray and conversion electrons), α decay, and fission, and discuss what these studies reveal in terms of the stability of the heaviest elements.

1. Introduction

What is the heaviest element? This seemingly simple question still evades an answer. Determining γ which combinations of protons and neutrons can form a nucleus stable enough to acquire electrons and form a neutral atom (a process with a characteristic time scale on the order of femtoseconds) is one part of the solution, but the nucleus must also survive long enough to be observed. Typical experimental set-ups of today, using large-acceptance separators, require that the nucleus survive around a microsecond for there to be a chance of observation.

Since superheavy elements owe their existence to the additional stability that arises from gaps in the quantal shell structure of the nucleon levels, it is particularly important to learn about the single-particle states of the heaviest nuclei. There has been a great deal of recent work involving both prompt and delayed γ-ray spectroscopy, measuring the properties of excited states in nuclei from Fm ($Z = 100$) on up [1, 2]. One can learn not only about the single-particle structure but also about the role of pairing, and the different shapes and excitation modes of the heaviest nuclei. These fundamental properties can then be used to test the current models, the same models that are used to make predictions of the extent of the heaviest elements.

While the single-particle structure is vital to understanding the stability of the heaviest elements, the α decay and fission processes ultimately determine how long a nucleus will survive. Recent observations in the decay chains of ^{270}Ds [3, 4] suggest that high-K isomeric states can decay via α emission where the metastable state is longer lived than the ground state of the same nucleus. Moreover, recent results on ^{254}Rf [5] suggest that such high-K states can also have a significant hindrance against fission. Such results have tremendous implications for how far we may be able to push the studies of the heaviest elements and it is important to gain a better understanding of these decay mechanisms.

In this contribution I describe recent results from work done at Lawrence Berkeley National Laboratory, which address the issues of single-particle structure of the heaviest nuclei, and the α-decay and fission hindrances of K-isomers. First, I briefly discuss the latest γ-ray spectroscopy of the decay chains of ^{288}115 [6–8], and the implications that such studies have on the knowledge of single-particle states. This leads on to a more general discussion of the predictions of theoretical models, especially concerning the occurrence and extent of

---

[a] e-mail: rmclark@lbl.gov

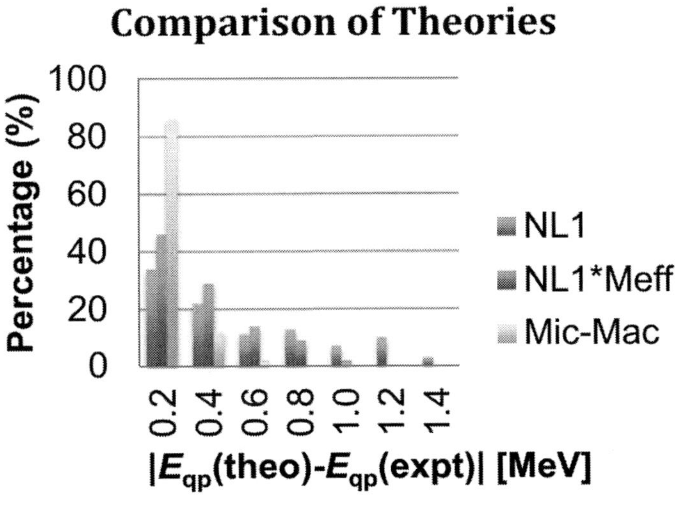

Figure 1. Distribution of the deviations of the calculated energies, E_{qp} (theo), of one-quasiparticle states from the experimental ones, E_{qp} (expt). The vertical axis shows the percentage of the states that deviate from experiment by the energy range specified on horizontal axis. NL1 refers to a particular interaction used in the microscopic calculation [20], while NL1*Meff indicates an *ad hoc* compression factor (see text). The Mic-Mac model refers to the calculations described in reference [18, 19].

spherical and deformed shell structure. Next, I describe efforts to understand the stability of metastable states, notably K-isomers in regions near the deformed shell gaps, against both α decay and fission. In the case of α decay, I describe the use of a simple superfluid-tunneling model to understand the role of pairing in enhancing the stability of the metastable states in superheavy nuclei, and use a hypothetical case in $^{294}118$ as an example [9]. For the fission of these states I attempt to answer the very simple question of how much one might expect the fission hindrance of a two-quasiparticle K-isomer to change relative to the corresponding ground state of an even-even nucleus.

2. Gamma-decay spectroscopy and single-particle structure

It is extraordinary to think that we are now able to perform γ-ray spectroscopy on nuclei in the decay chains of super-heavy isotopes. Two experiments were performed recently, one at GSI and a second at LBNL, using the $^{48}Ca+^{243}Am$ reaction to create isotopes of element-115, notably $^{288}115$ via the three-neutron evaporation channel. Combining the results from the two experiments, more than seventy chains of α decays were observed and assigned to the decay of this isotope [6–8, 10]. It was already known that there was a fine structure observed in the α decay [11] indicative of it passing through both the ground states and excited states in daughter isotopes belonging to these chains. It was hoped that the excited states might involve a competitive electromagnetic decay via some highly converted γ-ray transition, which would result in the observation of X-rays, confirming the assignment of the atomic number.

No clear-cut X-ray spectrum was observed, but multiple α-photon coincidences were seen, and discrete γ-ray transitions in the decay schemes of ^{276}Mt and ^{272}Bh were identified (see Fig. 1 of Ref. [7] and Fig. 2 of Ref. [8] to see the spectra and their interpretations). The fact that there is no apparent X-ray spectrum observed is important since it indicates that many transitions identified in these nuclei are not highly converted and most likely have electric dipole character. In turn, this indicates that the initial and final states involved in the decay have opposite parity. This is sufficient to place constraints on the models of

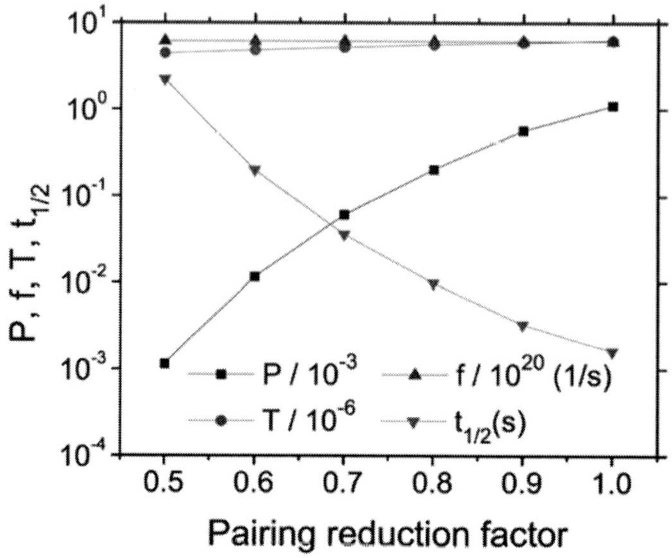

Figure 2. Contribution of the pairing reduction on the different factors f, P, and T, needed to calculate the half life, $t_{1/2}$, on the α decay of a hypothetical two-quasiparticle isomeric state in $^{294}118$. See text for more details.

single-particle structure for these nuclei [12]. These studies of decay chains of superheavy elements are extreme examples of γ-ray spectroscopy (involving cross-sections of only a few picobarns) and represent the first tentative steps toward more detailed studies, which should provide the most stringent tests of theory.

There are two broad categories of theory, which are making predictions of the single-particle structure of the heaviest elements. The first category involves microscopic-macroscopic (mic-mac) models [13–15]. These models involve describing the bulk properties of the nuclei through macroscopic approaches (such as the liquid drop model) and then superposing the microscopic (quantum) level structure. These approaches are often "tuned" to give the best reproduction of available experimental data and the issue is often raised of how well this category of model can be extrapolated to make predictions of the heaviest elements. The second category of models involves self-consistent microscopic approaches [20]; in today's parlance density functional theories (DFT) using effective nucleon-nucleon interactions. The interactions are usually fitted to the bulk properties of a few key nuclei (e.g., near ^{48}Ca and ^{208}Pb) rather than being fitted to local level structure, while the self-consistency leads to hopes that these models may be more reliably extrapolated to new regions such as the superheavy nuclei.

While these two approaches represent rather different philosophies, it is clear that any successful theory must be able to reproduce the available experimental data. Most microscopic-macroscopic treatments predict that there are major spherical shells at $Z = 114$ and $N = 172$, while the DFT approaches predict that the gaps could occur at $Z = 114$, $Z = 120$, or even $Z = 126$ and with neutron numbers $N = 172$ or $N = 184$. However, both categories of theory predict regions of low orbital degeneracy near these magic numbers, corresponding to extended regions of substantial shell stabilization, and the current experimental data is unable to differentiate between the models on this basis alone.

However, the differences between the theories should also be highlighted by how well they reproduce the known level structure of nuclei in the actinide and trans-actinide region, where there is more detailed experimental information. In Fig. 1, we show a comparison between theory and experiment. We have used the available experimental data on quasi-particle energies (differing from single-particle energies since the measured level energy determined from experiment is not a pure single-particle state and the effect of pairing must be taken into account) twenty-one odd-N and twenty-three odd-Z nuclei in the region with $Z = 89$ to $Z = 100$. The experimental data is compared against two models representative of microscopic-macroscopic [18] and the fully microscopic [20] approaches, respectively.

One clearly sees that the microscopic approach can only reproduce about one-third of the experimental level energies to within 200 keV and differences of around 1 MeV, or more, are quite common. Even after introducing an *ad-hoc* term, which "compresses" the theoretical level structure, less than half the experimental level energies are reproduced to within 200 keV, indicating that the level ordering is often incorrectly predicted. By contrast, the microscopic-macroscopic model reproduces over 85% of the known experimental levels to within 200 keV. This suggests that, from a purely pragmatic point-of-view, the mic-mac models better reproduce experiment and provide the most useful predictions. While the DFT approaches may be more readily extrapolated, they are not able to reliably reproduce the known experimental information, and their predictions of single-particle structure near the major spherical gaps have limited value in guiding experiment.

3. Alpha decay of ground and isomeric states

One consequence of the single-(quasi) particle structure is the possibility of the occurrence of isomers in specific regions of nuclei. Of particular interest are regions of high-K isomerism in deformed superheavy nuclei. These isomers arise because they involve configurations of nucleon orbitals, which yield high-K values, where K is the projection of the total angular momentum onto the axis of symmetry defined by the nuclear shape. Electromagnetic decays from these states may involve an energetically unfavorable re-orientation of nucleons to states with much lower K values and such transitions can become hindered, leading to the metastability. The decay of these high-K isomers have proven to be a useful tool for γ-ray spectroscopy in nuclei near $Z = 100$ and $N = 152$ (^{252}Fm) [1, 2]. However, it is the recent observation of α decay from high-K isomers in ^{270}Ds ($Z = 110$) and in the daughter ^{266}Hs that is of interest here [3, 4]. Not only do these isomers α decay, but the lifetimes of the isomeric states are longer than the corresponding ground state. This has profound implications for the survival of superheavy nuclei and the possibility of studying these states experimentally.

There are three major factors which influence the α decay of multi-quasiparticle states: i) the energies of the states involved – the larger the Q-value of the α decay the shorter the lifetime will be; ii) a large difference in angular momentum will give rise to a larger centrifugal barrier resulting in a longer lifetimes; iii) pairing enhances the decay through a barrier and so a reduction in pairing again implies a longer lifetime. Determining the relative influences of these competing effects is important for our understanding of the α decay of the excited states and the survival probability for superheavy elements.

In order to address this issue we have used the superfluid-tunneling model [21–23] to estimate these effects and apply it to the case of a superheavy nucleus [9]. The decay constant, λ, may be expressed as $\lambda = P \times f \times T$, where P is the preformation factor of the α particle, f is the frequency of the α particle hitting the barrier (the "knocking" frequency), and T is the transmission coefficient through the barrier. The methodology for calculating each term is explained in detail in [24]. We first applied the model to known ground-state-to-ground-state α decays and it reproduced the experimental data very well. For example, the calculated half-lives of the decays were found to be within a factor of 2.5 of the experimental values, comparable to the best semi-empirical models (see Fig. 1 of Ref. [9]). Next we used the same approach to try to reproduce the lifetimes for the α decays of multi-QP isomers in the neutron-deficient rare-earth and light actinide regions. By including a common pairing reduction factor of 0.6 we are able to reproduce the experimental data to within a factor of 3.5 (see Fig. 4 of Ref. [9]).

The ability to reproduce the known experimental data quite accurately gives confidence that the model contains the physical ingredients that are essential to understanding the decay process. We then turned to calculating a hypothetical case of relevance to the superheavy nuclei: we assume a two-quasiparticle $K^\pi = 10^+$ isomeric state at $E = 1.05$ MeV in the nucleus $^{294}118$. By modifying the pairing gap we calculated the effect on factors P, f, T, and $t_{1/2}$, as shown in Fig. 2. To form an α particle, a pair of neutrons, and a pair of protons, need to be coupled together. Therefore, the reduction in the pairing interaction has a remarkable effect on the α-particle preformation factor. The measured half-life of the ground-state of element $^{294}118$ is $t_{1/2} = 0.69$ ms, which is one to two orders of magnitude shorter than the theoretical isomeric state when the pairing reduction factor of 0.6 is used as before. Therefore, in regions where a stiff axial deformation persists, we can expect the heaviest elements to have isomeric states with longer half-lives than their ground states.

4. Fission decay of high-*K* states

Recent experiments on high-K isomers in the transfermium region have suggested that these metastable states may also have "unprecedented fission hindrance relative to the ground state" [5]. Since fission is the process that will ultimately dictate the survival of the heaviest elements it is important to understand this claim. A case of topical interest is that of ^{254}Rf [5]. Note, a compilation of all prior known data on K-isomer fission hindrances can be found in [25]. The ground state decay of ^{254}Rf is dominated by fission and has a half-life of 23.2(11) μs [5]. An exploratory run at Argonne National Laboratory using the Fragment Mass Analyzer (FMA) at the ATLAS accelerator facility, provided evidence for the occurrence of at least one high-K isomer in this nucleus. A second experiment, using the Berkeley Gas-filled Separator at the 88-Inch Cyclotron of Lawrence Berkeley National Laboratory (LBNL) was performed in order to obtain detailed information on the decay of such high-K states. In particular, it seemed reasonable to think that since the ground state decay was dominated by fission, then one might expect that any isomers in this nucleus could also have a significant fission branch. Definitive identification of fission decay of high-K states could provide a new probe of the fission process revealing information about the role of pairing and the influence of K conservation on the process.

The results of the second experiment revealed the presence of two isomers, one feeding the other, which decay by electromagnetic transitions with half-lives of 4.7(11) μs and 247(73) μs. The isomer with the shorter lifetime was interpreted as a two quasi-neutron state based on the $K^\pi = 8^- \{\nu^2(7/2^+[624], 9/2^-[734])\}$ configuration. The longer-lived isomer was interpreted as a four quasi-particle state with the $K^\pi = 16^+ \{\nu^2(7/2^+[624], 9/2^-[734])\pi^2(7/2^-[514], 9/2^+[624])\}$ configuration. We were able to estimate that the fission partial lifetimes are more than 2 (25) times longer for the two-(four-)quasi-particle isomers, respectively. The absence of any sizeable fission branch from either of the isomers was suggested as evidence of the claim of "unprecedented fission hindrance".

While the experimental results are exciting it is unclear what one *a priori* expects in terms of the fission hindrance of a high-K state. The excitation energy of a high-K two-quasiparticle isomer in this region is at least 1 MeV (roughly two times the pairing gap). In a simplified picture of a single parabolic barrier one could equate this with a lowering of the barrier height by 1 MeV. This would result in a shortening of the fission half-life. However, a high-K state involves broken pairs and in some sense it "looks" more like the configuration of the ground state of an odd-odd neighbour. One would then expect the half-life for fission to increase due to effects from the odd particles (often loosely described in terms of a "specialization energy" which accounts for the angular momentum, K value, and reduced pairing correlations).

In order to estimate the effect of the higher excitation energy of the high-K state we draw on the analysis of Vandenbosch and Huizenga for a simple parabolic fission barrier [26]. The fission half-life can be expressed as: $t_{1/2} = 2.77 \times 10^{-21} \exp[2\pi(B_f)/\hbar\omega]$. Assuming a constant barrier curvature of $\hbar\omega = 0.5$ MeV and changing the fission barrier height, B_f, from 6 MeV to 5 MeV one finds that the isomeric state would decay with a half-life that is $\sim 3 \times 10^5$ times faster than the ground state.

To estimate the effect of the "odd-odd character" of the high-K isomer on the fission hindrance we turn to an empirical estimate by looking at the hindrance factors of fission half lives of the ground states of odd-Z and odd-N neighbours of even-even nuclei. It is well known that the hindrance factor [27] arising from the odd particle, whether proton or neutron, is typically of the order of 10^3 to 10^5. There are very few cases where the odd-odd neighbour has a useful value, or limit, placed on the fission hindrance arising from the two odd particles, but it seems reasonable to expect that the effect of each odd particle is to contribute such that the odd-odd character would result in hindrances of $\sim 10^8$. One then can suggest that it would be reasonable to expect a fission hindrance for a two quasiparticle high-K state to be about 10^3 relative to the ground state. Limits, such as those established for the high-K isomers in ^{254}Rf, are significantly lower than this expectation. Indeed, looking at all the available data on cases where there are claims of fission from high-K isomeric states [25], several of the reports appear suspect. It is likely that experiments must improve their sensitivity by several orders of magnitude if we hope to understand fission from isomeric states and to use such observations as a new probe of the fission process.

5. Summary

In this contribution I have looked at some aspects of the electromagnetic, α, and fission decay of the heaviest elements. These processes can provide information on the stability of superheavy nuclei and the likelihood that we will be able to study them in experiment. We have yet to answer the very basic question of "what is the heaviest element?", but we are continuing to learn about the fundamental processes that dictate the answer.

This work was supported in part by the U.S. Department of Energy, Office of Science, Office of Nuclear Physics, under Contract No. DE-AC02-05CH11231(LBNL).

References

[1] R.-D. Herzberg and P.T. Greenlees, Prog. Part. Nucl. Phys. **61**, 674 (2008).
[2] M. Asai, F.P. Heßberger, A. Lopez-Martens, Nucl. Phys. **A944**, 308 (2015).
[3] S. Hofmann *et al.*, Eur. Phys. J. **A10**, 5 (2001).
[4] D. Ackermann, Nucl. Phys. **A944**, 376 (2015).
[5] H.M. David *et al.*, Phys. Rev. Lett. **115**, 132502 (2015).
[6] D. Rudolph *et al.*, Phys. Rev. Lett. **111**, 112502 (2013).
[7] J.M. Gates *et al.*, Phys. Rev. C **92**, 021301(R) (2015).
[8] D. Rudolph, L.G. Sarmiento, U. Forsberg, AIP Conf. Proc. **1681**, 030015 (2015).
[9] J. Rissanen *et al.*, Phys. Rev. C **90**, 044324 (2014).
[10] U. Forsberg *et al.*, Nucl. Phys. **A953**, 117 (2016).

[11] Yu.Ts. Oganessian *et al.*, Phys. Rev. C **87**, 014302 (2013).
[12] Yue Shi *et al.*, Phys. Rev. C **90**, 014308 (2014).
[13] H. Meldner, Ark. Phys. **36**, 593 (1967).
[14] R.R. Chasman *et al.*, Rev. Mod. Phys. **49**, 833 (1977).
[15] S. Ćwiok *et al.*, Nucl. Phys. **A573**, 356 (1994).
[16] A.V. Afansjev *et al.*, Phys. Rev. C **67**, 024309 (2003).
[17] M. Bender *et al.*, Nucl. Phys. **A723**, 354 (2003).
[18] A. Parkhomenko and A. Sobiczewski, Acta. Phys. Pol. B **35**, 2447 (2004).
[19] A. Parkhomenko and A. Sobiczewski, Acta. Phys. Pol. B **36**, 3115 (2005).
[20] A. Afanasjev and S. Shawaqfeh, Phys. Lett. B **706**, 177 (2011).
[21] F. Barranco, R.A. Broglia, G.F. Bertsch, Phys. Rev. Lett. **60**, 507 (1988).
[22] F. Barranco, E. Viogezzi, R.A. Broglia, Phys. Rev. C **39**, 2101 (1989).
[23] F. Barranco, G.F. Bertsch, R.A. Barranco, E. Vigezzi, Nucl. Phys. **A512**, 253 (1990).
[24] D.M. Brink and R.A. Broglia, *Nuclear Superfluidity* (Cambridge University Press, Cambridge, 2005).
[25] F.G. Kondev, G.D. Dracoulis, T. Kibedi, At. Data and Nucl. Data Tables **103–104**, 50 (2015), and references therein.
[26] R. Vandenbosch and J.R. Huizenga, *Nuclear Fission* (Academic Press, 1973).
[27] D.C. Hoffman and M.R. Lane, Radiochimica Acta **70/71**, 135 (1995).

EPJ Web of Conferences **131**, 02003 (2016) DOI: 10.1051/epjconf/201613102003
Nobel Symposium NS160 – Chemistry and Physics of Heavy and Superheavy Elements

Congruence of decay chains of elements 113, 115, and 117

Ulrika Forsberg[a], *Claes* Fahlander, and *Dirk* Rudolph

Department of Physics, Lund University, 22100 Lund, Sweden

Abstract. Superheavy element research relies heavily on the detection of correlated decay chains. When several chains have been detected, it is natural to ask whether they form one congruent set, or whether they should be further divided into subsets. To statistically assess the congruence between the chains in a data set, we developed a procedure to calculate a Figure-of-Merit for low-number statistics. In this work, the method is applied to the world data set of odd-Z superheavy nuclei with $Z \geq 113$. Some of the postulated sets of chains are found to be congruent, while others are not. The findings are of relevance for the IUPAC and IUPAP approvals of the new elements $Z = 113$, 115, and 117.

1. Introduction

Superheavy nuclei assumed to have atomic numbers $Z = 113$, 115, and 117, have been produced in heavy-ion fusion-evaporation reactions, and implanted into charged-particle detectors [1–10]. Superheavy elements often decay by subsequent α emission and thus form decay chains. These chains are often terminated by spontaneous fission, and they are found by searching for correlations in the data sets. A chain is characterized by its length and the types of decays that are observed, together with the decay energy and correlation time measured in each of the steps. Commonly, several different isotopes are expected to be produced in the same experiment. The chains can sometimes be divided into groups according to their characteristics, and then these groups can eventually be assigned to different isotopes. It is difficult to assess whether groups of chains are congruent, i.e., whether the chains are correctly grouped, or not. The advantages and disadvantages of various characteristics of decay chains led us to develop a low-statistics method, in which a Figure-of-Merit based on experimentally measured lifetimes is compared with the Figure-of-Merit of Monte-Carlo generated chains [11]. In this manuscript, the method is further motivated and applied to groups of the heaviest odd-Z superheavy nuclei. The results are of relevance to the IUPAC and IUPAP decisions to accept the new elements with $Z = 113$, 115, and 117.

2. Decay chain characteristics

Each of the characteristics of chains (length, decay modes, energies, and times) can be used to classify and divide them into different groups. However, any of these characteristics poses its own difficulty. The chain length and the decay modes are not necessarily good descriptors, since various decay modes can be present in the same nucleus. Spontaneous fission, α decay,

[a] Corresponding author: `ulrika.forsberg@nuclear.lu.se`

and electron capture (EC) decay branches are predicted to co-exist in many superheavy nuclei (see, for example, Ref. [12]). Here, it is also of interest to note that none of the detector setups employed in experiments on elements $Z = 113$, 115, and 117 have been sensitive to EC.

The α-decay energies can provide good characterisation of a chain if the decays always proceed to the same state in the daughter nucleus. This is usually the case for even-even nuclei. For odd-even, even-odd, or odd-odd nuclei, decays from one nucleus often proceed to several different states in the daughter. Such α decays give rise to a variety of different measured α energies. Hence, differing energies do not necessarily imply that the chains should be assigned to different isotopes.

In the cases where the α decays proceed to excited states in the daughter nucleus, there are additional aspects to keep in mind. If the excited state de-excites to a lower state by emission of electrons from internal conversion (IC), a partial (or full) summing of the electron and α energies may occur. This leads to a smearing of the otherwise discrete α peaks [13]. If, on the other hand, the excited state does not de-excite but rather undergoes α decay directly, other difficulties arise. The average lifetime for α decays starting from an excited state is not the same as the average lifetime for decays starting from the ground state. Therefore, a particular decay step associated with one isotope can decay with several different lifetimes. Furthermore, nuclear structure calculations (see, e.g., Ref. [11]) suggest that in some isotopes, α decays from different states in one nucleus lead to the population of different states also in the daughter nucleus. Therefore, decays starting from the same isotope can produce different-looking decay sequences.

Although there is no certain correspondence between different-looking decay sequences and different isotopes, the basic question of how we can test the congruence within a group remains. The measured times for decays proceeding from a particular state in a chain are observations of a random variable described by an exponential distribution. Thus, the decay times are not necessarily similar to each other. If the spread of the times in a data set is too small, the data has probably been divided into too many subsets. If, on the other hand, the spread is too large, the data probably originates from more than one exponential decay, and the data set must be further divided into subsets. In Ref. [14], K.-H. Schmidt characterized the spread as the standard deviation of the times, and used sets of Monte-Carlo generated exponentially distributed times to calculate 1-σ confidence limits for the standard deviation. This test applies to one decay step at the time. In Ref. [15], we proposed a generalisation of this method, where the entire chains can be taken into account. A similar approach was presented in Ref. [11], and it is this version of the method that we will use in this work. In short, a probability density function, based on the average lifetime and its uncertainty, is constructed for each decay step. The function is evaluated for each measured lifetime, and the result is taken as a Figure-of-Merit (FoM). The FoM for a chain is defined as the geometrical average over the individual FoMs in this chain. The FoM for a group of chains is defined as the arithmetic average of the FoMs of all the chains. The FoM for the data set is compared with the expected distribution for the FoM. The expected distribution is found by calculating the FoM for sets of Monte-Carlo generated chains that do fulfill the hypothesis of a common origin.

3. Available data sets for superheavy nuclei with $Z = 113$, 115 and 117

To date, odd-Z superheavy elements with $Z \geq 113$ have been directly produced in the reaction $_{30}Zn + {}_{83}Bi$ and in $^{48}_{20}Ca$-induced reactions on $_{93}Np$, $_{95}Am$, and $_{97}Bk$. The number

Table 1. Chains of superheavy nuclei with $Z = 113$, 115, and 117. The chains from each element have been divided into subsets according primarily to their lengths. In some cases, an isotope assignment is given. The chains are also sorted according to the laboratory where they were observed: JINR (Russia), GSI (Germany), LBNL (USA), and RIKEN (Japan).

	JINR	GSI	LBNL	RIKEN	total
"long" E117 ($^{294}117$)	4	2	–	–	6
"short" E117 ($^{293}117$)	16	–	–	–	16
$^{287}115$	2	1	–	–	3
$^{288}115$	31	22	43	–	96
"short" 115	4	7	3	–	14
$^{278}113$	–	–	–	3	3
$^{282}113$	2	–	–	–	2

of chains are summarized in Table 1. Of element 113, three chains have been produced at RIKEN [2–4] and two chains have been produced at JINR [5]. The RIKEN chains have been assigned to the isotope $^{278}113$ and the JINR chains to $^{282}113$. Of element 115, a total of 113 decay chains has been observed. The chains have coarsely been divided into three different groups – three long chains (five-α) with rather short lifetimes in all steps, 96 long chains (five-α) with relatively longer lifetimes, and fourteen "short" one-α and two-α decay chains. These three groups of chains have been assigned to the isotopes $^{287}115$, $^{288}115$, and $^{289}115$ in, e.g., Ref. [6]. Of element 117, twenty-two chains have been observed. Six of them consisted of six α decays and will be referred to as "long". Sixteen of the chains consisted of three or four α decays and will be referred to as "short". In, e.g., Ref. [7], the long chains have been assigned to the isotope $^{294}117$ and the short chains to $^{293}117$. Hence, the short element 117 chains are supposed to feed the isotope $^{289}115$, and it is of interest to also evaluate the congruence in the joint set of these chains.

4. Statistical assessments of element 113 decay chains

The proposed statistical test has been applied to the three α-decay chains assigned to start from $^{278}113$ [2–4]. Among these chains, two consist of five decay steps, and one chain consists of six decay steps. There are no missing members in the chains, i.e., all the lifetimes have been measured. The test is performed on the first five decay steps only, as there is only one chain which has more members. The FoM value for the experimental data set was 0.155. To obtain a reference distribution for the FoM, 10^5 sets – each consisting of three chains with five decay steps – were Monte-Carlo generated. The uncertainty in the lifetime in each step was taken into account by generating the lifetimes from different exponential distributions, where the inverse decay constant τ was selected from the appropriate likelihood function (given the average lifetime and the number of available lifetimes). The FoM was calculated for each of the 10^5 generated data sets, and the distribution was characterized by the 90% and 98% double-sided confidence intervals. These intervals are shown in Table 2. The FoM for the experimental data set (0.155) fits within the 90% confidence interval. Hence, from this perspective, there is no reason to question the common origin of these three chains.

It is interesting to note that the original Schmidt test [14] would yield the result that decay steps one, two, and four have standard deviations (1.13, 0.601, and 0.57, respectively) that are all within the respective 1-σ interval [0.19,1.91], while decay steps three and five have standard deviations (2.31, 2.18, respectively) that are outside this interval on the upper side. However, when all steps are considered in the new test, the conclusion is that the data set is congruent. These two views can still be in agreement: The low level of confidence associated with one σ means that the standard deviations should be outside the confidence intervals

Table 2. Experimental and reference values for the FoM. The experimental data set which is modeled by the Monte-Carlo generated chains, the FoM for this data set, the number of chains, the number of decay steps that are considered, and the double-sided 90% and 98% confidence intervals, are presented. The limits have been obtained from FoM distributions from 10^5 Monte-Carlo generated chains. See text and references [11, 16, 17] for details.

	Exp. FoM	No. chains	No. steps	90%	98%
$^{278}113$	0.155	3	5	[0.115, 0.226]	[0.094, 0.244]
$^{288}115$	0.224	96	6	[0.195, 0.221]	[0.189, 0.226]
$^{287}115$	0.225	3	6	[0.111, 0.214]	[0.090, 0.231]
"long" E117 ($^{294}117$)	0.226	6	6	[0.144, 0.234]	[0.126, 0.250]
"short" E117 ($^{293}117$)	0.173	16	4	[0.173, 0.244]	[0.158, 0.257]
all "short" E117 and E115	0.153	30	3	[0.187, 0.244]	[0.175, 0.254]

in some of the decay steps even in a congruent data set. Moreover, the new test is slightly more "generous" due to the inclusion of lifetime uncertainties when the reference chains are generated, and the use of an average lifetime that is not tailored to each of the generated data sets (see Ref. [16]). This translates to a situation where experimental sets of decay chains are rather accepted as congruent than discarded as consisting of more than one radioactive source.

5. Statistical assessments of element 115 decay chains

The statistical method has been applied to the data set containing 96 five-α long decay chains assigned to the isotope $^{288}115$ in Ref. [6, 8, 9]. 10^5 sets of 96 chains were Monte-Carlo generated. To mimic the detailed structure of the chains, i.e. to take into account that some members in the chains were missing, the Monte-Carlo generated sets of numbers where pruned accordingly. The FoM for the data set is 0.224. The different confidence intervals are presented in Table 2. The experimental FoM falls within the 98% confidence interval, but outside the 90% interval, on the upper side. This suggests that the measured lifetimes fit almost too well. However, as pointed out in the previous section, the test has been constructed such that congruence is rather exaggerated than understated. Therefore, the high FoM does not mean that the lifetimes are necessarily too similar. Although the test is inconclusive in this case, the reasons for a high FoM could be that the measurements have been insensitive to either (or both) very short or very long lifetimes, that some of the remaining chains observed in the ^{48}Ca on ^{243} Am reaction should be included in this group rather than in the other two groups associated with element 115, and/or that uncertain identification of chain members have led to the exclusion of the member rather than tentative inclusion (see, e.g., Table 1 in the Supplemental Material of Ref. [8], where it is indicated that there are "several options").

When the test is applied to the three five-α long chains assigned to start from the isotope $^{287}115$, the resulting FoM is 0.225. This value falls within the 98% confidence interval, but outside the 90% interval, on the upper side (see Table 2), which indicates that the similarities between the times are almost too good also in this case.

In Ref. [11], the test was applied to the fourteen "short" (one-α and two-α) element 115 chains. In brief, the set of all fourteen chains is not congruent. The FoM is well below the 98% double-sided confidence interval. Hence, the conclusion that the data set is not congruent has an error probability of less than 1%. However, several different subsets of the chains are internally congruent. It is suggested that some of the short chains should rather be assigned to the isotope $^{288}115$, which would lead to spontaneous fission branches in $^{284}113$ and $^{280}_{111}$Rg – or, as discussed in detail in Ref. [11], to undetected electron capture branches with subsequent spontaneous fission from the even-even daughters ^{284}Cn and ^{280}Ds, respectively.

6. Statistical assessments of element 117 decay chains

The test was applied to the six "long" element 117 chains. The experimental FoM of 0.226 is within the 90% confidence interval (see Table 2). Hence, this is a congruent data set. The test was also applied to the sixteen "short" element 117 chains. Although two of the chains have one more member, only four decay steps were considered. The resulting FoM is 0.173 – just on the lower limit of the 90% confidence interval (see Table 2). In Ref. [17], the congruence of a subset of ten of the sixteen "short" (three-α and four-α) element 117 chains was investigated. The subset was selected in accordance with Ref. [18], and contains those chains in which there are no missing members. The resulting FoM of 0.165 was between the 1% and 5% lower confidence limits. This indicates that the measured lifetimes, in both the full and the restricted data set, are too different from each other and that the data sets most likely should be divided into subsets.

7. Statistical assessments of the alleged link between element 115 and element 117

Assuming that the short element 117 chains originate from the isotope $^{293}117$ and that the short element 115 chains originate from the isotope $^{289}115$, the α decay of the former should feed the latter. In such case, these data sets could be combined into one. The statistical test was applied in Ref. [17] to the three decay steps E115→E113→E111 in such a combined data set. From element 117, the ten element 117 chains which had no missing members were used, and from element 115, the four chains observed at JINR. This set had a FoM which was below both the 90% and the 98% confidence interval, i.e., the error probability is less than 1% for the conclusion that this is not a congruent data set. This lack of congruence is important because the IUPAC and IUPAP approvals of the new elements 115 and 117 are based on Ref. [18], in which the postulated congruence of this data set poses the *only* argument for the discovery of these elements.

Further, the full data set on short element 115 and 117 chains – all fourteen short element 115 decay chains and all sixteen short element 117 decay chains – was investigated. The resulting FoM is 0.153, which is well below the 98% confidence interval (see Table 2). This means that inclusion of the subsequently acquired data from GSI and LBNL, and of element 117 chains which had missing members, does not improve the likeliness of this link.

8. Further comments on Ref. [18]

The approval of elements 115 and element 117 rely on a link between these elements that is, as shown by our statistical test, with a very high probability non-existent. This is problematic. Furthermore, there are other scientific inaccuracies in Ref. [18]. Some of these have already been discussed in Ref. [17]. Here, a selection of further aspects of the report are discussed in some detail.

As pointed out correctly in Ref. [18], α decays in odd-odd, odd-even, and even-odd nuclei tend to proceed to various states in the daughter nucleus (see Fig. 1). Hence, extreme care should be taken before attempting to calculate average α energies. If the measured energies correspond to different transitions, the average of these energies is a meaningless quantity. Possibly, it could be used to check consistency between subsequent measurements, in cases where the number of events is large enough that the branching ratios to different excited states are accurately reproduced. This is hardly the case for the reported measurements on elements 113, 115, and 117. Nonetheless – although data for single chains are readily available, and

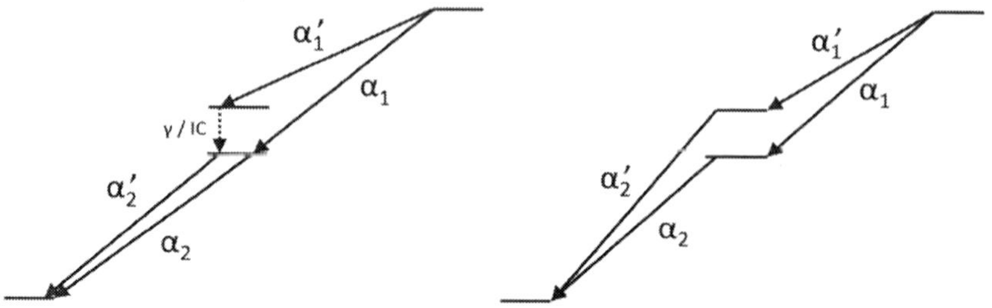

Figure 1. Schematic picture over two possible decay scenarios of a nuclide. The α decays proceed to different states in the intermediate nucleus. The populated excited states de-excites via emission of γ rays or internal conversion, and the subsequent α decays proceed from ground state to ground state (left). The populated excited state undergoes α decay to the ground state in the grand daughter (right).

although the authors of Ref. [18] acknowledge that the presence of transitions to different states has been demonstrated – such averages are used throughout the article.

Furthermore, the sum of α energies along decay chains is presented as a quantity that can shed light on the congruence between different chains. The anticipated physical background is shown in Fig. 1 (right) for chains consisting of two steps. The α decays from the mother isotope proceed to either the ground state or an excited state in the daughter nucleus. Then, the next α decays occur from the respective state to the ground state in the grand-daughter nucleus. Although the decay energies α_1 and α_1' are not the same, and also α_2 and α_2' are not the same, the summed energies $\alpha_1+\alpha_2$ and $\alpha_1'+\alpha_2'$ are equal. It is argued that if the sums are similar, it indicates that the chains start and end in the same nuclear states and, thus, that the chains do share a common origin.

This argumentation is valid as long as no γ rays or internal conversion electrons are emitted. Such de-excitations are most likely present in most odd-nuclei decays, and their presence has been firmly established in chains assigned to start from the isotope $^{288}115$ [8, 9]. If the sum-α energies agree for different chains although γ rays and internal conversions have not been excluded, this might be due to pure chance. Several hundreds of keV could be missed in each step (see, e.g., the decay step ^{276}Mt $\rightarrow$ ^{272}Bh) in one of the chains, while the decays in another (different-origin) chain might proceed from ground state to ground state. In this way, the sums can agree although this does not say anything about the origin of the chains. Naturally, the chances for such unfortunate coincidences between different-origin chains decrease if the energies are very precisely measured: The chances that two decay sequences should have identical sums is small. However, the individual α energies in the decay chains assigned to the nuclide $^{278}113$ (where the sum-energy argument is proposed) are measured with an uncertainty of $\sim$ 40 keV at best, according to Ref. [18].

Another interesting observation is that if a decay proceeds as in Fig. 1 (right), the α decays from the intermediate nucleus start from different nuclear states. Decays from two different nuclear states are characterized by two different decay constants, i.e., they exhibit two different lifetimes. This means that for subsequent decay steps where such behaviour has been established, the similarities in lifetimes become meaningless quantities that must not be used to (artificially) reinforce an energy-summing argument.

Yet another aspect of identification of superheavy elements that becomes important especially for odd-nuclei, is that different states may be reached in a nucleus when it is created directly in a fusion-evaporation reaction and when it is created as a decay product. A prime example is ^{257}Rf, where different decays sequences result when the isotope is directly produced and when it is created as the daughter of ^{261}Sg [19, 20]. Hence, the lack of an apparent link between the presumed isotopes $^{289}115$ and $^{293}117$ does not necessarily imply that the observed chains cannot originate from these isotopes. The lack of clear cross-bombardment cases for odd nuclei is not unexpected.

9. Summary

In this manuscript, new applications of the statistical test introduced in Ref. [11] have been presented. Some data sets are found to be congruent in terms of decay times, while others have FoM values that indicate that a revision is needed. For some data sets the lifetimes fit "too well", which indicates that the selection of chains has maybe been too restrictive. In other cases the FoM value is low, which indicates that the chains do not share a common radioactive origin. In particular, the alleged cross-bombardment link between element 117 and element 115 is, on a more than 2-σ level, absent. It is further exemplified that methods and numerical operations in Ref. [18] are questionable.

This work is supported by the Swedish Research Council (VR 2013-4271).

References

[1] Yu. Oganessian *et al.*, Phys. Rev. C **69**, 021601(R) (2004).
[2] K. Morita *et al.*, J. Phys. Soc. Jpn. **73**, 2593 (2004).
[3] K. Morita *et al.*, J. Phys. Soc. Jpn. **76**, 045001 (2007).
[4] K. Morita *et al.*, J. Phys. Soc. Jpn. **81**, 103201 (2012).
[5] Yu.Ts. Oganessian *et al.*, Phys. Rev. C **76**, 011601(R) (2007).
[6] Yu.Ts. Oganessian *et al.*, Phys. Rev. C **87**, 014302 (2013).
[7] Yu.Ts. Oganessian *et al.*, Phys. Rev. C **87**, 054621(2013).
[8] D. Rudolph *et al.*, Phys. Rev. Lett. **111**, 112502 (2013).
[9] J.M. Gates *et al.*, Phys. Rev. C **92**, 021301(R) (2015).
[10] J. Khuyagbaatar *et al.*, Phys. Rev. Lett. **112**, 172501 (2014).
[11] U. Forsberg *et al.*, Nucl. Phys. **A953** 117 (2016).
[12] A.V. Karpov *et al.*, Int. J. Mod. Phys. E **21**, 1250013 (2012).
[13] F.P. Heßberger *et al.*, Nucl. Instr. Meth. **A274**, 522 (1989).
[14] K.-H. Schmidt, Eur. Phys. J. **A8**, 141 (2000).
[15] D. Rudolph *et al.*, EPJ Web of Conferences **117**, 010001 (2016).
[16] U. Forsberg, PhD thesis, Lund University, ISBN 978-91-7623-813-4 (2016).
[17] U. Forsberg *et al.*, Phys. Lett. **B760**, 293 (2016).
[18] P.J. Karol *et al.*, Pure Appl. Chem. **88**, 139 (2016).
[19] B. Streicher *et al.*, Eur. Phys. J. **A45**, 275 (2010).
[20] F.P. Heßberger, these proceedings.

EPJ Web of Conferences **131**, 02004 (2016) DOI: 10.1051/epjconf/201613102004
Nobel Symposium NS160 – Chemistry and Physics of Heavy and Superheavy Elements

In-beam spectroscopy of the heaviest elements

Rolf-Dietmar Herzberg[a]

University of Liverpool, UK

Abstract. In-beam spectroscopy provides many powerful tools for the detailed study of nuclear structure. Over the past two decades the coupling of sensitive in-beam spectrometers to recoil separators has allowed the study of weakly populated reaction channels, such as the fusion-evaporation reactions leading to nuclei beyond fermium ($Z = 100$). The methods, observables, and limitations of this approach are discussed.

1. Introduction

With the recent discovery and naming of elements 113 (nihonium), 115 (moscovium), 117 (tennessine), and 118 (oganesson) [1, 2] the seventh row of the chemical table of elements is complete, and the search for the first elements of the 8th row has begun [3–6]. The last element of a row is traditionally a noble gas, which owes its chemical inertness to the strong stabilisation conferred by a completely filled major electron shell. In a nucleus the situation is similar: the nucleus is made up of fermions forming shell structures much like electrons do. On the other hand there are important differences: protons and neutrons form mostly independent shell structures, leading to singly or doubly closed shells. In addition the level density becomes much larger and the separation between shells is not as clear cut as in the atomic case. This leads to the current expectation that the the region of shell-stabilised nuclei owing their relatively long half-lives to large extra binding energies caused by regions of low level density, i.e. the "island of stability", is a complex two-dimensional region on the nuclear landscape whose shores are extremely sensitive to the details of the underlying single particle structure of both protons and neutrons (see, e.g., Ref. [7]).

One of the challenges of nuclear structure physics is therefore to study the single-particle structure of the heaviest nuclei in great detail. The situation is further complicated by the fact that the majority of non-magic nuclei are deformed, which results in a further splitting of the single-particle shells and a dramatic increase in the level density. In 1999 the realisation that the nobelium isotope ^{254}No could be synthesized with relatively large production cross sections of 1–2 μb led to the first in-beam studies of a nucleus beyond $Z = 100$ [8, 9], proved that it is indeed a well deformed nucleus and made the whole arsenal of spectroscopic tools available to study nuclei in this region. The state of the art has regularly been reviewed (see references [10–12] and references therein).

In this contribution I will summarise the principle of the experimental techniques, discuss the observables, and look at the inherent limitations of the methods.

---

[a] e-mail: R.Herzberg@Liverpool.ac.uk

2. In-beam spectroscopy

A major experimental challenge is the production of nuclei with more than 100 protons in a fusion reaction. The dominant background through fission and quasi-fission takes an ever larger fraction of the total reaction cross section, with the fusion-evaporation channels of interest becoming (much) less than 10^{-6} of the total. Tagging techniques developed in the 1990s [13] allowing a selection of the fusion products from this overwhelming background are crucial. They work by placing an array of detectors around the target position to record the radiation emitted promptly from the reaction. The reaction products are then passed into a recoil separator where they are selected over the unwanted reaction products and unreacted primary beam. Products of interest are detected in the separator focal plane and a time correlation with the promptly emitted radiation at the target position can be carried out using the well known flight time through the separator. For an illustration of the operating principle see, e.g., references [10–12].

The prompt detector array has mostly consisted of Ge-detectors for γ-ray spectroscopy. The initial experiment used GAMMASPHERE at Argonne National Laboratory [8, 14] and a number of similar detector arrays have since been used, notably the JUROGAM I and II arrays at the accelerator laboratory of the University of Jyväskylä.

This general detection scheme can be easily expanded. A conversion electron detector can be used either with or without γ-ray detection capability. An example of the latter is the SACRED spectrometer [15, 16], an example of the former is the SAGE spectrometer [17], where a highly segmented Si pixel detector for electrons is combined with the JUROGAM II array of Ge-detectors.

A further breakthrough came in 2002 when a new calorimetric approach [18] allowed selection of isomeric states populated in the initial reaction. Provided that the isomeric half-lives were at least as long as the flight time through the separator, this opened the door to the study of isomers in one of the most fertile isomer hunting grounds in the nuclear chart [19–21].

To illustrate the scope and reach of in-beam spectroscopic methods in the heaviest nuclei Fig. 1 summarises the available data in elements from curium upwards. The figure has been updated from [11, 12] and modified to indicate those nuclei studied using in-beam spectroscopic methods to date.

2.1 Observables

In-beam spectroscopy has been the workhorse of nuclear spectroscopy for many years, and a large number of spectroscopic techniques are available (see, e.g., Ref. [22]). The most important is the ability to use γ-ray coincidence techniques to build up even complex level schemes, which are indispensable for the reliable assignment of configurations, spins, and parities. From the energy spacings in the ground-state rotational band one can deduce the moment of inertia as a function of the rotational frequency of the nucleus. The nucleons sitting in the single-particle orbitals with the highest angular momentum experience the largest Coriolis force and will be the first ones to break their paired state and align their spins with the rotational axis. This "backbending" behaviour is directly observable in the moment of inertia and can be compared to model calculations probing the location and occupation of the high-j single-particle orbitals and the pairing strength of the nucleus. This is especially important as the main difference between microscopic-macroscopic models and fully self-consistent models is the location of the high-j orbitals in relation to the other single-particle configurations [23, 24]. Figure 2 shows the dynamic moments of inertia for even-even isotopes with $N = 146$–154.

DOI: 10.1051/epjconf/201613102004

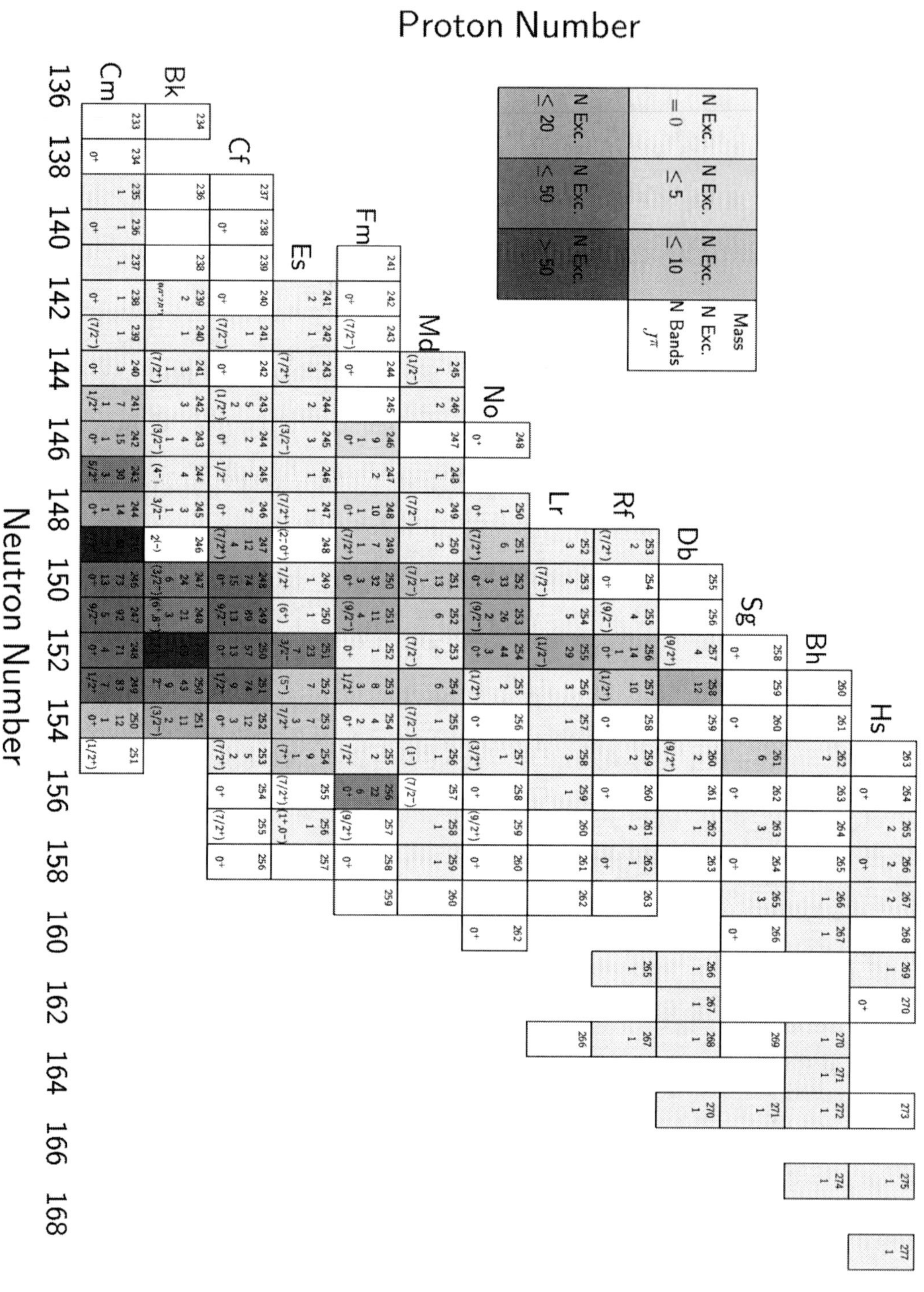

Figure 1. Schematic summary of evaluated nuclear data available in the ENSDF data base nuclei from Cm to Hs. We give for each isotope the mass number, the number of known excited levels, the number of assigned rotational bands, and the ground state spin. The blue boxes indicate those nuclei where in-beam spectroscopy has been carried out, the darker shades correspond to more levels identified. (Figure based on Fig. 1 in [11].)

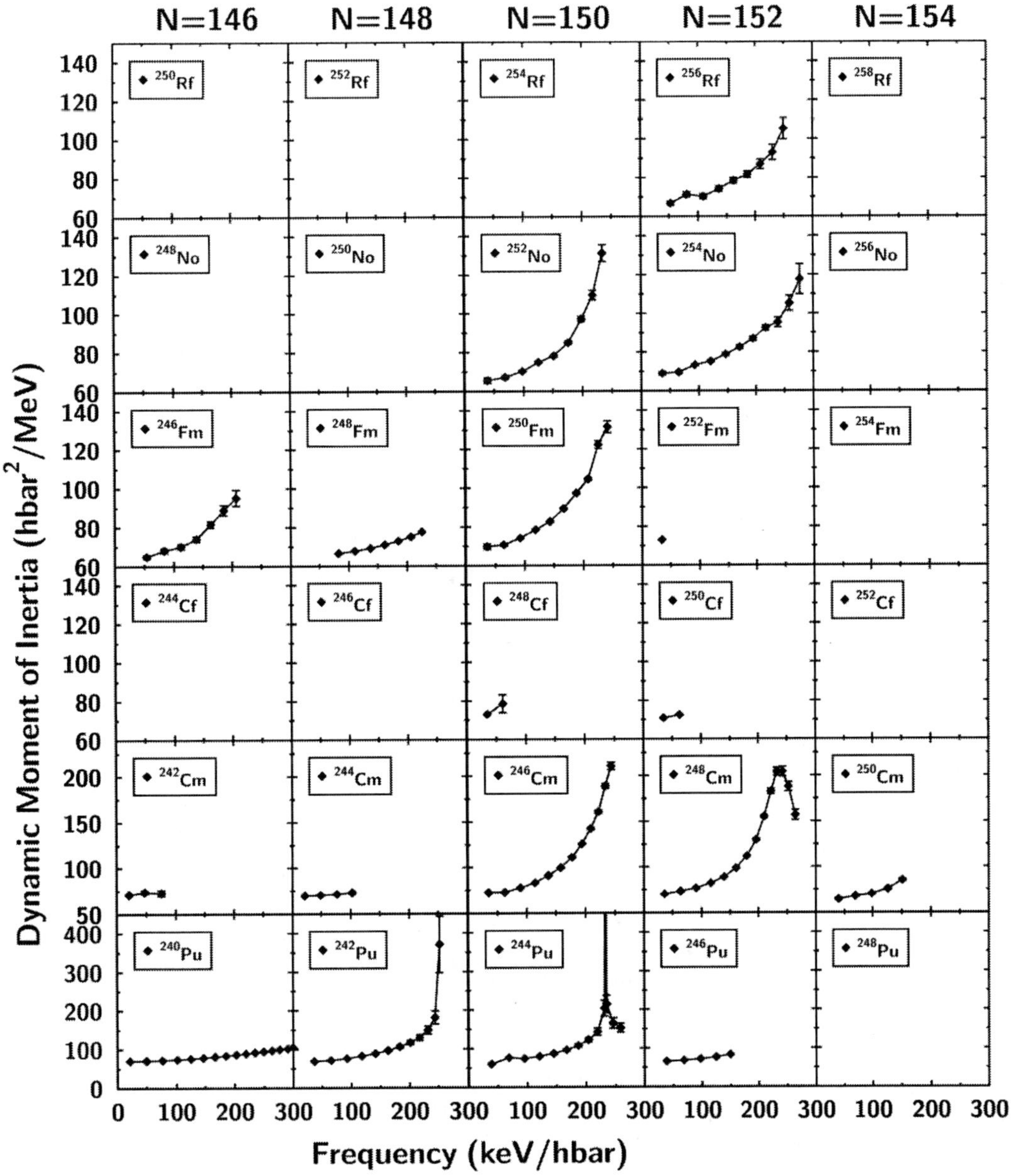

Figure 2. Dynamic moments of inertia for nuclei around ^{254}No plotted against angular frequency. The different upbending behaviours are clearly seen. (Adopted from Fig. 18 in [11], data for ^{256}Rf is taken from [25], data for ^{252}Fm is taken from [26].)

If one looks at the ground state rotational bands of ^{250}Fm, ^{252,254}No, and ^{256}Rf one can observe the difference in upbending behaviour between the $N = 150$ isotones and the $N = 152$ isotones [25]. In these nuclei the $i_{13/2}$ proton orbitals and the $j_{15/2}$ neutron orbitals are partially filled and are the first proton and neutron orbitals expected to align, respectively. Any difference in behaviour is therefore very sensitive to the relative position of these orbitals in relation to the Fermi surface.

In odd-mass isotopes the rotational band built on the ground state carries even more information. In general one will observe a strongly coupled system of two bands, one for each signature, connected with intraband $E2$ transitions and interband mixed $E2/M1$

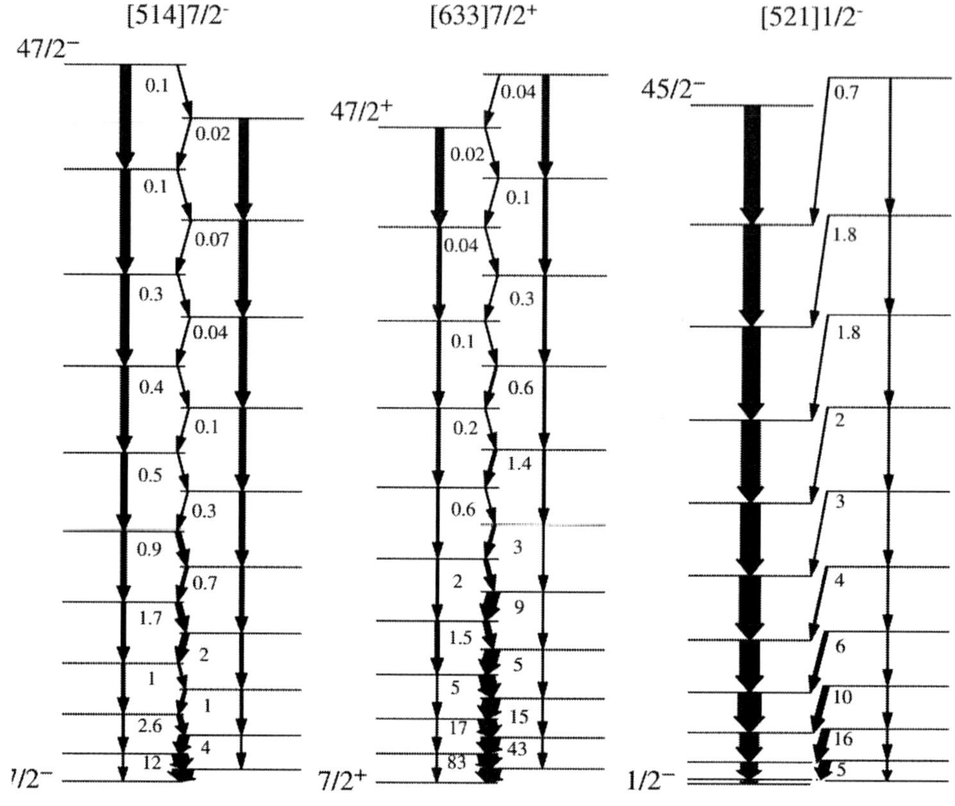

Figure 3. Schematic decay patterns for the three configuration candidates for rotational bands in ^{251}Md. The numbers labelling the states indicate the ratio $T(M1)/T(E2)$. (Taken from [27].)

transitions. The branching ratio between the interband/intraband transitions is sensitive to both the quadrupole moment of the band and the gyromagnetic factors of the ground-state configuration. Thus a measurement of the branching ratios in such bands can provide strong evidence to assign a specific configuration to the ground state, or, more generally, the band head. Such examples have been ^{251}Md [27] and ^{253}No [28–30]. Figure 3 illustrates the expected branching ratios in the case of ^{251}Md [27]. Here model calculations for the three configurations most likely to form the ground state, namely [514]7/2$^-$, [633]7/2$^+$, and [521]1/2$^-$ give very different and distinctive intensity patterns. On the basis of this pattern the ground-state configuration was assigned as built on the [521]1/2$^-$ single proton orbital [27].

This technique can equally well be applied to bands built on high-K isomeric states in even-even nuclei, such as the band built on the $K^\pi = 8^-$ isomers in ^{252}No [31] and ^{250}Fm [32, 33], resulting in an assignment of the two-quasineutron configuration $8^-([624]7/2^+{}_\nu \otimes [734]9/2^-{}_\nu)$ for the isomers. In ^{254}No the configuration of the 3^+ excited state populated in the decay of the lowest 8^- isomer was identified as $3^+([521]1/2^-{}_\pi \otimes [514]7/2^-{}_\pi)$. This is significant as the $[521]1/2^-{}_\pi$ proton orbital stems from the spherical $f_{5/2}$ single-particle level located *above* the spherical $Z = 114$ gap [19]. Although one starts to run into problems with statistics, it is possible to assign configurations in this way. Indeed, new methods to deal with extremely low statistics data have been developed for just this purpose [34].

2.2 Limitations

In-beam spectroscopy typically uses a recoil separator to pull the weak prompt signal from the dominant background by using a coincidence with the recoiling nucleus. However, the maximum beam current usable in these experiments is set by the maximum event rate observed at the target position. This is entirely dominated by unwanted background from fission and transfer channels, largely independent of the desired reaction. Thus the main limitation at cross sections below the nanobarn level comes from the total number of recoils produced on a target with a beam current limited by the prompt event rate. At present the lowest cross section for meaningful in-beam γ-ray spectroscopy in an even-even transfermium nucleus is 17 nb, reached in the reaction of ^{50}Ti on ^{208}Pb targets [25].

Modern γ-ray tracking arrays such as AGATA [35] or GRETA/GRETINA [36, 37] can push that limit down by at least one order of magnitude. They do this by tracking every γ-ray interaction in a full germanium shell and then reconstructing the number and energy of all simultaneously incoming γ rays using tracking algorithms. This is possible because the Compton scattering process gives a well defined relationship between the scattering angle and the deposited energy in each Compton scattering event. Thus these tracking arrays can make use of pile-up events which have to be discarded in non-tracking spectrometers. This allows the detectors to be operated at a much higher event rate, and thus allow a large beam intensity on the target.

References

[1] P.J. Karol, R.C. Barber, B.M. Sherrill, E. Vardaci, T. Yamazaki, Pure Appl. Chem. **88**, 139 (2016).
[2] P.J. Karol, R.C. Barber, B.M. Sherrill, E. Vardaci, T. Yamazaki, Pure Appl. Chem. **88**, 155 (2016).
[3] Ch. E. Düllmann *et al.*, GSI Scientific Report 2011, GSI Report 2012-1, p. 206.
[4] S. Hofmann *et al.*, GSI Scientific Report 2008, GSI Report 2009-1, p. 131.
[5] S. Hofmann *et al.*, GSI Report 2012-1, p. 205.
[6] Yu.Ts. Oganessian *et al.*, Phys. Rev. C **79**, 024603 (2009).
[7] M. Bender *et al.*, Phys. Lett. **B515** 42 (2001).
[8] P. Reiter *et al.*, Phys. Rev. Lett. **82**, 509 (1999).
[9] M. Leino *et al.*, Eur. Phys. J. **A 6**, 63 (1999).
[10] R.-D. Herzberg, J. Phys. **G30**, R123 (2004).
[11] R.-D. Herzberg and P.T. Greenlees, Prog. Part. Nucl. Phys. **61**, 674 (2008).
[12] Ch. Theisen *et al.*, Nucl. Phys. **A944**, 333 (2015).
[13] E.S. Paul *et al.*, Phys. Rev. C **51**, 78 (1995).
[14] I.-Y. Lee, Nucl. Phys. **A520**, 641c (1990).
[15] P.A. Butler *et al.*, Nucl. Instrum. Meth. **A381**, 433 (1996).
[16] H. Kankaanpää *et al.*, Nucl. Instrum. Meth. **A534**, 503 (2004).
[17] J. Pakarinen *et al.*, Eur. Phys. J. A **50**, 53 (2014).
[18] G.D. Jones, Nucl. Instrum. Meth. **A488**, 471 (2002).
[19] R.-D. Herzberg *et al.*, Nature **442**, 896 (2006).
[20] R.-D. Herzberg and D.M. Cox, Radiochim. Acta **99**, 441 (2011).
[21] G.D. Dracoulis *et al.*, Rep. Prog. Phys. **79**, 076301 (2016).
[22] H. Morinaga, *In-Beam Gamma-Ray Spectroscopy* (Elsevier Science Ltd, 1977).
[23] M. Bender *et al.*, Nucl. Phys. **A723**, 354 (2003).
[24] A.V. Afanasjev *et al.*, Phys. Rev. C **67**, 024309 (2003).
[25] P.T. Greenlees *et al.*, Phys. Rev. Lett. **109**, 012501 (2012).

[26] M. Asai, priv. comm. (2016).
[27] A. Chatillon, *et al.*, Phys. Rev. Lett. **98**, 132503 (2007).
[28] P. Reiter *et al.*, Phys. Rev. Lett. **95**, 032501 (2005).
[29] R.-D. Herzberg *et al.*, Eur. Phys. J. **A 42**, 333 (2009).
[30] A. Mistry *et al.*, to be published.
[31] B. Sulignano *et al.*, Phys. Rev. C **86**, 044318 (2012).
[32] J.E. Bastin *et al.*, Phys. Rev. C **73**, 024308 (2006).
[33] P.T. Greenlees *et al.*, Phys. Rev. C **78**, 021303(R) (2008).
[34] E. Parr *et al.*, Eur. Phys. J. A **48**, 134 (2012).
[35] S. Akkoyun *et al.*, Nucl. Instrum. Meth. **A668**, 26 (2012).
[36] I.Y. Lee *et al.*, Rep. Prog. Phys. **66**, 1095 (2003).
[37] I.Y. Lee, Nucl. Phys. **A746**, 255c (2004).

Nuclear structure of the transactinides – investigated by decay spectroscopy

Fritz Peter Heßberger[1,2,a]

[1] GSI Helmholtzzentrum für Schwerionenforschung GmbH, Planckstraße 1, 64291 Darmstadt, Germany
[2] Helmholtz-Institut Mainz, Johann-Joachim-Becherweg 45, 55128 Mainz, Germany

Abstract. Superheavy elements owe their stability due to a subtle balance between the disruptive Coulomb force and the attractive nuclear forces. Thus they represent an ideal laboratory to study basic interactions. The essential tools are detailed investigations of radioactive decay properties and nuclear structure of superheavy nuclei. The results of those studies will deliver valuable input to improve theoretical models. To fulfill this demand conclusive data of high quality are necessary, which is presently not so easy to meet due to small production cross sections and technical limitations (beam intensities, detection probabilities). Possibilities and problems concerning extraction of decay properties and nuclear structure information on the basis of a low number of observed decay events will be discussed for three illustrative examples, ^{257}Rf, ^{257}Lr, and ^{288}Fl.

1. Introduction

About fifty years ago first attempts to extent the nuclear shell model [1, 2] into regions far beyond the heaviest known doubly magic nucleus, ^{208}Pb ($Z = 82$, $N = 126$), were undertaken. They resulted in the prediction of spherical proton (Z) and neutron (N) shell closures at $Z = 114$ and $N = 184$ [3, 4], which came along with high shell-correction energies in the order of -8 MeV (see, e.g., Ref. [5]) leading to a high stabilization of nuclei in that region against prompt disruption. Early calculations [6] predicted, e.g. for ^{298}Fl ($Z = 114$, $N = 184$) a partial fission half-life of $2 \cdot 10^{19}$ years. The expected long-lived nuclei around $Z = 114$ and $N = 184$ were denoted "superheavy". These theoretical findings initiated tremendous efforts from experimental side to produce those "Superheavy Nuclei" (SHN) and to investigate their properties. Up to now synthesis of elements up to $Z = 118$ has been reported (see, e.g., Ref. [7]), and discovery of elements $Z > 112$ was recently approved by the International Union for Pure and Applied Chemistry (IUPAC) [8–10]. The decay properties of the reported isotopes indicate the existence of a region of shell stabilized nuclei in the transactinide region towards $N = 184$, although the "center" has not yet been reached. Information on the strength of the possible shells is still scarce.

From the theoretical side the situation is somewhat more complicated: while macroscopic-microscopic approaches based on the nuclear drop model [11] and the Strutinsky shell-correction method [12] agree in $Z = 114$ and $N = 184$ as proton and neutron shell closures (see, e.g., Refs. [13, 14]), calculations using self-consistent models like Skyrme-Hartree-Fock-Bogoliubov (SHFB) calculations or relativistic mean-field models

[a] e-mail: f.p.hessberger@gsi.de

(RMF) using NL3, NL-Z2 or NL-Z parametrizations [15, 16] disagree in locating the proton and neutron shells. Some predict $Z = 120$ as proton shell closure, while others predict $Z = 114$ or $Z = 126$. Skyrme-force based calculations agree in $N = 184$ as neutron shell closure, while the RMF calculations favour $N = 172$. Evidently, however, all these parametrizations result in a wide area of large shell effects. It has, however, already been discussed extensively by Bender *et al.* [17] that Z and N where the shells occur strongly depend on details in the description of the underlying forces, specifically on the values for the effective masses, m^*, and the strength of the spin-orbit interaction. It has been emphasized in [17] that the energy gap between the spin-orbit partners $2f_{5/2}$ and $2f_{7/2}$ determines whether the proton shell occurs at $Z = 114$ or $Z = 120$. Under this aspect prediction of different shell closures in different models may be regarded rather as a feature of "fine tuning" of the models than as a principle disagreement. More detailed knowledge of properties and structure of heaviest nuclei is thus undoubtedly decisive for improvements of theoretical models. Therefore such investigations will become in future even more important than synthesis of new elements.

The experimental success in the synthesis of isotopes up to $Z - 118$, and the recent and still ongoing theoretical attempts to use effective nucleon-nucleon interactions may be suited to change the point of view of SHN research. While so far the hunt for the spherical proton and neutron shells was in the foreground, it should be stressed out more in future that nucleon shells and enhanced stability are the result of a delicate balance between the Coulomb force and nuclear forces. Since SHN are unstable against prompt disruption using a macroscopic liquid drop picture, their existence and properties are essentially governed by subtleties of the nuclear forces. Thus they represent an ideal laboratory to study the latter. The resulting vision is a common description of phenomena in the "macro-cosmos" (cosmology, neutron stars) and the "micro-cosmos" (nuclei, sub-nuclear phenomena). Such ideas are not new, they were discussed already by, e.g., Chabanat *et al.* [18, 19]. Yet, it has to be stated, that presently we are far away from such a description. Nevertheless, it is suggested as a way to go, and detailed investigations of the properties of SHN will be a valuable tool to reach that goal.

To promote scientific progress, a critical debate about theoretical models is necessary. Experimental results have not just to be discussed with respect to confirm theoretical predictions, as it is often done. Rather, they have to be regarded as firm tests of theoretical predictions, and discrepancies (e.g. in Q_α-values, excitation energies and ordering of nuclear levels, fission half-lives, etc.) clearly point to deficiencies that have to be overcome. Nevertheless, the quality of theoretical predictions is often high enough for a classification of experimental data, e.g. tentative isotope assignments on the basis of decay Q-values, tentative nuclear level and spin assignments on the basis of predicted energies, and ordering of nuclear levels. The reason is that in general a sufficiently large amount of data of sufficient quality is available to achieve an impression of the predictive power of a model. Regarding this feature, in a first step the experimental assignments seem reasonable when in line with theoretical predictions within their limits of validity, while those leading to too extreme deviations seem rather improbable.

The discussions above lead to the trivial statement that developments of experiments and theories have to go hand-in-hand and to influence each other, but always having a critical view on the results.

From the experimental side refined techniques to deliver data of high quality are required, as it does not make any sense to draw any conclusions, also concerning the validity of the theoretical predictions, from data of low quality, which are not conclusive by themselves. Unfortunately, often only "low quality" data are available. In such cases, however, it would not be wise to discard the results, but they have to be presented critically and deficiencies

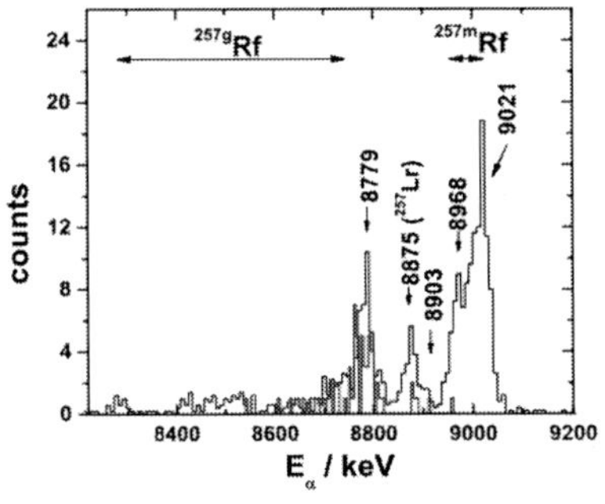

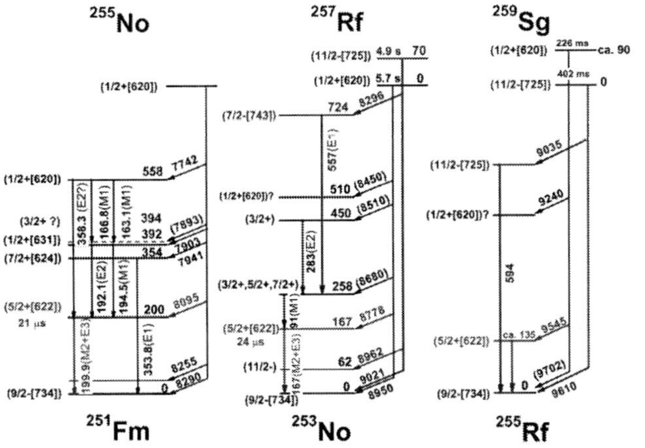

Figure 1. Left side: alpha spectra of ^{257}Rf; black line: direct production via ^{208}Pb(^{50}Ti,n)^{257}Rf; red line: indirect production via ^{208}Pb(^{58}Fe,n)^{265}Hs $\overset{\alpha}{\rightarrow}$ ^{261}Sg $\overset{\alpha}{\rightarrow}$ ^{257}Rf.

Figure 2. Right side: decay schemes of ^{255}No, ^{257}Rf, and ^{259}Sg.

and problems connected with their interpretation must not be concealed but have to be clearly pointed out. Only under this aspect they represent valuable information and challenge for improvements to obtain data of better quality. Insofar experimental work is a process of learning and gradual improvement of knowledge step by step. Sometimes numerous experiments spanning a couple of decades were performed, and despite large progress a final conclusion on decay properties and nuclear structure still has not yet been reached. An example are decay studies of ^{257}Rf which will be discussed in the following.

2. Alpha and EC decay studies of ^{257}Rf

The isotope ^{257}Rf was first synthesized by Ghiorso *et al.* [20] in the reaction ^{249}Cf(^{12}C,4n)^{257}Rf using the He-jet technique. They observed a complex α spectrum with energies in the range (8.5–9.0) MeV and reported a half-life of (4.5 ± 1.0) s. Some years later, Bemis *et al.* [21] confirmed the results. In addition, they also measured photons in coincidence with α particles and observed K X rays of energies agreeing with those expected for nobelium. It thus was the first direct Z-identification of a transactinide element. Part of the K X rays were observed in delayed coincidence with α particles; this was interpreted as due to populating an isomeric state in ^{253}No by the α decay. It was settled at an excitation energy of ≈ 300 keV and its half-life was given as $T_{1/2} = (31.1 \pm 4.1)$ μs [21]. At SHIP this isotope was produced in the reaction ^{208}Pb(^{50}Ti,n)^{257}Rf [22, 23] (direct production) and by α decay of ^{265}Hs ($\overset{\alpha}{\rightarrow}$ ^{261}Sg $\overset{\alpha}{\rightarrow}$ ^{257}Rf) [23, 24] (indirect production). Comparison of the α spectra from the direct and indirect production revealed that the α lines of highest energy (9021, 8968 keV) were missing in the indirect production (see Fig. 1). It was concluded that these α lines stem from the decay of an isomeric level, not populated by the α decay of ^{261}Sg. In addition α transitions at $E_\alpha < 8.5$ MeV could be identified, which were not accessible in the previous studies due to α background from lighter isotopes [20, 21]. In [23] also first decay schemes of $^{257\,g,257\,m}$Rf were presented. Due to interpretation of a weak α line at 8903 keV (see Fig. 1) as the ground-state to ground-state (gs-gs) transition, the $5/2^+$[622] Nilsson level was located at $E^* = 122$ keV, leading to a discrepancy overseen in [23]: the $5/2^+$[622] level was identified as an isomeric level in the lighter $N = 151$ isotones (^{249}Cf, ^{251}Fm) decaying to the ground state by highly converted $M2$ transitions (with $E3$ admixtures). Thus it could be also assumed in ^{253}No as the source of the delayed K X rays observed in [21]; its excitation energy of 122 keV, assigned in [23], however, is below the K binding energy in nobelium. Subsequently, more precise measurements clarified this problem and finally settled the excitation energy of the $5/2^+[622]$ state at 167 keV by measuring γ transitions $5/2^+[622] \rightarrow 9/2^-[734]$(gs) in delayed coincidence with α particles [25].

The long-lived isomeric state had been attributed as $11/2^-[725]$ in [23] on the basis of systematics of Nilsson levels in $N = 153$ isotones and calculated levels schemes [26], as it was the only predicted level that allowed for highly hindered internal transition into the ground state necessary to form an isomer decaying by α emission with a half-life of about 4 s.

Recent studies at SHIP [27] resulted in the observation of two α-γ-coincidences (8296 keV, 557 keV), which were attributed to the α decay of ^{257m}Rf, populating the $7/2^-[743]$ Nilsson level in ^{253}No and then decaying into the isomeric $5/2^+[622]$ level [27] as shown in Fig. 2.

Another feature of the recent study [27] was investigation of electron capture (EC) decay of ^{257}Rf by measuring delayed coincidences between conversion electrons (CE) from decay of excited levels in ^{257}Lr populated by the EC decay and α decays of ^{257}Lr. Comparison of the decay data of ^{257}Lr resulting from the direct and indirect [25] production of ^{257}Rf as well as from production of ^{257}Lr by α decay of ^{261}Db [25, 28] revealed significant differences in the energy and time distributions of the α decays of ^{257}Lr (see Fig. 3), which were interpreted as α decays from two low-lying levels in ^{257}Lr. The lower energy (8811 keV) was attributed to the decay of the isomeric state with $T_{1/2} = (0.20^{+0.16}_{-0.06})$ s and tentatively assigned as $1/2^-[521]$, the higher one (8878 keV) to the decay of the ground state having $T_{1/2} = (1.24^{+0.85}_{-0.36})$ s (obtained from the ^{261}Db-^{257}Lr-correlations) and tentatively assigned as $7/2^-[514]$ or $9/2^+[624]$ in Ref. [27] as shown in Fig. 4a.

The occurrence of two low-lying states decaying by α emission is not unexpected, as such states have been reported previously in the lighter odd-mass isotopes ^{253}Lr [30] and ^{255}Lr [31, 32]. The situation is shown in Fig. 5 where the experimental results are compared with theoretical predictions [29]. Of specific interest are here the relative positions of the $1/2^-[521]$ and the $7/2^-[514]$ states. The former originates from the $2f_{5/2}$ subshell, the latter from the $2f_{7/2}$. As pointed out in [17] (see above) the gap between the spin-orbit partners $2f_{5/2}$–$2f_{7/2}$ defines the location of the spherical proton shells with $Z = 114$ or $Z = 120$. Therefore identification and determination of their excitation energies in the lawrencium isotopes already delivers valuable data for determining the spin-orbit coupling and thus for prediction of the location of the spherical proton shell. Clearly more data are required.

Another result of the analysis of the delayed CE-α coincidences was the observation of some photons in prompt coincidence with CE. Evidently, at a low background rate, three photons fit to energies of lawrencium $K_{\alpha1}$ and $K_{\beta1}$ X rays [see Fig. 4b], supporting the EC decay of ^{257}Rf and also showing the possibility to use K X rays from EC decay for Z-identification of SHN, as recently demonstrated for EC decay of ^{258}Db [33].

3. Alpha-decay properties of ^{288}Fl

Nuclear structure investigations require clean measurements of nuclear transitions. In case of measuring α particles this feature might be difficult to fulfil sometimes, when the technique of implanting nuclei into Si-detectors, commonly used in SHN research, is applied. If α decay populates excited daughter levels that decay by internal conversion energy summing between α particles and CE will occur [34]. The result will be a rather broad distribution of α-particle

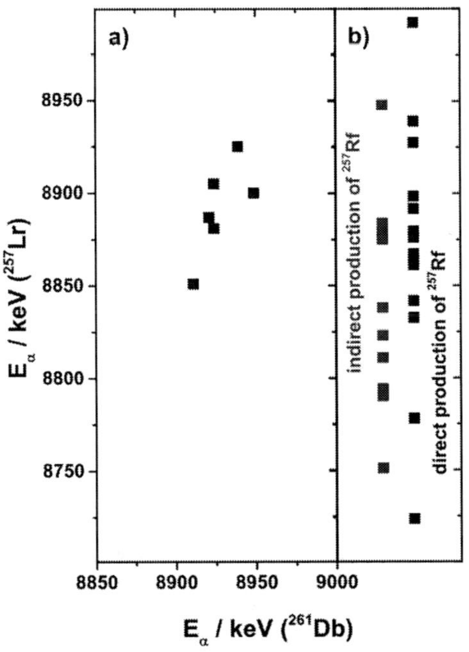

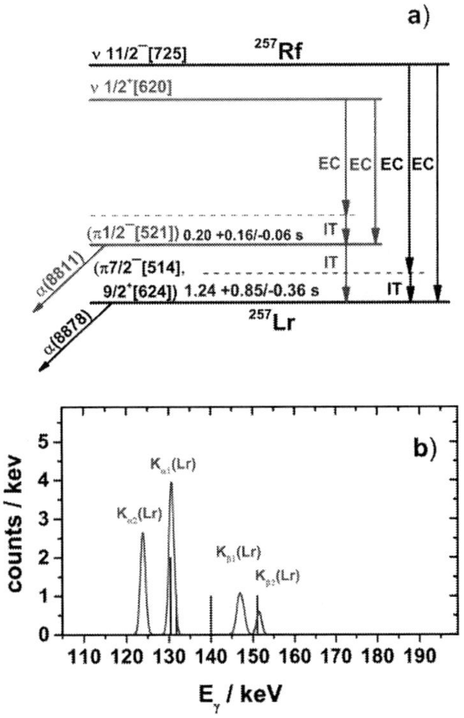

Figure 3. Left side: (a) two-dimensional plot of $\alpha(^{261}Db)$-$\alpha(^{257}Lr)$ correlations; (b) energy distributions of α decays of ^{257}Lr following EC decay of ^{257}Rf, either from direct production or indirect production.

Figure 4. Right side: (a) suggested EC decay schemes of $^{257g,257m}Rf$; (b) spectrum of photons measured in prompt coincidence with CE preceding α decays of ^{257}Lr within $\Delta t \leq 1.5$ s.

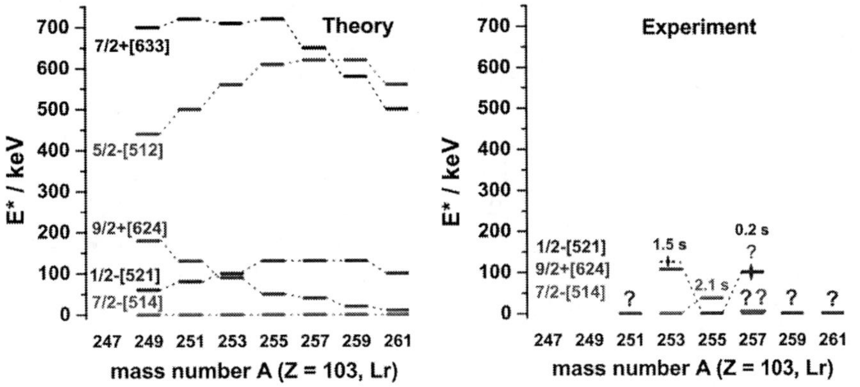

Figure 5. Predicted [29] and (tentative) experimental level schemes of odd-mass Lr isotopes.

energies, washing out peak structures, as shown in [35] for the case of ^{255}No. At low number of observed events in the extreme case just a broad distribution of α events might be observed. As a consequence one may obtain a superposition of α energy distributions of neighbouring isotopes or of α decays from different isomeric states in a specific nucleus.

In the range of $Z > 112$ it was recently shown [36, 37] that a postulated link [38] between the α-decay chain starting from $^{293}117$ (produced via $^{249}Bk(^{48}Ca,4n)^{293}117$) and from $^{289}115$

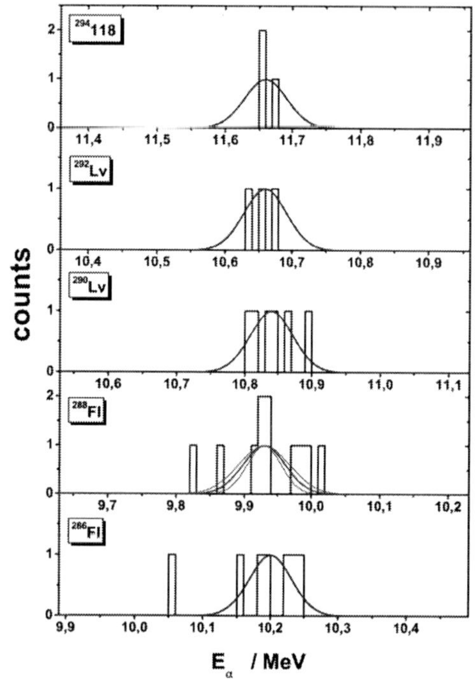

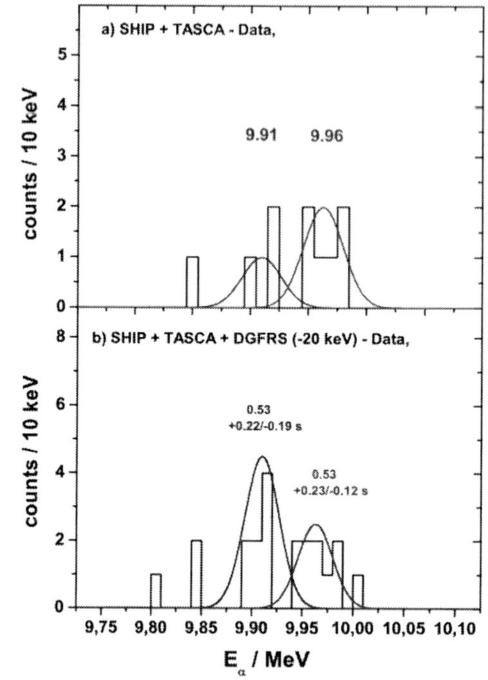

Figure 6. Left side: alpha spectra of even-even isotopes observed at the DGRFS (see text for details).

Figure 7. Right side: alpha spectrum of ^{288}Fl; (a) sum of SHIP and TASCA data; (b) sum of SHIP, TASCA, and DGFRS data (see text for details.)

(produced via ^{243}Am(^{48}Ca,2n)^{289}115) does not exist. This link was built on assigning all "short" decay chains observed in the latter reaction with levity to the decay of the same state in ^{289}115 (see [39] for details).

Another example for inconsistencies in the α spectra of isotopes in the region $Z > 112$ is the even-even nucleus ^{288}Fl. Alpha-spectra of even-even nuclei usually have a quite simple structure as the $0^+ \to 0^+$ transition between the ground-states dominates due to the highest Q-values and the absence of spin hindrance. In Fig. 6 the α spectra of all $Z > 112$ even-even nuclei observed in experiments at the Dubna gas-filled separator (DGFRS) are shown. To avoid distortions due to worse energy resolution of α particles registered as sum events in the focal plane (stop) detector and the box detector, only events with full energy release in the stop detector are considered. The red lines represent Gaussians calculated using a detector resolution of 75 keV (FWHM), the mean value given for the range (60–90) keV for the stop detector in [43]. The distributions for ^{294}118 [40, 41], ^{292,290}Lv [40–42], and also ^{286}Fl [40, 44] are justifiable with a single α line, while for ^{288}Fl [42, 43] this is questionable. To emphasize this feature also the Gaussians using detector resolutions of 60 keV and 90 keV are shown by the magenta lines. Indeed, a possible fine structure in the α decay of ^{288}Fl has also very recently mentioned by Oganessian and Utyonkov [7], however, without drawing definite conclusions. ^{288}Fl was also observed in experiments at the gas-filled separator TASCA [45] and the velocity filter SHIP [46] at GSI. The results are shown in Fig. 7a. The detector resolutions in these experiments were significantly better, 25 keV (FWHM) [45] and 24 keV (FWHM) [46]. The data also indicate the existence of two α lines. The existence of two α lines becomes even more evident when the results of all experiments are summed, as shown in Fig. 7b. (The results from DGFRS have been scaled by -20 keV to match apparent differences in the calibrations.) Mean energies and half-lives of (9.11 ± 0.01) MeV, $(0.53^{+0.22}_{-0.19})$ s and (9.96 ± 0.01) MeV, $(0.53^{+0.23}_{-0.12})$ s are obtained.

The α spectrum of ^{288}Fl presently cannot be explained satisfactorily: although the energy difference between the two lines is similar to typical differences for the 0^+ ground-state and the first 2^+ state in highly deformed heavy nuclei [47], this feature cannot explain the intensity ratio of the two lines. The intensities for decay into the 2^+ level are typically only 10–20% of those of decays into the 0^+ ground state, and the $2^+ \rightarrow 0^+$ $E2$-transitions are highly converted. Energy summing between α particles and CE will shift the energies towards the energy of transition into the ground state and thus the lines would not be resolved. Another possibility could be the existence of a very low lying 1^- vibrational state in the daughter nucleus ^{284}Cn, as existing, e.g., in ^{222}Ra ($E^* = 242.13$ keV) and ^{224}Ra ($E^* = 215.99$ keV) [47]; those decay by weakly converted $E1$ transitions into the ground state [47]; however, α-decay branchings into these levels are very low (< 0.015).

As ^{288}Fl is one member of a "mesh of isotopes" a simple assignment to neighbouring odd-mass nuclei seems not possible as it is followed by a short-lived fission activity. As the daughter products of the neighbouring isotopes (^{289}Fl, ^{287}Fl) are long-lived α emitters [7] this would imply a shorter fission half-life of an odd-mass nucleus compared to a neighbouring even-even nucleus, which contradicts our knowledge of fission properties of SHN. A third possibility would be existence of an isomeric state, but with respect to the existing data (including the same half-lives within the error bars) such an assumption is not well based. To summarize: one has to state a strange behavior of the α spectrum of ^{288}Fl, but no simple explanation for it can be given so far. Careful studies are required to draw final conclusions, and also to exclude that this behavior is not just an artifact, caused by a "random" scattering of the data and a rather low (≈ 20 events) number of observed α decays.

4. Conclusion

It has been shown for the case of ^{257}Rf-^{257}Lr that valuable nuclear structure information can be extracted by careful and critical analysis of the data and considering the complete physical enviroment even on the basis of a quite low number of observed events. Certainly they do not deliver a complete picture, but mark the road for further investigations. The case of ^{288}Fl, on the other hand, shows that a blinkered data analysis (see e.g. [48]) can lead to missing essential features. Indeed, the situation is sometimes ambivalent, as often "unexpected" results later turned out to be artifacts (see, e.g., discussion on ^{255}Rf in [23, 30]). These facts, however, simply corroborate the demand for high-quality data free of ambiguities before drawing conclusions on identification, decay properties, or nuclear structure of superheavy nuclei.

References

[1] M. Göppert-Mayer, Phys. Rev. **74**, 235 (1948).
[2] O. Haxel *et al.*, Phys. Rev. **75**, 1769 (1949).
[3] A. Sobiczewski *et al.*, Phys. Lett. **22**, 500 (1966).
[4] H. Meldner, Arkiv för fysik **36**, 593 (1967).
[5] S.G. Nilsson *et al.*, Phys. Lett. **28B**, 458 (1969).
[6] S.G. Nilsson *et al.*, Nucl. Phys. **A115**, 545 (1968).
[7] Yu.Ts. Oganessian and V.K. Utyonkov, Nucl. Phys. **A944**, 62 (2015).
[8] R.C. Barber *et al.*, Pure Appl. Chem. **83**, 1485 (2011).
[9] P.J. Karol *et al.*, Pure Appl. Chem. **88**, 139 (2016).

[10] P.J. Karol *et al.*, Pure Appl. Chem. **88**, 155 (2016).
[11] C.F. von Weizsäcker, Z. Phys. **96**, 461 (1935).
[12] V.M. Strutinsky, Nucl. Phys. **A95**, 442 (1967).
[13] R. Smolanczuk and A. Sobiczewski, Proc. EPS Conf. *Low Energy Nuclear Dynamics*, St. Petersburg 1995, eds. Yu.Ts. Oganessian *et al.* (World Scientific, Singapore, New Jersey, London, Hong Kong, 1995), p. 313.
[14] P. Möller *et al.*, At. Data and Nucl. Data Tab. **59**, 185 (1995).
[15] K. Rutz *et al.*, Proc. of the Second Int. Conf. *Fission and Properties of Neutron-Rich Nuclei*, St. Andrews 1999, eds. J.H. Hamilton *et al.* (World Scientific, Singapore, New Jersey, London, Hong Kong, 2000), p. 449.
[16] M. Bender *et al.*, Rev. Mod. Phys. **75**, 121 (2003).
[17] M. Bender *et al.*, Phys. Rev. C **60**, 034304 (1999).
[18] E. Chabanat *et al.*, Nucl. Phys. **A627**, 710 (1997).
[19] E. Chabanat *et al.*, Nucl. Phys. **A635**, 231 (1998).
[20] A. Ghiorso *et al.*, Phys. Rev. Lett. **22**, 1317 (1969).
[21] C.E. Bemis *et al.*, Phys. Rev. Lett **31**, 647 (1973).
[22] F.P. Heßberger *et al.*, Z. Phys. **A322**, 557 (1985).
[23] F.P. Heßberger *et al.*, Z. Phys. **A359**, 415 (1997).
[24] S. Hofmann *et al.*, Z. Phys. **A350**, 277 (1995).
[25] B. Streicher *et al.*, Eur. Phys. J. **A45**, 275 (2010).
[26] S. Cwiok *et al.*, Nucl. Phys. A **573**, 356 (1997).
[27] F.P. Heßberger *et al.*, Eur. Phys. J. **A52**, 192 (2016).
[28] Z. Gan *et al.*, Eur. Phys. J. **A20**, 385 (2004).
[29] A. Parkhomenko and A. Sobiczewski, Acta Phys. Pol. B **35**, 2447 (2004).
[30] F.P. Heßberger *et al.*, Eur. Phys. J. **A12**, 57 (2001).
[31] A. Chatillon *et al.*, Eur. Phys. J. **A30**, 397 (2006).
[32] S. Antalic *et al.*, Eur. Phys. J. **A38**, 219 (2008).
[33] F.P. Heßberger *et al.*, GSI Scientific Report 2014, MU-NUSTAR-SHE-P-09, 177 (2015).
[34] F.P. Heßberger *et al.*, Nucl. Instr. Meth. **A274**, 522 (1989).
[35] M. Asai, F.P. Heßberger, A. Lopez-Martens, Nucl. Phys. **A944**, 308 (2015).
[36] U. Forsberg *et al.*, Nucl. Phys. **A953**, 117 (2016).
[37] U. Forsberg *et al.*, Phys. Lett. **B760**, 293 (2016).
[38] Yu.Ts. Oganessian *et al.*, Phys. Rev. C **87**, 054621 (2013).
[39] U. Forsberg, PhD Thesis, Lund University, ISBN 978-91-7623-813-4 (2016).
[40] Yu.Ts. Oganessian *et al.*, Phys. Rev. C **74**, 044602 (2006).
[41] Yu.Ts. Oganessian *et al.*, Phys. Rev. Lett. **109**, 162501 (2012).
[42] Yu.Ts. Oganessian *et al.*, Preprint E7-2004-160 (2004).
[43] Yu.Ts. Oganessian *et al.*, Phys. Rev. C **69**, 054607 (2004).
[44] Yu.Ts. Oganessian *et al.*, Phys. Rev. C **70**, 064609 (2004).
[45] J.M. Gates *et al.*, Phys. Rev. C **83**, 054018 (2011).
[46] S. Hofmann *et al.*, Eur. Phys. J. **A48**, 62 (2012).
[47] R.B. Firestone *et al.*, Table of Isotopes, 8th Edition (Wiley & Sons, New York, 1996).
[48] S. Hofmann *et al.*, Eur. Phys. J. **A52**, 180 (2016).

EPJ Web of Conferences **131**, 03001 (2016) DOI: 10.1051/epjconf/201613103001
Nobel Symposium NS160 – Chemistry and Physics of Heavy and Superheavy Elements

Stability and synthesis of superheavy elements: Fighting the battle against fission – example of ^{254}No

A. Lopez-Martens[1,a], *G.* Henning[1,2,b], *T.L.* Khoo[2], *D.* Seweryniak[2], *M.* Alcorta[2], *M.* Asai[3], *B.B.* Back[2], *P.* Bertone[2], *D.* Boilley[4], *M.P.* Carpenter[2], *C.J.* Chiara[2,5], *P.* Chowdhury[6], *B.* Gall[7], *P.T.* Greenlees[8], *G.* Gurdal[6], *K.* Hauschild[1], *A.* Heinz[9], *C.R.* Hoffman[2], *R.V.F.* Janssens[2], *A.V.* Karpov[10], *B.P.* Kay[2], *F.G.* Kondev[2], *S.* Lakshmi[6], *T.* Lauristen[2], *C.J.* Lister[6], *E.A.* McCutchan[2], *C.* Nair[2], *J.* Piot[7,c], *D.* Potterveld[2], *P.* Reiter[11], *N.* Rowley[12], *A.M.* Rogers[2], and *S.* Zhu[2]

[1] CSNSM, IN2P3/CNRS and Université Paris Sud, France
[2] Argonne National Laboratory, USA
[3] Japan Atomic Energy Agency, Japan
[4] GANIL, CEA and IN2P3/CNRS and Normandie Université, France
[5] University of Maryland, USA
[6] Univeristy of Massachusetts Lowell, USA
[7] IPHC, IN2P3/CNRS and Université Louis Pasteur, France
[8] University of Jyväskylä, Finland
[9] Chalmers Tekniska Högskola, Sweden
[10] Flerov Laboratory of Nuclear Reactions, JINR, Russia
[11] Universität zu Köln, Germany
[12] IPN Orsay, IN2P3/CNRS and Université Paris Sud, France

Abstract. Superheavy nuclei exist solely due to quantum shell effects, which create a pocket in the potential-energy surface of the nucleus, thus providing a barrier against spontaneous fission. Determining the height of the fission barrier and its angular-momentum dependence is important to quantify the role that microscopic shell corrections play in enhancing and extending the limits of nuclear stability. In this talk, the first measurement of a fission barrier in the very heavy nucleus ^{254}No will be presented.

1. Introduction

The nucleus of interest, ^{254}No, is situated at the very top of the nuclear chart in the region of transfermium nuclei. These nuclei are characterized by a decreasing, and for the heaviest nuclei, a vanishing liquid-drop fission barrier. If ^{254}No was just as a liquid drop of incompressible fluid consisting of 102 positively-charged protons and 152 neutrons, its fission barrier would be of the order of 0.9 MeV [1]. This means that ^{254}No would fission with a lifetime of less than a nanosecond. Experimentally, however, we know that the half life of ^{254}No is 51 s and that its main mode of decay is α decay. Since its production cross section

[a] e-mail: `araceli.lopez-martens@csnsm.in2p3.fr`
[b] Current address: IPHC, Strasbourg
[c] Current address: GANIL, Caen

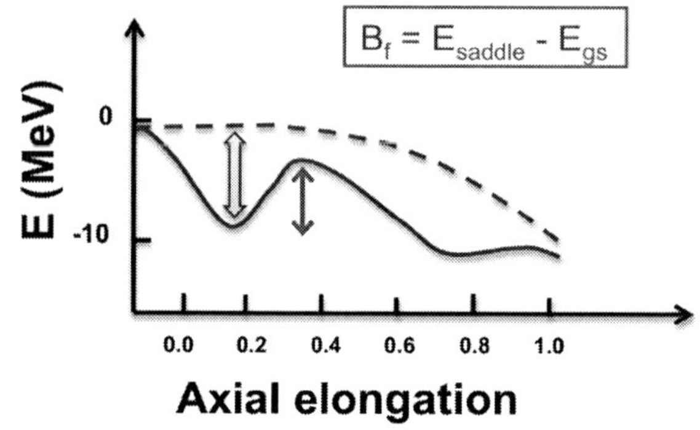

Figure 1. Potential energy of the nucleus as a function of axial deformation. The dashed line shows the liquid drop prediction for ^{264}Hs (Z=108). The solid line shows the effect of shell corrections on the potential energy.

is one of the highest in the mass region, it is also a nucleus that has been extensively studied via prompt spectroscopy [2–5]. These studies have revealed that ^{254}No is deformed and emits a spectrum of photons and conversion electrons, typical of regular rotational motion. ^{254}No is in fact very robust against fission and has been observed up to spin 24 $\hbar$.

The reason why ^{254}No has an enhanced stability against fission is due to the fermionic nature of the neutrons and protons in the nucleus, which occupy discrete energy levels grouped into shells. It is the regions of low level density, or gaps in the single-particle energy spectra, which give additional binding to the nucleus: the ground-state is lowered with respect to the liquid drop energy and a barrier against fission appears. As illustrated in Fig. 1, the height of the barrier B_f is defined as the energy difference between the ground state and the highest point on the path to fission: the saddle point. This barrier is what enables the existence of the heaviest nuclei. This makes it a valuable quantity to study – though technically it is not an observable one can measure directly.

Different theoretical models predict different barrier heights. So measuring B_f also provides a test of theories. In a macroscopic-microscopic approach where the liquid drop part of the potential is calculated using the Finite Range Liquid Drop Model (FRLDM) and the folded Yukawa potential is used for the microscopic shell correction, the barrier for ^{254}No is predicted to be 6.7 MeV [6]. In a fully microscopic approach, J.L. Egido and L.M. Robledo used the Gogny effective interaction and the Hartree Fock Bogoliubov approximation to compute the potential energy surface of the nucleus as a function of axial deformation for different values of total angular momentum [7]. A barrier of 8.66 MeV is obtained. What this model can also tell us is how the barrier evolves with spin. In accordance with experimental observations, the self-consistent calculations predict that the barrier will survive to quite high spin.

How the barrier behaves as a function of spin, in other words the moment of inertia of the saddle point configuration, is important to study because if the barrier were to disappear at a finite spin, the heaviest elements could not be synthesized. Indeed, one has to keep in mind that these heavy nuclei are produced in fusion-evaporation reactions, which bring in a substantial amount of angular momentum. The cross section to produce a given evaporation residue depends on three terms: (i) The probability that the colliding projectile and target will penetrate inside the entrance channel potential barrier and reach the contact point, (ii) the probability that this capture state evolves into a fused compound nucleus, and finally, (iii) the probability that the excited compound nucleus evaporates neutrons and decays to the ground state of the nucleus of interest while surviving fission every step of the way. This survival

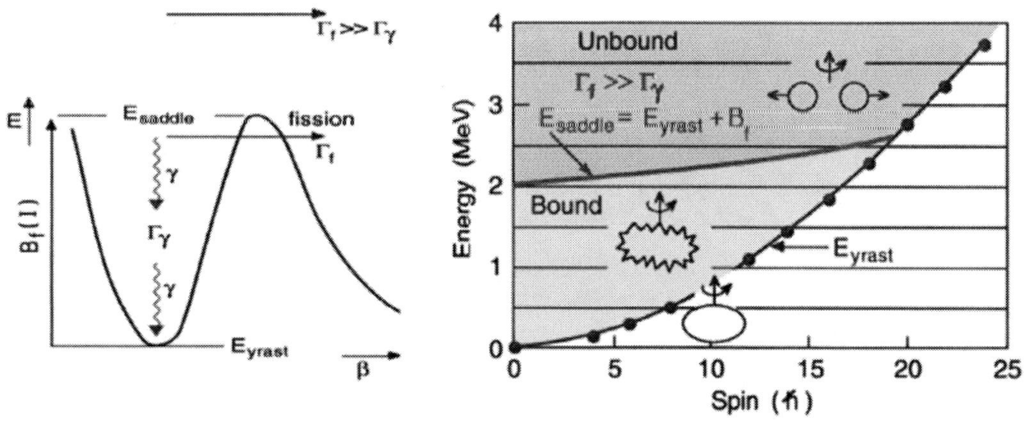

Figure 2. Left: schematic plot of the potential energy of the nucleus as a function of axial deformation illustrating how γ decay dominates below the saddle-point energy while fission takes over at energies above the barrier. Right: cartoon showing how the competition between γ emission and fission splits the excitation energy vs. spin plane into two regions.

probability reflects the competition between neutron emission, fission and γ decay and one of the key parameters is the fission barrier, which enters into the fission width Γ_f. In fact, the high fission barriers predicted around $Z = 114$ and $N = 184$ are thought to be responsible for reversing (at $Z = 112$) the trend of ever decreasing evaporation-residue cross sections as a function of atomic number Z, resulting in a maximum effect at $Z = 114 - 115$ [8]. One should also note that factors, such as dissipation, may also contribute to reduce Γ_f [9].

Experimentally, no fission barriers are known above Bk ($Z = 99$). The probability to fission has been measured in direct reactions on actinide targets [10, 11], by electromagnetic excitation of secondary beams following fragmentation reactions [12] and in a few cases the electron capture (EC) delayed fission branch was used to determine the height of the fission barrier [13]. Unfortunately, these methods are not easily applicable to the transfermium region. The two first methods cannot be used because of lack of suitable targets and primary beams. And the third method is limited to the cases where EC populates states around and above the fission barrier in the daughter nucleus.

2. Entry distribution method

A different approach for determining the fission barrier was proposed by P. Reiter more than ten years ago [14]. It is called the entry distribution method and has the advantage that it is the only way to study the spin dependence of the barrier. As is shown in Fig. 2, it relies on the fact that in the cases where the fission barrier lies below the neutron separation energy, fission is the dominant process above the barrier. This means that the excitation energy vs. spin plane can be divided into two regions above the yrast line: a region where the nucleus is bound and γ decays to the ground state and a region above the saddle point, where the nucleus fissions. In order to probe the saddle-point line separating the two regions, the idea is to measure the distribution of total energy and total spin of the ^{254}No nucleus right after neutron evaporation. As more energy is put into the system in the fusion reaction, one should be able to observe the distribution move up in energy and spin and be eventually cut off by fission. The distribution falls to half its maximum value at the energy at which the fission and γ-decay modes compete with equal probability. This energy is called the 1/2 energy and differs from the saddle point energy by a small factor $\Delta(I)$. A recent analysis has shown that the model dependency of Δ is small [15].

The entry distribution of ^{254}No was measured by Reiter *et al.* using the reaction ^{208}Pb(^{48}Ca,2n) at two different bombarding energies: 215 and 219 MeV. At the lowest energy, it was found that the distribution was limited by the maximum energy available in the reaction. At the higher bombarding energy, no cut off of the entry distribution was observed, and so only a lower limit of 5 MeV for the fission barrier at 12 $\hbar$ could be extracted [14]. The experiment, however, showed that the bulk of the cross section to produce ^{254}No occurs at high spin.

3. Experiment

The experiment was repeated in order to probe the entry distribution at higher beam energy: 219 and 223 MeV. The 10 pnA ^{48}Ca was delivered by the Argonne Tandem Linac Accelerator System. ^{254}No was produced at the target position of the Fragment Mass Analyzer (FMA) [16] which separated the evaporation residues from the other unwanted reaction products. The recoiling ^{254}No ions were identified at the focal plane of the FMA while γ rays emitted by the excited ^{254}No were detected by GAMMASPHERE [17], which was used in its calorimetric mode. In such a mode, the anti-Compton BGO shields surrounding the Ge detectors are also used to detect photons allowing for an efficiency of $\sim$78% to be reached [18]. The response function of GAMMASPHERE was measured with radioactive sources so that the distribution of emitted photon energies and multiplicities could be reconstituted from the measured total energies and number of firing detector modules. This is achieved via an iterative unfolding procedure [19, 20] starting with an initial guess at the emitted distribution. This trial distribution is then passed through the measured response function and compared to the measured distribution. In doing so, an improved guess at the true distribution can be made and the whole procedure can be repeated. To get to the total energy vs. spin distribution, one has to know a little bit more about the nucleus of interest, in particular the contributions of conversion electrons to the total spin and energy. These parameters can be obtained and/or extracted from the known level scheme or from experimental spectra. In the case of ^{254}No, this was made possible by extensive prompt and decay spectroscopy performed by many groups worldwide [2, 3, 5, 21–24]. The average spin and energy removed by conversion electrons obtained from the known level decay scheme of ^{254}No and tabulated in reference [15] is in excellent agreement with what was measured by Butler *et al.* [4] before the level scheme was known, giving confidence in our understanding of the average decay cascade in ^{254}No. The entry distributions obtained at 219 and 223 MeV bombarding energies are shown in Fig. 3. The crosses indicate the points where the distributions fall to half of their maximum value. What is immediately noticeable is that even though more spin and energy (4 MeV) are put into the system between the left and right panels of Fig. 3, the crosses between 12 and 20 $\hbar$ remain at constant excitation energy, E^*. This saturation is attributed to the effect of fission depleting the γ-emitting states. The saddle point energies extracted from the saturated half-maximum points yield a moment of inertia of the saddle of 125(60) MeV^{-1}. This is in good agreement with what can be extracted from the self-consistent potential energy surfaces of Ref. [7]. The value of the barrier height at spin 15 $\hbar$ is found to be $B_f = 6.0(5)$ MeV. Extrapolating down to zero spin using the saddle point moment of inertia, one finds $B_f(0) = 6.6(9)$ MeV. This value of the excitation energy above which fission dominates the decay process is in good agreement with the predictions of Möller *et al.* [6]. An interesting thing to note is that a 6.6 MeV barrier is larger than what one would naively obtain by adding the liquid drop barrier height and the amplitude of the microscopic shell correction in ^{254}No [25]: $B_{f_0} = 0.9 + 3.97 = 4.9$ MeV. This implies a positive shell effect of more than 1.5 MeV at the saddle point. Another point to mention is that the measured average spin distribution of

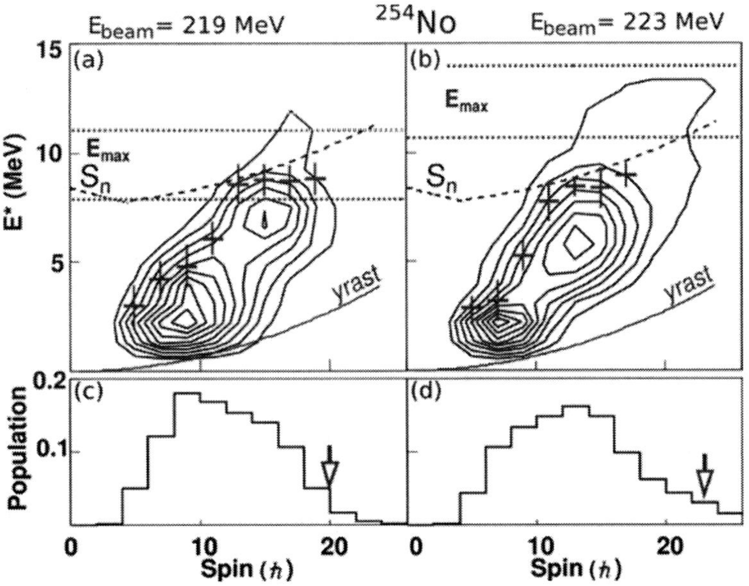

Figure 3. Plot of the entry distribution of ^{254}No measured at 219 (left) and 223 (right) MeV bombarding energies and their corresponding projections onto the spin axis. The yrast line and neutron separation energy are shown together with the interval of maximal energy available in the reaction. The crosses represent the half-maximum energies at a given spin value. The points at high spin have been discarded as they originate from events resulting from random summing. The red arrows point to the spins at which the distribution falls by one tenth of its maximum value.

the surviving ^{254}No is shifted towards higher spins by nearly $\sim 5\ \hbar$ compared to calculations (e.g., see Fig. 6 of Ref. [26]).

4. Conclusions and perspectives

This is the first time that the excitation energy at which fission starts to dominate the decay process has been determined in a transfermium nucleus. The results obtained and extrapolated to spin zero are in agreement with macroscopic-microscopic calculations of the barrier height. This may not be so surprising since these calculations are the only ones which reproduce the single-particle energy spectra in this mass region [27].

Extending the method to other nuclei requires the knowledge of their average decay schemes. This should be made possible soon by the coupling of a new generation γ-ray spectrometers, such as AGATA (Advanced GAmma Tracking Array) [28], GRETINA (Gamma Ray Energy Tracking In beam Nuclear Array) [29] and also digital GAMMASPHERE to powerful spectrometers such as VAMOS (VAriable MOde Spectrometer) gas-filled [30] and AGFA (Argonne Gas Fille Analyzer) [31]. Decay schemes of nuclei around ^{254}No will also be established and/or improved by decay spectroscopy studies performed with increased beam intensities and more efficient decay stations. These technical developments should allow the investigation of the average fission barrier height in odd systems and also the magnitude of the barrier associated to particular configurations of the nucleus (e.g. high-K states) for which the hindrance towards fission is expected to be large.

Finally, measuring the entry distribution of evaporation residues is useful in itself as it can be used as a starting point to Monte Carlo simulations of the decay flow to the ground state and compared to experiment in order to get a handle on level densities and γ strength functions in super heavy nuclei.

References

[1] A.J. Sierk, Phys. Rev. C **33**, 2039 (1986).
[2] M. Leino *et al.*, Eur. Phys. J. A **6**, 63 (1999).
[3] P. Reiter *et al.*, Phys. Rev. Lett. **82**, 509 (1999).
[4] P.A. Butler *et al.* Phys. Rev. Lett. **89**, 202501(2002).
[5] S. Eeckhaudt *et al.*, Eur. Phys. J. A **26**, 227 (2005).
[6] P. Möller *et al.*, Phys. Rev. C **79**, 064304 (2009).
[7] J.L. Egido and L.M. Robledo, Phys. Rev. Lett. **85**, 1198 (2000).
[8] Y. Oganessian, Nucl. Phys. News Vol. **23**, 15 (2013).
[9] R. Yanez *et al.*, Phys. Rev. Lett. **112**, 152702 (2014).
[10] B.B. Back *et al.*, Phys. Rev. C **9**, 1924 (1974).
[11] B.B. Back *et al.*, Phys. Rev. C **10**, 1948 (1974).
[12] A. Grewe *et al.*, Nucl. Phys. **A614**, 400 (1997).
[13] D. Habs *et al.*, Z. Phys. **A285**, 53 (1978).
[14] P. Reiter *et al.*, Phys. Rev. Lett. **84**, 3542 (2000).
[15] G. Henning *et al.*, Phys. Rev. Lett. **113**, 262505 (2014).
[16] C.N. Davids *et al.*, Nucl. Instr. Meth. **B70**, 358 (1992).
[17] I.Y. Lee, Nucl. Phys. **A 520**, 641c (1990).
[18] T. Lauritsen *et al.*, Phys. Rev. C **75**, 064309 (2007).
[19] M. Jääskeläinen *et al.*, Nucl. Instr. Meth. **204**, 385 (1983).
[20] P. Benet, PhD thesis, Université Louis Pasteur, Strasbourg (1988).
[21] R.-D. Herzberg *et al.*, Nature **442**, 896 (2006).
[22] S. Tandel *et al.*, Phys. Rev. Lett. **97**, 082502 (2006).
[23] F.P. Heßberger *et al.*, Eur. Phys. J. A **43**, 55 (2010).
[24] R.M. Clark *et al.*, Phys. Lett. B **690**, 19 (2010).
[25] P. Möller, J. Nix, W. Myers, W. Swiatecki, At. Data Nucl. Data Tables **59**, 185 (1995).
[26] V.I. Zagrebaev *et al.*, Phys. Rev. C **65**, 014607 (2001).
[27] D. Seweryniak *et al.* Nucl. Phys. **A834**, 357c (2010).
[28] S. Akkoyun, *et al.*, Nucl. Instr. Meth. **A668**, 26 (2012).
[29] S. Paschalis, *et al.*, Nucl. Instr. Meth. **A709**, 44 (2013).
[30] Ch. Schmitt *et al.*, Nucl. Instr. Meth. **A621**, 558 (2010).
[31] B.B. Back *et al.*, Proposal for the Argonne Gas Filled Analyzer - AGFA (2013).

EPJ Web of Conferences **131**, 03002 (2016) DOI: 10.1051/epjconf/201613103002
Nobel Symposium NS160 – Chemistry and Physics of Heavy and Superheavy Elements

The limits of the nuclear chart set by fission and alpha decay

Peter Möller[a]

P. Moller Scientific Computing and Graphics, Inc., PO Box 1440, Los Alamos, NM 87544, USA

Abstract. I will review how our picture of heavy-element nuclear structure has evolved through remarkably simple ideas and related models. It is well known that the Bethe-Weizsäcker semi-empirical mass model had an important role in unraveling radioactive decay and element transmutation in the heavy-element region in the 1930s. A remarkable aspect is that this model could immediately after the discovery of fission be generalized to explain this phenomenon through the consideration of deformation of a charged liquid drop. Bethe and Bacher already raised the possibility that shell structure (by them calculated in terms of a single-particle oscillator potential) could give rise to noticeable deviations between results of the macroscopic mass model and experiment but limited data prevented firm conclusions. In the 1950s the single-particle models took a realistic form and also included deformation. The possibility of the existence of a relatively stable "island" of superheavy elements was raised already then. But it was not until the work by Strutinsky in the mid 1960s that a quantitative model for the nuclear potential-energy emerged in the form of the macroscopic-microscopic model. Although new elements have been discovered at an almost steady pace since 1940, theory indicates that we are close to the end of this era: repulsive Coulomb effects will set the limit of observable elements to near $Z = 120$.

1. Introduction

Most insights into and understanding of nuclear properties have historically and are currently obtained in terms of surprisingly simple models. In systems which depend mainly on electron behaviour the system properties can be obtained from solving the Schrödinger equation with realistic, known Coulomb potentials. The nuclear system with its up to a few hundred nucleons leads to a many-body Schrödinger equation which cannot be solved, because the potential is based on the much more complicated nuclear forces. Therefore other types of models are used. They are often referred to in a condescending manner as phenomenological, but, actually all "models" are phenomenological. We will here focus on the remarkable insights into nuclear properties that have been obtained by the "liquid-drop model" and the single-particle model and a combination of those that still today are providing much insight about the properties of superheavy nuclei.

[a] e-mail: `mollerinla@gmail.com`

2. The liquid-drop model and limits to nuclear stability

The liquid-drop model has its origin in the semi-empirical mass model, usually attributed to von Weizsäcker [1] and Bethe and Bacher [2]. The nuclear mass is

$$M(Z, N, \text{shape}) = M_{\text{H}}Z + M_{\text{n}}N - B(N, Z). \tag{1}$$

The first two terms are obviously giving the mass of all the nucleons and electrons (hydrogen mass is the sum of the a proton and an electron mass). The last term is the binding energy and accounts for the mass deficit of a nucleus relative to the sum of the masses of the separated particles. In the semi-empirical mass model it is given by

$$B(N, Z) = +a_{\text{v}}A - a_{\text{s}}A^{2/3} - a_{\text{C}}\frac{Z^2}{A^{1/3}} - a_I\frac{(N-Z)^2}{A} - \delta(A). \tag{2}$$

This simple expression gives the nuclear binding energy, which for a heavy nucleus is roughly 1600 MeV (8 MeV per nucleon) to an accuracy of one percent. The terms represent in order volume energy, surface energy, Coulomb energy, symmetry energy and a pairing correction. The notation Z, N, and A stand for proton number, neutron number and total nucleon number, respectively. The model was highly useful in interpreting the decay chains following element transmutations. However, in some experiments in which uranium was bombarded with neutrons a confusing number of radioactive decay products, which were difficult to reconcile with properties given by the semi-empirical mass model, were observed. This all was explained when Hahn and Strassmann [4] identified barium in the products following neutron irradiation of uranium and Meitner and Frisch [5] suggested one could think of the nucleus as a deformable charged liquid drop that had split into two smaller nuclei of about equal size in fission. It then turned out that the simple semi-empirical mass model could be generalized to describe this division and, just a few months later, Bohr and Wheeler suggested the following generalization [6]

$$B(N, Z, \alpha) = +a_{\text{v}}A - a_{\text{s}}A^{2/3}B_{\text{s}}(\alpha) - a_{\text{C}}\frac{Z^2}{A^{1/3}}B_{\text{C}}(\alpha) - a_I\frac{(N-Z)^2}{A} - \delta(A). \tag{3}$$

Only the Coulomb and surface energies depend on deformation. $B_{\text{s}}(\alpha)$ and $B_{\text{C}}(\alpha)$ are the ratios of the surface and Coulomb energies at deformation α to that for spherical shape. With this simple model one can show that stability with respect to fission decreases with increasing proton number and is completely lost at proton numbers somewhat above $Z = 100$. Bohr and Wheeler showed this by deriving Taylor expansions for the Coulomb and surface shape dependencies. If the surface shape as function of deformation (to lowest order in α) is given by

$$r(\theta) = R_0\left[1 + \alpha_2 P_2(\cos\theta)\right]$$

then surface and Coulomb energy to lowest order Taylor expansion are given by

$$E_{\text{s}} = E_{\text{s}}^0\left(1 + \frac{2}{5}{\alpha_2}^2\right) \qquad \text{and} \qquad E_{\text{C}} = E_{\text{C}}^0\left(1 - \frac{1}{5}{\alpha_2}^2\right).$$

The energy at deformation α_2 relative to spherical shape can be written as

$$E_{\text{def}}(\alpha_2) = E_{\text{C}}(\alpha_2) + E_{\text{s}}(\alpha_2) - (E_{\text{C}}^0 + E_{\text{s}}^0).$$

If E_{def} is negative then the spherical system has no barrier with respect to fission, i.e.

$$E_{\text{def}}(\alpha_2) = \frac{2}{5}{\alpha_2}^2E_{\text{s}}^0 - \frac{1}{5}{\alpha_2}^2E_{\text{C}}^0 < 0$$

Table 1. Calculated values of the fissility parameter x for some sample nuclei.

Z	50	82	92	100	114	125	130
A	124	208	238	252	298	328	335
x	0.402	0.645	0.709	0.792	0.870	0.950	1.006

This condition is usually written as

$$1 < \frac{E_{\rm C}^0}{2E_{\rm s}^0} = x.$$

With common values of the constants ($E_{\rm s}^0 = 17.80A^{2/3}$ and $E_{\rm c}^0 = 0.7103Z^2/A^{1/3}$), the fissility parameter x is then obtained as

$$x = \frac{Z^2}{50.13A}.$$

In Table 1 we calculate the fissility x for a few nuclides throughout the periodic system. Clearly there are no fission barriers at $Z = 130$. However, already slightly above $x = 0.80$ the barrier in the liquid-drop model becomes so low that one has to expect that the spontaneous-fission half-lives become sufficiently short to make such a system unobservable. It is usually taken to be around $Z = 104$. Thus, already in 1939 there were credible theories that showed that that there is an end to the nuclear chart and that the end is due to fission. These insights are still pretty realistic today, but have been refined because methods have been developed to model microscopic "quantum" effects and how they can be combined with the liquid drop model to allow more accurate descriptions of nuclear masses, fission barriers, and many other nuclear properties. We will now discuss this macroscopic-microscopic model in which microscopic effects are obtained from a single-particle model.

3. Macroscopic-microscopic model and single-particle shell effects and their benchmarks

Already Bethe and Bacher [2] considered the possibility that microscopic effects might give rise to deviations from the semi-empirical mass model. They noted that in an oscillator potential there are large gaps in the corresponding single-particle level spectrum and investigated if there were unusually large deviations between their theory and measured masses at $Z = 20$, $N = 20$, namely ^{40}Ca. They found none, but concluded that the masses were not measured sufficiently accurately to observe the effect they expected. In Fig. 1 we compare masses calculated in a modern liquid-drop model [3] to measured masses [13]. We observe that there are no particularly large deviations at these nucleon numbers. But for heavier nuclei there are large deviations at the by now well-known magic numbers. Around 1950 a modern single-particle model was developed [14–16]. It showed large gaps at the nucleon numbers where there were large deviations between the semi-empirical mass model and measured masses and it also agreed with some observed ground-state spins. But it soon became clear that for large regions of the nuclear chart spins were not well explained. This observation together with the experimental observations that nuclei in these regions had large intrinsic quadrupole moments led to the suggestion that these nuclei were not spherical but deformed [17–19]. Motivated by these developments, Nilsson in 1955 generalized the spherical single-particle model to describe levels in deformed nuclei for any prescribed spheroidal shape. A large part of the work was to devise a model specification that could be implemented on computers of the day. Nilsson solved this by introducing coordinate transformations that resulted in matrices of low dimensionality about ten or so [20].

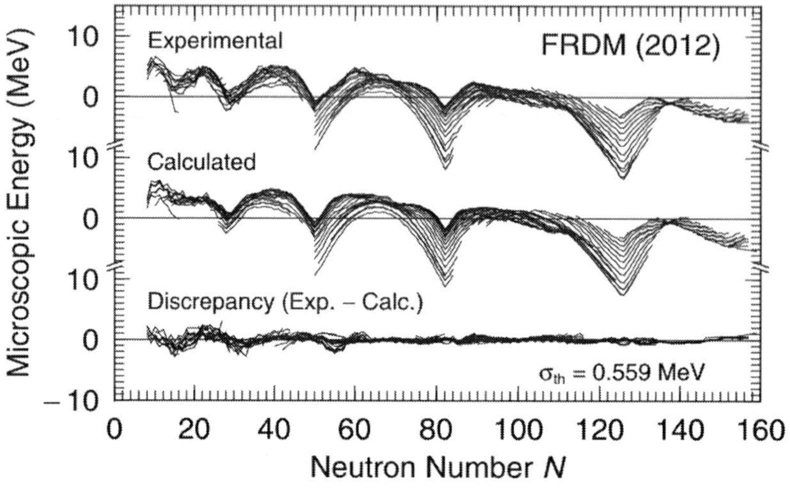

Figure 1. Top panel: difference between experimental masses and masses calculated in a modern macroscopic liquid-drop model. These are also called "experimental" shell corrections, but do depend somewhat on the theoretical macroscopic model used. Middle panel: microscopic corrections *calculated* by use of Strutinsky's method. Bottom panel: difference between the top and middle panel. This difference is also equivalent to the difference between experimental and calculated masses. Lines connect isotopes, full details are found in Ref. [3].

The model results for ground-state spins and low-lying level structure were shown to agree well with experimental observations [21]. To calculate the level order it was necessary to know for what deformation or nuclear shape to calculate the levels. This was done by calculating what nuclear shape has the lowest energy. The potential energy for a particular shape was calculated by summing the energy of the levels that are occupied at that shape. This was done for a succession of shapes and the ground-state shape is the shape corresponding to the minimum energy. It is the level structure here that is compared to experiment.

However, it turned out that apart from having some issues from fundamental grounds, this method of obtaining the potential energy versus shape, also did not work well in practice when the method was extended to larger deformations or additional multipole distortions. But ten years after Nilsson introduced the deformed single-particle model, Strutinsky introduced a method that combined physics from the macroscopic liquid-drop model and the single-particle model [22, 23]. Briefly, the potential energy for any prescribed deformation is calculated in the following manner. The energy for the charged liquid drop is calculated for this shape, somewhat simplified it is the sum of a surface and electrostatic Coulomb energy. The single-particle energies are calculated. A shell-plus-pairing correction is then obtained from the calculated single-particle spectrum. In Strutinsky's method the correction is about -6 MeV when there is a large gap in the spectrum such as in the neutron spectrum at $N = 126$ for ^{208}Pb. Since there is also a gap in the proton levels at $Z = 82$ the total shell-plus-pairing correction at spherical shape is about -12 MeV. In situations where there is a high density of levels the shell-plus-pairing correction sum is larger than 10 MeV.

By use of the method, potential energies versus shape are calculated, the shape and energy at the minimum of this function are obtained. The energy at the minimum corresponds to the ground-state mass. In Fig. 1 we compare a modern macroscopic-microscopic mass model to experimental data, see discussion in the caption. One can gain much insight from calculated single-particles levels versus deformation, a so-called Nilsson diagram. An example is shown and discussed in Fig. 2.

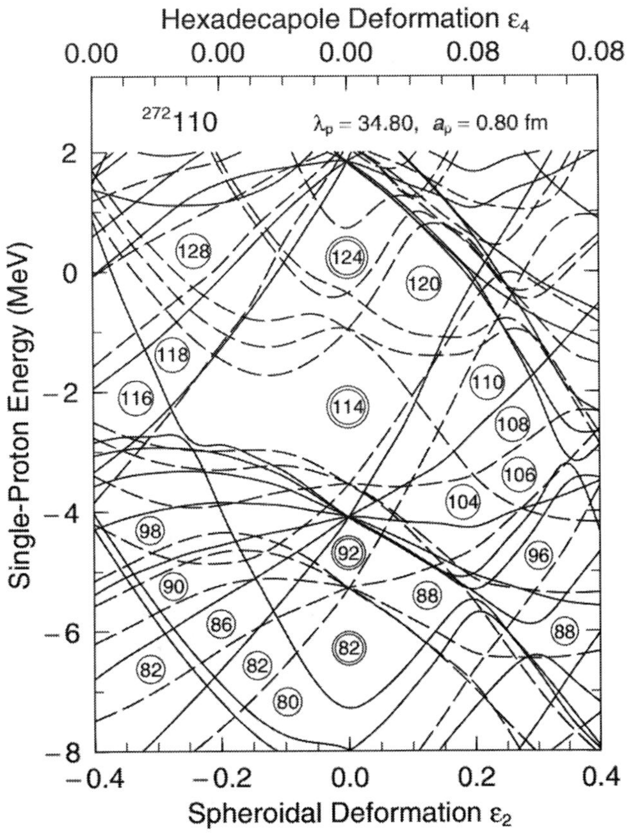

Figure 2. Calculated single-particle levels versus nuclear shape [7]. The higher multipole ϵ_4 values are those appropriate to the nucleus ^{272}Ds ($Z = 110$). For spherical shape we see the well-known magic proton number 82 but also a predicted gap at $Z = 114$. For some time it was thought that no observable elements would exist beyond Rf ($Z = 104$) and that one would need to "jump" or sail across a sea of instability to reach nuclei in the vicinity of the next predicted magic proton number 114. But by now all elements up to $Z = 118$ have been observed and named. Thus, there are elements in the sea of instability. The origin of their stability are the gaps in the single-particle level diagrams for *deformed* shapes. The large negative shell corrections associated with these were present in early calculations [8–10] but their significance was not realized until later [11], which led to the more detailed visualization of the previous theoretical calculations [12].

No matter how well a model agrees with experimental results known at the time of publication one of several requirements to qualify as a bona fide model is that it can predict new experimental results. We have tested our FRDM (1992) mass model finalized in 1992 and published in 1995 [30] in this respect. It was adjusted to a 1989 data base of measured masses [31]. It is compared to masses in Ref. [32] that are not in the 1989 evaluation that is "new" masses in the right part of Fig. 3. The accuracy 0.642 MeV is about the same as for the masses to which the model was adjusted and with no increase in the deviations with distance from stability. In the left part of the figure we compare to a larger data base of new masses; those that are new in the AME2003 evaluation relative to the 1989 evaluation. Now the deviation is actually substantially smaller than in the region the model was adjusted, only 0.462 MeV. An interesting feature is that many dots on the right representing large mass deviations are not present on the left, most easily noticeable on the neutron-deficient side. Since the theory did not change, this means that masses in the 1993 evaluation were either

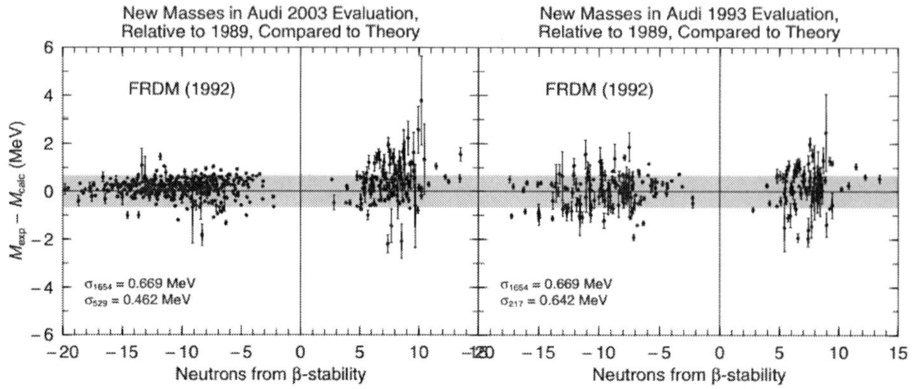

Figure 3. Comparison between masses calculated in the FRDM (1992) to measured masses that were not known when the mass table was finalized, see text for further discussion.

removed (so they are not in the 2003 evaluation) or revised by more than the error bars. One can also note that in the comparison to the right, theory and experiment did not know of each other whereas in the 2003 evaluation the evaluators knew of the FRDM(1992) since it was published in 1995 [30]. The right figure represents a double-blind result whereas the left only a single-blind result.

The calculation of fission barriers is much more complicated than the calculation of masses so we cannot review it here. Suffice to say a state-of-the-art calculation needs to calculate the potential energy for at least all major shape types that might be involved in the evolution from a single ground-state shape to separated fragments. We argue that a minimum of five shape variables are needed: elongation, left and right emerging fragment deformation (we limit our treatment to ellipsoidal shapes), neck diameter and mass division between the emerging fragments. A reasonable spacing of the shape grid points leads to calculations of potential energies for five million shapes for each nuclide considered. A slightly simplified macroscopic model, the FRLDM is used in fission calculations. Extensive details and benchmarks are given in Refs. [33–35].

4. Superheavy nuclei

It is unclear who first suggested the existence of a region of nuclei stabilized by magic numbers, beyond the region of increasingly unstable nuclei at the end of the nuclear chart. Gertrud Scharff-Goldhaber mentions the possibility of a stable region centered at $Z = 126$ in a rather subdued fashion in Ref. [36] in 1957, so we conclude it was generally discussed already at that time. Later it was concluded that $Z = 114$ was more likely [37] the next magic proton number. We give in Fig. 4 some examples of how the FRDM(1992) and FRLDM(1992) have *predicted* α-decay properties of some superheavy nuclei. Recall that the heaviest nuclei to which the model was adjusted were $A \approx 260$. Note that the kinks in the experimental data are very similar in the calculated results. For $Z = 113$ the origin of the kink is the gaps at $Z = 108$ and $N = 162$ in the *deformed* single-particle level spectra, see Fig. 2 for the corresponding neutron diagram from Ref. [7]. For $Z = 117$ the kinks are related to changes in deformation.

It is α-decay that usually serves to identify the observation of new elements or isotopes, but it is fission that sets the limit to what is the heaviest element that can eventually be reached. We show in Fig. 5 3282 calculated fission barrier heights. The calculations reproduce

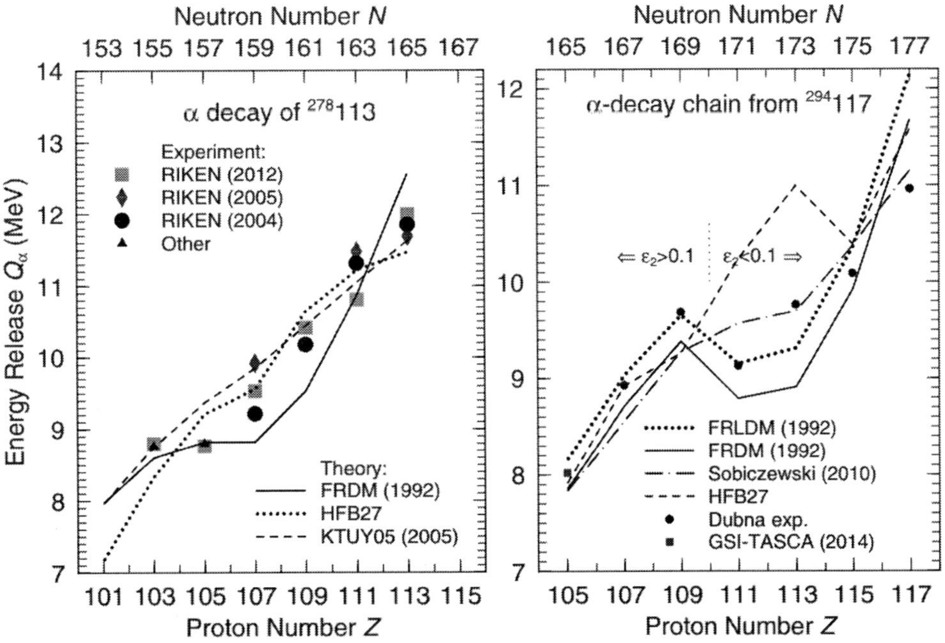

Figure 4. Measured α-decay Q values compared to calculated ground-state to ground-state Q values. For $Z = 113$ different colors indicate different decay chains. The measured values may not correspond to ground-state to ground-state transitions as evidenced by the differences between $Z = 113$ chains. In the left figure the experimental data are from Ref. [24], HFB27 from Ref. [25] and KTUY05 from Ref. [26]. In the right figure the experimental data are from Refs. [27] and [28], Sobiczewski from Ref. [29].

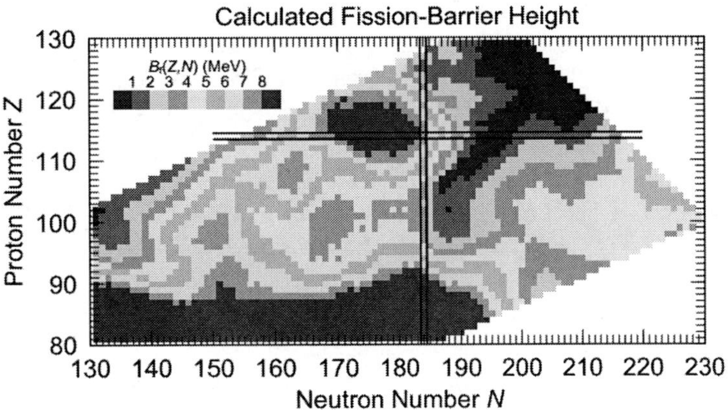

Figure 5. Calculated fission-barrier heights in the heavy region. See text and Ref. [35] for extensive discussions of barriers and the r-process.

the well-known regions of locally enhanced stability centered at ^{252}Fm ($Z = 100$) and ^{270}Hs ($Z = 108$), evidenced by the calculated high barriers. For still heavier nuclei we see a larger region of high barriers, the spherical superheavy region. It is slightly offset to the north-west from the predicted doubly-magic numbers $Z = 114$ and $N = 184$. These calculated results and considerations suggest that it is unlikely that elements much above $Z = 120$ or isotopes much beyond $N = 184$ will be observed.

References

[1] C.F. von Weizsäcker, Z. Phys. **96**, 431 (1935).
[2] H.A. Bethe and R.F. Bacher, Rev. Mod. Phys. **8**, 82 (1936).
[3] P. Möller, A.J. Sierk, T. Ichikawa, H. Sagawa, At. Data Nucl. Data Tables **109–110**, 1 (2016).
[4] O. Hahn and F. Strassmann, Naturwiss. **27**, 11 (1939).
[5] L. Meitner and O.R. Frisch, Nature **143**, 239 (1939).
[6] N. Bohr and J.A. Wheeler, Phys. Rev. **56**, 426 (1939).
[7] P. Möller, J.R. Nix, K.-L. Kratz, At. Data Nucl. Data Tables **66**, 131 (1997).
[8] P. Möller, S.G. Nilsson, J.R. Nix, Nucl. Phys. **A229**, 292 (1974).
[9] P. Möller and J.R. Nix, Nucl. Phys. **A361**, 117 (1981).
[10] P. Möller and J.R. Nix, At. Data Nucl. Data Tables **26**, 165 (1981).
[11] P. Armbruster, priv. comm., LBL Cafteria, summer 1982.
[12] R. Bengtsson, P. Möller, J.R. Nix, Jing-ye Zhang, Phys. Scr. **29**, 402 (1984).
[13] M. Wang *et al.*, Chin. Phys. **C36**, 1603 (2012).
[14] M. Mayer, Phys. Rev. **75**, 1969 (1949).
[15] M. Mayer, Phys. Rev. **78**, 16 (1950).
[16] M. Mayer, Phys. Rev. **78**, 22 (1950).
[17] R. Rainwater, Phys. Rev. **79**, 432 (1950).
[18] A. Bohr, Kgl. Danske Videnskab Selskab, Mat. Fys. Medd. **26** No. 14 (1952).
[19] A. Bohr and B.R. Mottelson, Kgl. Danske Videnskab Selskab, Mat. Fys. Medd. **27** No. 16 (1953).
[20] S.G. Nilsson, Kgl. Danske Videnskab Selskab, Mat. Fys. Medd. **29** No. 16 (1955).
[21] B.R. Mottelson and S.G. Nilsson, Kgl. Danske Videnskab Selskab, Mat. Fys. Skr. **1** No. 8 (1959).
[22] V.M. Strutinsky, Nucl. Phys. **A95**, 420 (1967).
[23] V.M. Strutinsky, Nucl. Phys. **A122**, 1 (1968).
[24] K. Morita *et al.* J. Phys. Soc. Jpn. **81**, 103201 (2012).
[25] S. Goriely, N. Chamel, J.M. Pearson, Phys. Rev. C **88**, 061302(R) (2013).
[26] H. Koura, T. Tachibana, M. Uno, Yamada, Prog. Theor. Phys. **113**, 305 (2005).
[27] Yu.Ts. Oganessian, J. Phys. G **34**, R165 (2007).
[28] J. Khuyagbaatar *et al.*, Phys. Rev. Lett. **112**, 172501 (2014).
[29] A. Sobiczewski, P. Jachimowicz, M. Kowal, Int. J. Mod. Phys. E **19**, 493 (2010).
[30] P. Möller, J.R. Nix, W.D. Myers, W.J. Swiatecki, At. Data Nucl. Data Tables **59**, 185 (1995).
[31] G. Audi, Midstream atomic mass evaluation, priv. comm. (1989), with four revisions.
[32] G. Audi and A.H. Wapstra, Nucl. Phys. **A565**, 1 (1993).
[33] P. Möller, D.G. Madland, A.J. Sierk, A. Iwamoto, Nature **409**, 785 (2001).
[34] P. Möller *et al.*, Phys. Rev. C **79**, 064304 (2009).
[35] P. Möller *et al.*, Phys. Rev. C **91**, 024310 (2015).
[36] G. Scharff-Goldhaber, Nucleonics **15**, 122 (1957).
[37] A. Sobiczewski, F.A. Gareev, B.N. Kalinkin, Phys. Lett. **22**, 500 (1966).

EPJ Web of Conferences **131**, 03003 (2016) DOI: 10.1051/epjconf/201613103003
Nobel Symposium NS160 – Chemistry and Physics of Heavy and Superheavy Elements

Fission in the landscape of heaviest elements: Some recent examples

J. Khuyagbaatar[1,2,a], *A.* Yakushev[2], *Ch.E.* Düllmann[1,2,3], *D.* Ackermann[2,b], *L.-L.* Andersson[1], *M.* Block[1,2,3], *H.* Brand[2], *J.* Even[1,3,c], *U.* Forsberg[4], *W.* Hartmann[2], *R.-D.* Herzberg[5], *F.P.* Heßberger[1,2], *J.* Hoffmann[2], *A.* Hübner[2], *E.* Jäger[2], *J.* Jeppsson[4], *B.* Kindler[2], *J.V.* Kratz[3], *J.* Krier[2], *N.* Kurz[2], *B.* Lommel[2], *M.* Maiti[6,d], *S.* Minami[2], *D.* Rudolph[4], *J.* Runke[2], *L.G.* Sarmiento[4], *M.* Schädel[2], *B.* Schausten[2], *J.* Steiner[2], *T.* Torres De Heidenreich[2], *J.* Uusitalo[7], *N.* Wiehl[1,3], and *V.* Yakusheva[2]

[1] Helmholtz Institute Mainz, 55099 Mainz, Germany
[2] GSI Helmholtzzentrum für Schwerionenforschung, 64291 Darmstadt, Germany
[3] Johannes Gutenberg-Universität Mainz, 55099 Mainz, Germany
[4] Lund University, 22100 Lund, Sweden
[5] University of Liverpool, Liverpool L69 7ZE, UK
[6] Saha Institute of Nuclear Physics, Kolkata 700064, India
[7] University of Jyväskylä, 40351 Jyväskylä, Finland

Abstract. The fission process still remains a main factor that determines the stability of the atomic nucleus of heaviest elements. Fission half-lives vary over a wide range, $10^{-19}-10^{24}$ s. Present experimental techniques for the synthesis of the superheavy elements that usually measure α-decay chains are sensitive only in a limited range of half-lives, often $10^{-5}-10^{3}$ s. In the past years, measurement techniques for very short-lived and very long-lived nuclei were significantly improved at the gas-filled recoil separator TASCA at GSI Darmstadt. Recently, several experimental studies of fission-related phenomena have successfully been performed. In this paper, results on $^{254-256}$Rf and ^{266}Lr are presented and corresponding factors for retarding the fission process are discussed.

1. Introduction

The stability, i.e. the existence of an atomic nucleus that consists of Z protons and N neutrons, strongly depends on the interplay between the attractive nuclear and repulsive Coulomb forces. In heavy nuclei the latter force eventually becomes dominant and spontaneous fission (SF) limits their stability [1]. Some observables of the fission process were successfully explained within the liquid drop model, which also predicted that nuclei with too many protons cannot exist due to the absence of a (macroscopic) fission barrier (e.g. $Z \gtrsim 112$ [2]). In the 1960's the shell-structure influence on the mass of the atomic nucleus has been investigated and accordingly the fission barrier was represented as a superposition of

[a] Corresponding author: J.Khuyagbaatar@gsi.de
[b] Present address: GANIL, CEA/DSM-CNRS/IN2P3, Bd. Henri Becquerel, BP. 55027, 14076 Caen Cedex 5, France
[c] Present address: KVI-Center for Advanced Radiation Technology University of Groningen, 9747 AA Groningen, The Netherlands
[d] Present address: Indian Institute of Technology Roorkee, Roorkee 247667, India

the liquid-drop and shell-correction energy terms [3, 4]. Within these earlier versions of the macro-microscopic models the next spherical magic numbers $Z = 114$ and $N = 184$ have been predicted and nuclei at around these numbers are inferred to be surprisingly stable against spontaneous fission. A possible next magic proton number has several times alternatively been allocated to $Z = 120$ or 126 within modern self-consistent theoretical calculations [5, 6].

To date many superheavy nuclei (SHN) with Z and N up to 118 and 176, respectively, have been synthesized. The heaviest ones have been produced in ^{48}Ca-induced fusion reactions with actinide targets and identified by their subsequent α-decay chains, which always ended with fission of relatively neutron-rich unknown isotopes of elements down to Db [7].

One way for probing Z and A assignments of SHN is testing the systematics of observables like α-decay energies and α-decay (T_α) and SF (T_{SF}) half-lives. Especially T_{SF} of the relatively neutron-rich isotopes at the end of the $Z = 114-118$ chains can be compared with systematics based on the well known neutron-deficient ones, where shell-structure effects are well established.

The shell structure of the nucleus can strongly affect the stability against fission, primarily on its mass surface. Such an influence can be studied using the even-even nuclei as it has been observed for $N = 152$, where enhanced fission stabilities were found compared to neighbouring isotopes of Fm ($Z = 100$) and No ($Z = 102$) elements. Single and multi-particle configurations also hinder SF, and the strength of the hindrance depends strongly on the corresponding nuclear orbitals, e.g. quantum numbers associated with intrinsic angular momentum.

Present experimental techniques for the detection and identification of SHN are limited: they are sensitive for fissioning nuclei with half-lives roughly between tens of μs up to a few hours at most. The identification of nuclei with $T_{1/2}$ outside these limits is challenging at best using standard experimental data acquisition (DAQ) methods. DAQ systems based on fast sampling ADCs have been in use for about a decade for detection of charged particles [8]. Recently, such a method has successfully been implemented at the gas-filled recoil separator TASCA and has been used for the first time in experiments dedicated to syntheses of SHN [9–11]. Its advantages for the detection of α particles have already been described in Ref. [12]. In the present paper its benefits for the detection of SF-decaying nuclei will be demonstrated with examples of both short-lived ($^{254-256}$Rf [13]) and long-lived (^{266}Lr [11]) isotopes. Each case corresponds to a particular effect of the shell structure on the fission stability. Accumulated data and analyses procedures of fission activities will also be presented: (i) Spontaneous fission of even-even ^{254,256}Rf will demonstrate the influence of the shell structure on the nuclear mass. (ii) An influence of the single-particle configuration on the fission stability will be shown with even-odd ^{255}Rf. (iii) Fission hindrances due to multi-particle configurations will be demonstrated with even-even ^{254}Rf (K-isomeric state) and odd-odd ^{266}Lr.

2. Experimental setup

Since 2009 the gas-filled recoil separator TASCA is available for experiments aiming at SHN and has successfully been used for the studies of the elements flerovium (Fl, $Z = 114$) [14–16]), moscovium (Mc, $Z = 115$) [9], and Tennessine[1] (Ts, $Z = 117$) [11]. TASCA consists of a deflecting dipole and a focusing quadrupole doublet and can be operated in

[1] Here we adopt the suggested IUPAC names for elements 115 and 117.

different ion-optical modes [17], and with various fill gases [18], depending on experimental requirements.

Nuclei considered in this work were produced as evaporation residues (ERs) of excited compound nuclei formed in nuclear fusion reactions between ^{48}Ca and ^{50}Ti projectiles (beam) and solid ^{204,206,208}Pb [13] and ^{249}Bk [11, 19] targets. Pulsed beams (pulse length: 5 ms, repetition frequency: 50 Hz) were accelerated by the UNIversal Linear ACcelerator (UNILAC) to energies around the Coulomb barrier. Typical target thicknesses were 0.5 mg/cm^2. ERs leaving the target with relatively high charge states with broad distributions enter the gas-filled volume of TASCA, where charge-state equilibrating charge-exchange processes occur. After sufficiently many collisions the charge-state distribution becomes narrow at around an average one (often $q < 10$), before reaching the dipole magnetic field. Consequently, the trajectory of the ERs through TASCA can be defined by adjusting the dipole magnetic field to their magnetic rigidity [18, 20]. For the present studies TASCA was filled with helium gas at 0.8 mbar pressure. At the focal plane of TASCA, ERs were implanted into a double-sided silicon strip detector (DSSSD) comprising 144 vertical (X) and 48 horizontal (Y) strips on the front and back sides, respectively.

TASCA was significantly upgraded in 2011 by improving its performance including the assembling of a fast DAQ system for the implantation detectors [21]. All pre-amplifier signals from the Y-strips of the DSSSD were digitized by 60 MHz-sampling ADCs and stored in 50 μs-long traces. Signals from the X-strips of were processed with standard analog electronics and peak-sensing ADCs with a dead-time of $\leqslant 35\,\mu$s. These analog and digital branches, which independently initialized data storage, allowed determining the origin of events in two independent ways [11, 12, 21].

3. Experimental results

3.1 SF of ^{256}Rf, ^{255}Rf and ^{254}Rf

The isotope ^{256}Rf was produced in the ^{50}Ti(^{208}Pb,2n) reaction. A typical energy spectrum of fission fragments detected in the DSSSD by the analog part of the DAQ is shown in Fig. 1a. These high-energetic events were in anti-coincidence with signals from the MWPC. Spatial and time correlation analysis between them and preceding evaporation residues (hereafter noted as ER-SF) allowed to determine the half-life of ^{256}Rf [see Fig. 2a] according to a method described in Ref. [22]. Traces of these fission events were stored as saturated signals in the digital part of the data. The saturation is due to the limited dynamic input range of the sampling ADCs. The saturation time can be used as a variable for the determination of the energy deposited in the DSSSD by the fission fragments in the digital data, as shown in Fig. 1b. The time distribution of fission events from decay of even-odd ^{255}Rf produced via ^{50}Ti(^{206}Pb,1n) is shown in Fig. 2b. Due to the long half-life, 1.9(4) s, all fission traces were detected as single saturated signals similar to ^{256}Rf.

^{254}Rf was produced in the ^{50}Ti(^{206}Pb,2n) reaction. Only a few events with saturated single signal, like the ones shown in Fig. 1a, were detected. They were found to be correlated with ERs within the short time as shown in Fig. 2c. The correlation time distribution of these events was terminated at about 50 μs, because most of the fission events from ^{254}Rf were recorded as a second signal in ER-traces. Thus, these fission events were lost in the analog part of the data as being an individual fission-like event! One of them is exemplified in Fig. 1d. Energies of the first signal in such traces were corresponding to ER energies similar to that observed from the decay of ^{255,256}Rf. After extracting such short correlated ER-SF signals from the ER-traces, the full correlation time distribution was reconstructed; the corresponding half-life of ^{254}Rf was determined to $T_{SF} = 20(3)\,\mu$s.

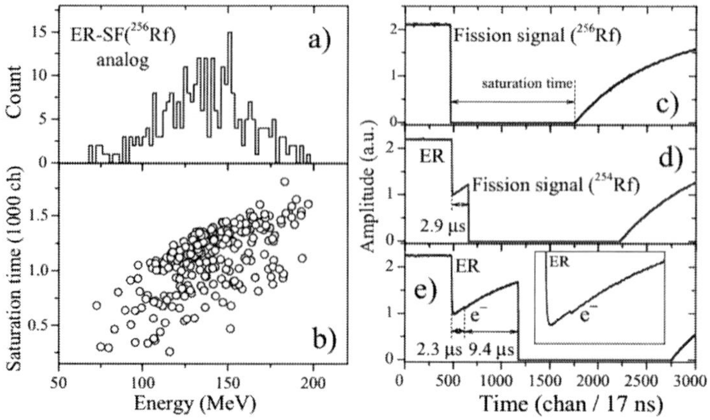

Figure 1. (a) Energy spectrum and (b) saturation times of traces versus energies of fission events from the decay of ^{256}Rf measured in the analog part of the data. (c)–(e) Examples of different types of traces containing fission events: (c) single-signal trace corresponding to ^{256}Rf. (d) Double-signal trace where the implantation signal (noted by ER) of ^{254}Rf was followed by fission. (e) Triple-signal trace where ^{254m}Rf was implanted and after 2.3 μs the isomeric state de-excited to the ground state which decays by SF after 9.4 μs. The insert highlights the region of the electron signal.

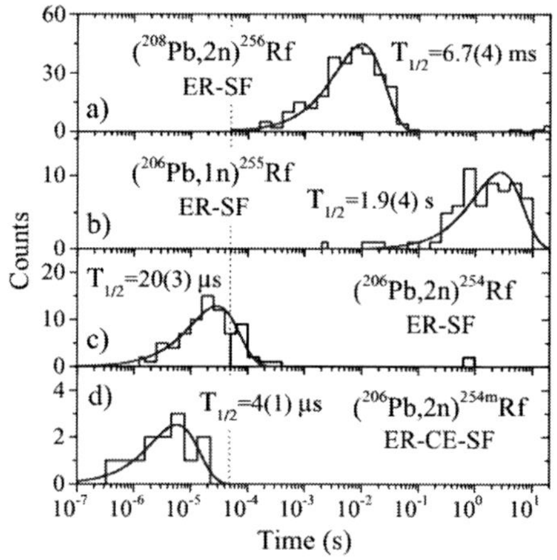

Figure 2. Correlation-time distribution of fission events relative to preceding ERs (a)–(c). The distribution of single-signal fission traces which terminated at ≈ 50 μs (marked by the dotted line) is also shown in panel (c). (d) Time distribution of electron-like signals relative to ERs correlated with fissions from ^{254}Rf. Lines represent the calculated density distribution of the fission events on a logarithmic time scale according to Ref. [22]. The center of gravity of the distribution determines the lifetime ($\tau = 1.443 \cdot T_{1/2}$) of the fissioning nucleus. Half-lives of the corresponding isotopes with uncertainties (in the bracket) and reactions in which they were produced are given. See text for details.

Furthermore, in several traces of ER-like events a small signal was detected in between the ER and fission signals [23]. These signals with amplitudes not exceeding 500 keV were attributed to originate from the registration of electrons. Time differences between the ER and these electron-like signals are shown in Fig. 1e. They represent an activity with a half-life of 4(1) μs. Therefore, all observed events containing ER-electron-SF signals were attributed to implantation of ^{254}Rf in an isomeric state associated with the emission of conversion electrons (CEs).

3.2 Fission from ^{266}Lr and α-decay assignment for ^{270}Db

The nucleus ^{266}Lr has not been produced directly so far. In 2012 at TASCA, this isotope was assigned to be produced in α-decay chains associated with ^{294}117 [11]. Its safe observation was possible thanks to several experimental conditions, e.g., the use of a highly pixelized DSSSD [15], fast electronics [11], the UNILAC pulsed beam, and well-tuned TASCA [20] that allowed to assign a long-lived α decay for the mother nucleus ^{270}Db in both α-decay chains stemming from element 117.

As an example, energies of events measured in the pixel X=103 and Y=41 [11] without a coincident signal from the MWPC are shown in Fig. 3 as a function of time counting from the beginning of the experiment. Events occurring in the DSSSD during beam off periods are marked by additional horizontal lines. Within the energy range of 7–12 MeV where most of the known α decays from heavy nuclei should occur, 55 beam-off events were detected. From the even-by-event analysis, 32 of them were identified as originating from known isotopes around ^{208}Pb produced in transfer reactions (see Fig. 3).

Alpha decays in the region "northeast" of ^{208}Pb occur with short half-lives. If the α-particle energy is above 8 MeV and it originates from nuclei with $Z \approx 82-90$ and $A \approx 210-230$ then the corresponding ERs can mostly be found in the data within 1 s [24]. Therefore, depending on the counting rate of the individual DSSSD pixel, the predominant amount of beam-off events can be identified.

Several events were detected within a short time window (cf. Fig. 3, cluster). They were unambiguously attributed to originate from element $Z = 117$ based on the similarity to findings reported from the Dubna Gas-Filled Recoil Separator (DGFRS) [25–27]. Only one fission-like event was observed 5.1 hours after the last α from the cluster-event. In between, four α-like events with energies of 6.82, 7.89, 11.53 and 6.29 MeV were measured. Only the 7.89-MeV event could not be assigned to any known isotope. Furthermore, this was the only one event left without an assignment from in total seven events detected during the entire experiment within the energy range of 7.7–8.2 MeV, where a possible decay of ^{270}Db has been discussed in Ref. [27]. A similar situation has occurred in the second α-decay chain, where the energy of the corresponding α event was 7.90 MeV [11]. Therefore, these events were attributed to ^{270}Db and the fission events that terminate the decay chains to the decay of ^{266}Lr [11].

4. Discussion and summary

The influence of the shell structure on the spontaneous fission of heavy nuclei is well visible in systematics of T_{SF}, as shown in Fig. 4. The results on isotopes presented in this work [11, 18] are marked by solid symbols. Measured T_{SF} of $^{254-256}$Rf were in agreement with literature data [24]. Maxima of T_{SF} of Fm ($Z = 100$) and No ($Z = 102$) isotopes observed at $N = 152$ show the enhanced fission stability due to the shell closure. However, even-even Rf ($Z = 104$) isotopes do not anymore show a well pronounced peak at $N = 152$ compared to lighter elements. This might be related to a weakened influence of the shell closure on the fission barrier. These nuclei are deformed, and thus the gaps between those nuclear orbitals responsible for the shell closure are strongly affected by interactions between the nucleons, i.e. both numbers of protons and neutrons. This is one of the distinct evidences of differences on the "magic" numbers ($Z = 100$, 108 and $N = 152$, 162) occurring in the deformed nuclei from those at, for example, $Z = 82$ and $N = 126$, where deformation is absent.

In case of nuclei with an unpaired proton and/or neutron (odd-A Rf, Lr and Db), T_{SF} are significantly longer than in the neighbouring even-even ones, showing the effect of single and

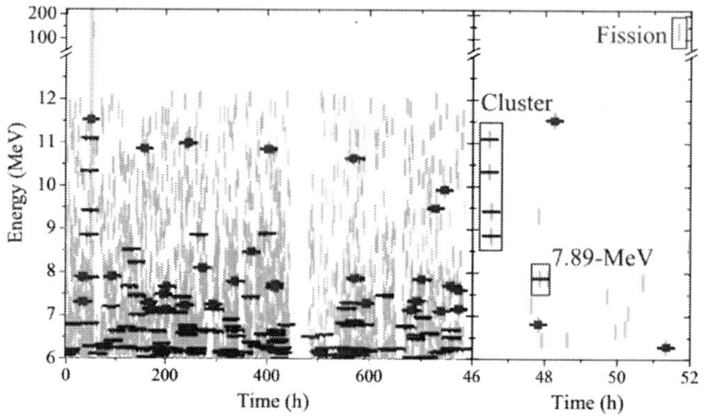

Figure 3. Energy versus time plot for events detected in pixel X=103 and Y=41 of the DSSSD where one of the α-decay chains associated with $^{294}117$ was observed [11]. The left panel shows the full time period of the experiment. The right panel shows a detailed view of the period the chain was observed. Events without a coincident signal from the MWPC and occurring during beam-off periods are marked by horizontal and vertical lines, respectively. Events with energies in the range of 7–12 MeV (an additional two low-energy events in the right panel), which were identified to originate from the products of transfer reactions, are marked by additional solid squares.

multi-particle configurations on the fission process. We note that experimentally observed SF branches from odd-A and/or odd-odd nuclei do not necessarily always correspond to the ground-state decay. Though T_{SF} of odd-A Rf, Lr and Db isotopes show a maximum at $N = 153$ associated with the influence of both the $N = 152$ shell closure and single and/or multi-particle configurations. An increase in T_{SF} towards $N = 162$ was observed in Rf-Hs, and in the case of Db a peak-like trend is observed at $N = 163$ similar to the situation $N = 153$, which indicates a shell closure.

An intriguing finding is the assignment of the long-lived α decay of ^{270}Db. From the experimental data analysis' point of view, two identical cases of 7.9 -MeV α's were clearly found to be member of the decay chains. However, in Ref. [28], the non-observation of analogous α-branching in the odd-N neighbour ^{268}Db ($N = 163$) was pointed out, questioning the assignment of α decay in ^{270}Db. According to Ref. [28] α-decay properties of ^{268}Db could be expected to be similar to those of ^{270}Db. However, no α decay was identified and assigned for ^{268}Db in several experiments performed at DFGRS [28], TASCA [9] and Berkeley Gas-filled separator (BGS) [29]. However, we note that the DGFRS and BGS experiments were significantly less sensitive for measuring long-lived α decay because of the use of a DC beam where no (background-free) beam-off periods occur. The absence of an α-decay branch in ^{268}Db would mean that it has a longer T_α compared to $\approx$ 1 hour of the neighbouring ^{270}Db ($N = 165$). Such cases are well known to occur in odd-odd and even-odd nuclei located above closed shells: for instance, T_α of ^{256}Lr ($N = 153$) is about eight times longer than ^{258}Lr ($N = 155$). In cases of spherically shaped nuclei such T_α differences between $N = 127$ and 129 isotopes of odd-Z elements vary up to $6{\cdot}10^5$. Thus, it would not be surprising to expect a longer T_α and different α-decay energy for ^{268}Db than ^{270}Db. For the future, more dedicated experiments to search for the long-lived α decay of ^{268}Db and more data on ^{270}Db need to be performed. Their results will help to shed light on the shell structure around $N = 162$. We also note that in the case of odd-odd nuclei where both α decay and SF are hindered, other radioactive decay modes (β decay and electron capture) can become preferable. Recently, statistical analyses of the α-decay chains attributed to start at $Z = 115$ checking for the presence of such decays, have been carried out [31]. Their results show that

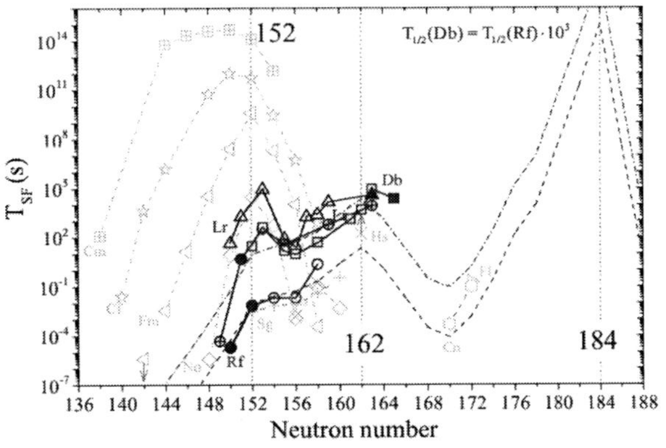

Figure 4. Partial SF half-lives of even-even isotopes of the heaviest elements including odd-A Rf, Lr, and Db isotopes, as function of neutron number. Solid symbols mark experimental data provided in this paper, open symbols are from literature [24]. Theoretically calculated T_{SF} of even-even Rf isotopes are also given as dashed lines [30]. Dashed-dotted lines represent the same calculation but multiplied by 10^3, which simulates a hindrance factor due to an unpaired proton and/or neutron. Location of known and predicted neutron shell closures are marked by vertical dotted lines.

different scenarios for the observed α-decay chains can be proposed, assuming that one or more of the chain-members underwent electron capture decay.

Apparently, the identification of long-lived nuclei around $N = 162$ (expected also at $N = 184$) turned out to be a major task for further SHN studies. Suitable experimental techniques have to be implemented.

Similarly, very short-lived fissioning nuclei can also be studied in the future as demonstrated in the case of ^{254}Rf and its isomeric state. The latter observation made through electron detection without requesting a particular trigger was possible thanks to a fast DAQ, whose further advantage for the solution of fission related topics still needs to be explored. For instance, the observed ER-CE-SF sequence for ^{254}Rf indicates that an isomeric state is de-exciting into the ground state instead of direct fission. When such situation occurs in the strongly spontaneously fissioning nucleus then the most probable origin of the isomer is an excited state with high K-quantum number. K represents the projection of the total nuclear spin along the symmetry axis of the nucleus [32, 33]. The same isomeric state in ^{254}Rf has recently been identified and half-lives are in agreement [34]. A two-quasi-neutron configuration $\nu 7/2^+[624] \otimes \nu 9/2^-[734]$ forming $K = 8$ has been suggested for this 4-μs isomeric state in Ref. [34]. An additional 247-μs isomeric state has also been observed. In the present experiment a threshold for the detection of signals in the DSSSD were not sensitive enough for measurements of electrons, thus the 247-μs-isomeric decays reported in Ref. [34] could not be observed.

K-hindered fission stability might become an important issue for further studies of ultra short-lived nuclei with extreme numbers of Z and N. Very neutron-deficient nuclei with T_{SF} much shorter than 1 μs, which is approximately the flight-time of ERs through short in-flight separators for SHN, are just one example. Such an experimental scenario has been applied at TASCA for the hitherto unknown ^{252}Rf with a predicted $T_{SF} = 0.6$ μs [30]; these results will be presented elsewhere [23].

We are grateful for GSI's ECR ion-source and UNILAC staff, and the Experimental Electronics department for their continuous support of the experiments.

References

[1] R. Vandenbosch and J.R. Huizenga, *Nuclear Fission* (Academic Press, New York, 1973).
[2] A. Sobiczewski and I. Muntian, Nucl. Phys. **A734**, 176 (2004).
[3] A. Sobiczewski *et al.*, Phys. Lett. **22**, 500 (1966).
[4] W.D. Myers and W.J. Swiatecki, Nucl. Phys. **81**, 1 (1966).
[5] K.M. Rutz *et al.*, Phys. Rev. C **56**, 238 (1997).
[6] S. Čwiok *et al.*, Nucl. Phys. **A611**, 211 (1996).
[7] J.H. Hamilton, S. Hofmann, Yu.Ts. Oganessian, Ann. Rev. Nucl. Part. Sc. **63**, 383 (2013).
[8] S.N. Liddic *et al.*, Phys. Rev. Lett. **97**, 082501 (2006).
[9] D. Rudolph *et al.*, Phys. Rev. Lett. **111**, 112502 (2013).
[10] J. Khuyagbaatar *et al.*, GSI Scientific Report-2012, 131 (2013).
[11] J. Khuyagbaatar *et al.*, Phys. Rev. Lett. **112**, 172501 (2014).
[12] J. Khuyagbaatar *et al.*, Phys. Rev. Lett. **115**, 242502 (2015).
[13] J. Khuyagbaatar *et al.*, GSI Scientific Report-2012, 133 (2013).
[14] Ch.E. Düllmann *et al.*, Phys. Rev. Lett. **104**, 252701 (2010).
[15] J.M. Gates *et al.*, Phys. Rev. C **83**, 054618 (2011).
[16] A. Yakushev *et al.*, Inorg. Chem. **53**, 1624 (2014).
[17] A. Semchenkov *et al.*, Nucl. Instr. Meth. **B266**, 4153 (2008).
[18] J. Khuyagbaatar *et al.*, Nucl. Instr. Meth. **A689**, 40 (2012).
[19] J. Runke *et al.*, J. Radioanal. Nucl. Chem. **299**, 1081 (2014).
[20] J. Khuyagbaatar *et al.*, Phys. Rev. A. **88**, 042703 (2013).
[21] A. Yakushev *et al.*, to be published.
[22] K.H. Schmidt *et al.*, Z. Phys. **A316**, 19 (1984).
[23] J. Khuyagbaatar *et al.*, to be published.
[24] `http://www.nndc.bnl.gov/ensdf/`.
[25] Yu.Ts. Oganessian *et al.*, Phys. Rev. Lett. **104**, 142502 (2010).
[26] Yu.Ts. Oganessian *et al.*, Phys. Rev. Lett. **109**, 162501 (2012).
[27] Yu.Ts. Oganessian *et al.*, Phys. Rev. C. **87**, 054621 (2013).
[28] Yu.Ts. Oganessian and V.K. Utyonkov, Rep. Prog. Phys. **78**, 036301 (2015).
[29] J.M. Gates *et al.*, Phys. Rev. C **92**, 021301(R) (2015).
[30] R. Smolanczuk, J. Skalski, A. Sobiczewski, Phys. Rev. C **52**, 1871 (1995).
[31] U. Forsberg *et al.*, Nucl. Phys. **A953**, 117 (2016).
[32] D. Peterson *et al.*, Phys. Rev. C **74**, 014316 (2006).
[33] H.B. Jeppesen *et al.*, Phys. Rev. C **79**, 031303 (2009).
[34] H.M. David *et al.*, Phys. Rev. Lett. **115**, 132502 (2015).

EPJ Web of Conferences **131**, 04001 (2016) DOI: 10.1051/epjconf/201613104001
Nobel Symposium NS160 – Chemistry and Physics of Heavy and Superheavy Elements

Prospects of heavy and superheavy element production via inelastic nucleus-nucleus collisions – from ^{238}U + ^{238}U to ^{18}O + ^{254}Es

Matthias Schädel[a]

GSI Helmholtz Centre for Heavy Ion Research, Planckstr. 1, 64291 Darmstadt, Germany

Abstract. Multi-nucleon transfer reactions, frequently termed deep-inelastic, between heavy-ion projectiles and actinide targets provide prospects to synthesize unknown isotopes of heavy actinides and superheavy elements with neutron numbers beyond present limits. The ^{238}U on ^{238}U reaction, which revealed essential aspects of those nuclear reactions leading to surviving heavy nuclides, mainly produced in 3n and 4n evaporation channels, is discussed in detail. Positions and widths of isotope distributions are compared. It is shown, as a general rule, that cross sections peak at irradiation energies about 10% above the Coulomb barrier. Heavy target nuclei are essential for maximizing cross sections. Experimental results from the ^{238}U on ^{248}Cm reaction, including empirical extrapolations, are compared with theoretical model calculations predicting relatively high cross sections for neutron-rich nuclei. Experiments to test the validity of such predictions are proposed. Comparisons between rather symmetric heavy-ion reactions like ^{238}U on ^{248}Cm (or heavier targets up to ^{254}Es) with very asymmetric ones like ^{18}O on ^{254}Es reveal that the ones with ^{238}U as a projectile have the highest potential in the superheavy element region while the latter ones can be advantageous for the synthesis of heavy actinide isotopes. Concepts for highly efficient recoil separators designed for transfer products are presented.

1. Introduction

The quest for superheavy elements (SHE), chemical elements with atomic number, Z, larger than 103 [1], is driven by the desire to determine and understand one of the extreme limits of existence of matter. After an early phase of discussing the possible existence of SHE [2], the year 1966 marked a breakthrough with a seminal meeting in Lysekil, Sweden, [3] and additional theoretical work [4, 5] both shifting the focus from element 126 to 114. A maximum of stability was, and still is, expected at the neutron number $N = 184$ [6].

Early on, possible ways to reach this magical "island of stability" were discussed and were fostering developments and installations of heavy-ion accelerators. Fully exhausted are the two fusion paths; the Pb and Bi target based "cold fusion" that at element 113 has reached the sensitivity limit of today's experiments [7] and, with element 118, the actinide target based reactions with ^{48}Ca as the projectile [8]. While the latter one gives access to the presently known most (relative) neutron-rich nuclei, like ^{289}114 [8, 9], alternative paths are essential to reach and explore the exciting unknown territory towards $N = 184$. As the next generation radioactive ion beam facilities, which will become operational worldwide within

[a] e-mail: m.schaedel@gsi.de

the next decade, will not provide sufficient beam intensity for SHE research with potentially interesting more neutron-rich projectiles like ^{46}Ar or ^{50,52}Ca, this leaves scientists with a path that is often termed multi-nucleon transfer. These reactions were already proposed between the mid-60s and mid-70s [10, 11]. However, it was only after ^{238}U beams with sufficiently high energy and intensity became available at the Gesellschaft für Schwerionenforschung (GSI), Darmstadt, that the entire range of multi-nucleon transfer or deep-inelastic reactions up to ^{238}U on ^{238}U and ^{248}Cm were explored [12, 13]. These reactions, their results on the synthesis of heavy actinide isotopes, the understanding of the processes leading to these products and the potential for the synthesis of new, neutron-rich isotopes will be discussed in this article. To a large extent it includes information available in [12–14] and, more recently, summarized and expanded in [15].

2. Experimental techniques

All studies on the production of heavy actinides discussed in this article were performed applying radiochemical methods. Because of the very high yield and efficiency, which includes a close to 100% collection efficiency of all products and a high-yield chemical separation of individual elements together with the nearly background-free off-line measurement of samples, this technique is best adapted to cross-section measurements of individual isotopes of heavy actinides. Discontinuous, off-line techniques, as applied in the ^{238}U on ^{238}U reaction [14, 16] and, thereafter, the ^{238}U on ^{248}Cm one [13] are most sensitive for isotopes with half-lives between about one hour and a few months that undergo α decay or spontaneous fission (SF) as their characteristic nuclear decay mode. This is the case for most isotopes in the investigated region from the target to the Fm and Md region allowing for the determination of complete isotope distributions.

Beam times from a few hours up to two days with cumulated numbers of 0.5 to $1.2{\cdot}10^{16}$ ^{238}U particles enabled cross-section measurements for individual isotopes in the millibarn to nanobarn range. In these experiments, thick ^{238}U-targets (300 and 500 mg/cm^2 depleted U metal explosion-cladded onto 1.5-mm thick Cu sheet-metal) simultaneously served as an infinitely thick target and catcher material regarding the nuclear reaction and the collection of reaction products, respectively. Irradiations were performed with different incident beam energies and calculated effective target thicknesses were used for cross-section calculations. This also allowed to determine the energy dependence of the actinide production. Chemical separations, followed by sample preparation and measurement of the characteristic radioactive decay, started immediately after end of irradiation; for further details see [15].

Up to 7.3-mg/cm^2 thick metallic ^{248}Cm targets [17] on Mo foils were on particular thick but thin enough to let reaction products recoil out of the target. They were collected in a Cu catcher that was dissolved and chemically processed after end of irradiation to determine cross sections for actinide isotopes up to element 101 [13].

3. Results and discussion

3.1 The ^{238}U on ^{238}U reaction

In the reaction of $\leq$7.5 MeV/u ^{238}U ions on ^{238}U, cross sections for a very large number of isotopes for chemical elements from $Z = 23$ to 100 were measured. This provided the basis not only to establish a complete cross-section surface but also to decompose it into individual reaction channels. An essential information obtained in view of the synthesis of SHE was that primary fragments of element 70 are produced with a cross section of 0.1 mb.

This means that in the symmetric and binary $_{92}U + _{92}U$ reaction also element 114 primary fragments are synthesized with the same high cross section of 0.1 mb [12, 15]. As the search for SHE, however, remained unsuccessful [18], it became especially interesting to investigate and better understand the synthesis of highly fissile actinide products formed in multi-nucleon transfer reactions. This provided the basis to clarify perspectives for the synthesis of SHE.

At $\leq$7.5 MeV/u ^{238}U on ^{238}U all isotope distributions for elements beyond the target are Gaussian in shape with a common width (FWHM) of 3.3 mass units, u. They peak at about the following mass number, A, (in brackets) and cross section: Pu (239) 1 mb, Am (241.5) 300 μb, Cm (244) 50 μb, Bk (247.5) 3 μb, Cf (249.3) 250 nb, Es (252) 10 nb and Fm (254) 1 nb. Consecutive experiments at higher and lower beam energy showed that the Gaussian shape with a width of 3.2(5) u is a common feature for all actinide isotope distributions in ^{238}U on ^{238}U reactions [15].

Experiments performed with incident ^{238}U energies of 6.49, 6.84, 7.50 and 8.65 MeV/u, corresponding to an energy, E, of 1.07, 1.13, 1.24 and 1.43 MeV/u above the Coulomb barrier, B, respectively, show an increase in cross section up to 1.13 E/B followed by a slow decrease at higher energies. As an example, peak cross sections for the Cf isotope distributions, all centred at $A = 249.3$, first increased from about 140 to 250 nb and fell off to 200 and 110 nb for the two increasingly higher energies. It is remarkable that shapes and centroids of all actinide isotope distributions remained unchanged and are energy independent.

A comparison of the energy dependence for the synthesis of $_{98}Cf$ as the heavy partner and $_{86}Rn$ as the light one in the symmetric ^{238}U on ^{238}U reaction has been carried out. After a small correction for minor fission losses, the cross-section increase for $_{86}Rn$ represents the general trend of increasing primary yield with increasing energy for a $\Delta Z = 6$ charge (and associated mass) transfer. The rise and fall of the complementary $_{98}Cf$ cross section reveals the projectile energy dependence in the low-energy part of the broad excitation energy distribution [15]; see below for more details. The Cf/Rn cross-section ratio gives the survival probability of Cf. It has an exponentially decreasing slope with increasing beam energy at a rate of about one order of magnitude per 0.5 MeV/u. This slope, which depends on the number of neutrons evaporated in the deexcitation of the primary fragments, and by applying empirical systematics for the survival probability, $\Gamma_n/(\Gamma_n + \Gamma_f)$, [19] one obtains an average number between three and four neutrons evaporated for Cf isotopes that survive the fission competition in the deexcitation process of the primary formed, hot fragments. Independently, a comparison of the centroid of the experimentally observed Cf isotope distribution, $\bar{A}_{Cf} = 249.3$, with the most probable primary fragment mass, $\bar{A}'_{Cf} = 251.9$, derived from a generalized systematics of ground-state Q-values, Q_{gg}, [20] again yields an average of 3.5 neutrons evaporated [15].

To obtain deeper and more quantitative insights into the reaction mechanism leading to surviving heavy actinides and to SHE, calculations were performed to reproduce measured isotope distributions of $_{98}Cf$, $_{99}Es$ and $_{100}Fm$ in the reaction of $\leq$7.5 MeV/u ^{238}U on ^{38}U; for details see [12, 15]. These calculations started with measured isotope distributions of the complementary light elements $_{86}Rn$, $_{85}At$ and $_{84}Po$, assuming that these distributions were not strongly altered by prompt fission. Furthermore, it was assumed that (i) for a given element, the most probable primary fragment mass, $\bar{A}'_Z$, can be deduced from a generalized Q_{gg}-systematics including corrections for pair breaking of the transferred neutrons [20] and (ii) the variance of an isotope distribution, $\sigma^2(A)$, can be described as a superposition of the mass dispersion of the primary fragments, $\sigma^2(A')$, the excitation energy dispersion, $\sigma^2(E^*)$, and the dispersion in the number of evaporated neutrons, $\sigma^2(\nu)$.

For the $_{84}Po$-$_{100}Fm$ pair, discussed here as the prime example, evaporation code calculations show that the latter one is a minor contribution with $\sigma^2(\nu) \leq 0.5$ u^2. The largest part, $\approx$4.5 u^2, comes from the wide excitation energy dispersion, which has an estimated FWHM of about 100 MeV. For the primary fragment distribution a dispersion of $\sigma^2(A') \approx$ 2.1 u^2 was estimated [12, 15].

A comparison of $\bar{A}'_Z$ with the measured mean product mass number, $\bar{A}_Z$, gives the average number of neutrons evaporated from the primary light fragment and from this, the mean excitation energy in the lighter element fragments. With the conventional assumption that the mean total excitation energy of the combined light and heavy fragment is shared between the two fragments according to their initial masses, the mean excitation energy in the heavy fragments was calculated.

Using these ingredients as input for calculating isotope distributions, and by comparing these with the measured distributions, the following was found [12, 15]:

(i) The most abundant isotopes of the complementary *light elements* are predominantly formed in collisions where energies close to the mean value are present in the system, e.g. $\approx$168 MeV for the combined di-nuclear system associated with a $\Delta Z = 8$ charge split leading to an excitation energy of $\approx$77 MeV and, consecutively, an average number of 8.6 neutron evaporated for Po isotopes.

(ii) The *surviving heavy actinides* originate exclusively from the low-energy tail (!) of the excitation energy distribution. Heavy element fragments with excitation energies above $\approx$45 MeV have such a low probability to survive the fission competition that they do not significantly contribute to the product distribution.

(iii) To reproduce the shape and centroid, e.g. of the measured Fm isotope distribution, a cut-off of 35 MeV needed to be introduced on the low-energy side of the excitation energy distribution. This reflects the very steep fall-off and extremely low probability for the formation of fragments with a given number of transferred number of protons and neutrons. A steep fall-off was also obtained in a diffusion model calculation [21].

(iv) As a result of (ii) and (iii), heavy actinide isotopes are produced in a narrow excitation energy bin; e.g. for Fm from approximately 35 to 45 MeV, which corresponds to a deexcitation with typically three and four-neutron evaporated.

(v) While shape and centroids of the measured Cf, Es and Fm distributions are excellently described by the calculation, it overpredicts their absolute cross section by about one order of magnitude. As discussed in [15], this discrepancy may have its origin in shell effects, higher fission probabilities originating from high angular momenta, nonstatistical processes leading to nonequilibrium states or large fragment deformations in the exit channel.

3.2 The ^{238}U on ^{248}Cm reaction

Inelastic collisions between ^{238}U and ^{248}Cm were believed to provide a closer and more widespread approach to synthesize SHE in multi-nucleon transfer reactions. Aqueous and gas-phase radiochemical experiments, with a sensitivity down to $\approx$50 pb for isotopes with half-lives in the range from about an hour to several years, only resulted in cross-section upper-limits [22]. However, simultaneously performed studies on the production of heavy actinides up to Md revealed a strong enhancement in their synthesis [13]. In the reaction of $\leq$7.4 MeV/u ^{238}U ions on ^{248}Cm cross sections for Fm isotopes increase by about three orders of magnitude as compared with ^{238}U targets [13, 15]. This distinctly underlines the importance of the highest possible target-Z for the synthesis of heavy actinides and SHE.

Maximum cross sections for Cf, Es, Fm and Md were about 2 mb, 20 μb, 1 μb and 100 nb, respectively. For ^{259}No, an upper cross-section limit of 30 nb was determined.

Despite orders of magnitude different cross sections, it is remarkable that the Cf, Es and Fm isotope distributions originating from ^{248}Cm do not only have the same shape and width as the ones from ^{238}U but their centroids are also located at the same mass number; e.g. $\bar{A}_Z = 254$ for Fm. As the measured cross sections for the lighter fragments $_{84}$Po and $_{85}$At were nearly the same in reactions of ^{238}U with ^{238}U and ^{248}Cm, it was assumed that also the primary actinide fragment yield before fission are about the same for a given (ΔZ, ΔN) transfer channel in both reactions. If this is the case, then cross-section differences for the same (ΔZ, ΔN) products should simply reflect the different survival probability of the different product nuclei. With an assumption of unchanged excitation energies and angular momenta, a calculation was performed, which took into account the different survival probabilities. For a given number of neutrons evaporated, cross sections were calculated for Es, Fm and Md. Very good agreement between calculated and experimentally determined isotope distributions were obtained for an average of three to four neutrons evaporated. As in the reactions with ^{238}U, this again implies a very similar reaction mechanism and average fragment excitation energies between about 35 and 45 MeV for surviving products. In addition, differences between experimentally determined most probable masses, $\bar{A}_Z$, and calculated ones for primary fragments, $\bar{A}'_Z$, also indicate the synthesis predominantly in 3n and 4n evaporation channels; for details see [13, 15]. While heavy element products from ^{238}U based reactions are produced in 3n or 4n evaporations channels [12, 13, 15], results from experiments with ^{136}Xe indicate a production in 2n to 3n channels and for ^{48}Ca and lighter projectiles it is closer to the 2n and 1n channel [23, 24]. The higher survival probability of products formed in the latter reactions is in reactions with ^{238}U overcompensated by a much larger charge and mass transfer, giving rise to much large production of primary fragments. This makes the ^{238}U reaction the most favourable one for synthesis of neutron-rich isotopes further away from the target, i.e. for superheavy elements.

While excitation functions for the synthesis of heavy actinides are not available for the ^{238}U on ^{248}Cm reaction, it can safely be assumed that it will also have its maximum at about 10% above the Coulomb barrier. This expectation does not solely rely on the ^{238}U on ^{238}U data and the above mentioned successful reproduction of isotope distributions from the ^{238}U on ^{248}Cm reaction but is most strongly supported from results obtained with ^{136}Xe, ^{86}Kr, ^{48}Ca and ^{18}O projectiles on ^{248}Cm and ^{249}Cf targets [23–27] all clearly showing a cross-section maximum at $E/B \approx 1.1$.

Measured heavy actinide cross sections in the ^{238}U on ^{248}Cm reaction provide pivotal data challenging results and predictions of recent two-center shell-model calculations; see [28] and references therein. On the one hand, they exhibit excellent agreement with experimentally determined cross sections [13]. On the other hand, these theory results [28] show a most highly desired but unexpected extension of all isotope distributions towards the neutron rich side, which was never seen before. These isotope distributions have an enormous width of about 10 u (FWHM) and they are – in contrast to all experimental findings so far – not Gaussian in shape anymore. If these theoretical results reflect what can be achieved experimentally, this would dramatically open up intriguing perspectives to synthesize and investigate more than 20 new, neutron-rich isotopes from $_{100}$Fm to $_{108}$Hs with cross sections ranging from hundreds of microbarn to the picobarn level. Among these are fascinating isotopes like the ones between $^{262,\,264}$Fm and $^{266,\,268}$Rf with cross section of more than 1 nb, or ^{272}Rf and ^{274}Sg accessible with more than 1 pb. However, as the authors of [28] point out, these cross section essentially depend on the strength of shell effects. Fortunately, the validity of these calculations on the neutron-rich side can be tested in the Fm, Md and No region, where cross section between 1 μb and 1 nb are predicted. As previous experiment did not anticipate such high cross section, they did not look for the decays of these isotopes with unknown α energies or spontaneous fission half-lives. New and dedicated experiments are required to answer this pressing question.

4. Summary and perspectives

Multi-nucleon transfer reactions with ^{238}U as a projectile and targets with the highest Z and N/Z-ratio as possible, are the prime, and sole, candidates for the synthesis of new neutron-rich transactinides or superheavy elements. Presently, no other nuclear reaction, including radioactive ion beams, provides a competing alternative. Based on current theoretical predictions [28], ^{248}Cm as a target can provide access to most interesting new isotopes in the heavy actinide and light transactinide region, like the $N = 162$ nuclei ^{262}Fm, ^{264}No and ^{266}Rf. Moreover, it is essential for putting these theoretical predictions on the synthesis of new isotopes for heavy and superheavy elements on the test bench. Only after having verified the predicted high cross sections for neutron-rich isotopes, which show that shell effects are strongly enhancing the synthesis of these isotopes, trustable predictions for the synthesis of SHE isotopes approaching $Z = 114$ and $N = 184$ can be made. However, the use of such an elusive target material like ^{251}Cf, or much better ^{254}Es, would increase the scope of studies into the SHE region by orders of magnitude over the use of ^{248}Cm. So far, the enormous potential of ^{254}Es as a target material in multi-nucleon transfer reactions has only been demonstrated in reactions with light projectiles like ^{18}O, ^{22}Ne [29] and ^{48}Ca [30]. Subsequently, results of ^{18}O and ^{22}Ne reactions with ^{254}Es led to the discovery of ^{260}Md and enabled extended studies of "bimodal fission" [31, 32].

The tools to embark on the endeavour to discover and investigate new neutron-rich isotopes of the heavies actinides and SHE are readily at hand or in an advanced conceptual state. The former ones are discontinuous off-line and continuous on-line radiochemical techniques, which are optimized and highly sensitive to study isotopes with half-lives in the range from hours to years and from seconds to hours, respectively. Over the last decades, these techniques were developed and applied to chemically investigate heavy actinides and transactinides; for aspects of automated separations in the aqueous phase see, e.g. [16, 33] and [34] for a comprehensive review focusing on techniques applied in SHE studies.

In the $_{100}$Fm to $_{103}$Lr region, half-lives from seconds to hours, even up to days, are expected for all new isotopes up to $N \leq 165$ and in some specific cases, like ^{274,275}Db, ^{269,273}Db and ^{279,280}Hs, also for transactinides [35]. Generally, beginning with $_{104}$Rf, half-lives of less than one second are estimated for all SHE. This clearly calls for in-flight separation of SHE in dedicated recoil separators specially designed for multi-nucleon transfer reactions. While today's recoil separators are often excellently suited to separate fusion-evaporation products emitted in a very narrow angular cone around 0°, their limited aperture and only partial coverage of the angular spread limits their use for transfer products.

Despite such severe limitations, multi-nucleon transfer studies were performed, e.g. at the SHIP in collisions of ^{58,64}Ni on ^{207}Pb at E/B values from about 0.9 to 1.2 [36]. With an angular range acceptance of $(0 \pm 2)°$, products with atomic number 84 to 89 were measured. They are relatively stable against fission and are not representative for highly fissile SHE. As this experiment inherently could not measure products emitted at scattering angles larger than two degree, it was not sensitive to products from typical quasi-elastic or grazing collisions for all experiments performed above the Coulomb barrier. Any data analysis that requires efficiencies, e.g. cross-section evaluations, are therefore strongly model depended. This is also true for any conclusion drawn on the synthesis of the heaviest elements in transfer reactions from such an evaluation. However, experiments performed at or slightly below the Coulomb barrier, were the grazing angle for heavy transfer products is 0°, can be very useful and can provide important information.

None of the presently operated recoil separators, not even advanced versions like VAMOS, a recently for multi-nucleon transfer reactions applied separator with a relatively high angular aperture of ±5% and the possibility to position it at a suitable angle [37], provide sufficiently high efficiency for sensitive and meaningful searches for new isotopes of the heaviest actinides and SHE below the microbarn level.

What is required for an optimized investigation are separators and peripheral components which are best adapted to the following necessities, characteristics and kinematic features of the emission of heavy actinides and SHE:

(i) Indispensable is the use of often highly radioactive, typically about 0.5-mg/cm^2 thick actinide targets; generally as rotating wheels to accept heavy-ion beam intensities beyond 100 nA$_{part}$, ideally in the range of particle microampere.

(ii) As discussed, all excitation functions for the production of heavy elements in multi-nucleon transfer reactions peak at about 10% above the Coulomb barrier. Based on experimental evidence collected over a long period of time, e.g. [38], and preliminary data from a recent dedicated experiment [39], it can be safely assumed that transfer products are formed in reaction with kinematics very much resembling quasi-elastic transfer reactions. Therefore, recoil separators should accept in forward direction all products, which are emitted within a conical angular range of (minimum) of ±5°(ideally ±10°or more) and the centre of the cone located at the grazing angle, Θ_{gr}. For target-like heavy fragments, the corresponding angle in the laboratory frame is at about 32°relative to the beam direction. It should be noted, that a theoretical calculation including modelling of the kinematics of neutron-rich platinum isotopes formed in collisions of ^{136}Xe with ^{208}Pb targets shows the emission of theses products predominately under 0° [40]. From this calculation it was concluded that "head-on" collisions mainly contribute to the synthesis of heavy target-like transfer products. As Pt isotopes are much less fissile and can survive high excitation energies, this system cannot be seen as a suitable model for SHE. However, extending their model calculations to the ^{238}U on ^{248}Cm system, again the authors concluded that heavy actinides and SHE do not reveal any "grazing feature" and their wide-spread angular distribution (±25°in the laboratory frame) peaks at 0°; not only for reactions at the Coulomb barrier but also about 6% higher [40]. The discrepancy between results of this theory, with no grazing feature seen, and the assumption about typical grazing features for the kinematics of heavy and superheavy element products, as derived from the experimental observations outlined above, provides an interesting challenge.

(iii) Following the arguments based on experimental observations, as outlined in this work, products will be emitted with typical "grazing" velocities, which span a very wide kinetic energy range from about 20 MeV to ≥1000 MeV when going from ^{22}Ne to ^{238}U induced reactions, respectively. Consequently, when designing a separator, a decision has to be made to build (i) a universal one with a very large velocity acceptance or (ii) one which is dedicated either to studies of the heaviest actinides synthesized with light to medium heavy ($A \leq 50$) projectiles or to SHE where ^{238}U beams are advantageous.

(iv) Highly desirable is a separator design, which shows flexibility in the usual trade-off between separation efficiency and collection efficiency. While the former is required to reach an acceptable data rate in a focal-plane detector the latter one is needed when a second-stage in-flight separation is foreseen and for devices to further investigate chemical and physical properties. Automated chemistry set-ups, ion-traps, remote low-background nuclear spectroscopy devices or lasers for atomic excitation and ionization are just examples. All of them use rate insensitive buffer gas cells as an interface.

Within the IRiS project, a first step design study, it was shown that various separator options can provide the required high efficiency, selectivity and flexibility [41]. As the primary choice, a solenoid with 2-m length, a bore radius of 0.3 m and a maximum magnetic field of $B_{max} \approx 2.3$ T was studied. Adjustable positions of the target outside the solenoid and of the beam dump in its centre are valuable features. For ^{48}Ca-based reactions and ^{260}Fm as a typical transfer product, collection efficiencies between about 40% and 56% can been achieved depending on the desired separation efficiency from unwanted species, which ranged from >99.9% to 98%, respectively. For the ^{238}U on ^{248}Cm reaction collection efficiencies range from 12% to 30%, again with associated separation efficiencies of >99.9% to 98%, respectively. More options, including gas-filled magnets, are conceivable which may also include new designs of very large aperture quadrupoles in combination with a dipole. Obviously, not only the understanding of nuclear reactions leading to surviving heavy elements in multi-nucleon transfer reactions has made a breakthrough but also the technical options to perform such experiments and significantly move into the fascinating and yet uncharted territory of neutron-rich actinides and superheavy elements.

The time is ripe to revive multi-nucleon transfer reactions to discover and investigate neutron-rich isotopes of heavy actinides and of SHE with readily available, highly sensitive radiochemical techniques and dedicated, highly-efficient and "tailor-made" recoil separators down to today's experimentally achievable picobarn level.

It is my great pleasure and honour to acknowledge advice, help, suggestions and wonderful and most fruitful collaborations with G. Herrmann, N. Trautmann, J.V. Kratz, W. Brüchle, G. Wirth, H.W. Gäggeler, A. Türler, Ch.E. Düllmann, E.K. Hulet, R.W. Lougheed, J.F. Wild, D.C. Hoffman, D. Lee, K.E. Gregorich, G.T. Seaborg and K.J. Moody at various stages of my work; several named here as representatives for entire groups. I like to thank M. Götz for preliminary results (2016) on kinematic studies of the ^{48}Ca + ^{248}Cm reaction and J. Dvorak for private communications (2011) in the context of design options for dedicated recoil separators for multi-nucleon transfer products.

References

[1] M. Schädel, Phil. Trans. R. Soc. A **373**, 20140191 (2015).

[2] G. Herrmann, Historical Reminiscences: The Pioneering Years of Superheavy Element Research. In *The Chemistry of Superheavy Elements,* 2nd edn. (eds. M. Schädel, D. Shaughnessy), p. 485 (Springer, Berlin, 2014).

[3] Proceedings International Symposium. *Why and How Should We Investigate Nuclides far off the Stability Line.* Lysekil, 1966. Arkiv Fysik **36** (1967).

[4] W.D. Myers and W.J. Swiatecki, Nucl. Phys. **81**, 1 (1966).

[5] A. Sobiczewski, F.A. Gareev, B.N. Kalinkin, Phys. Lett. **22**, 500 (1966).

[6] R.-D. Herzberg, Nuclear Structure of Superheavy Elements. In *The Chemistry of Superheavy Elements*, 2nd edn. (eds. M. Schädel, D. Shaughnessy), p. 83 (Springer, Berlin, 2014).

[7] K. Morita *et al.*, J. Phys. Soc. Jpn. **81**, 103201 (2012).

[8] Yu.Ts. Oganessian and V.K. Utyonkov, Nucl. Phys. **A944**, 62 (2015).
[9] Ch.E. Düllmann *et al.*, Phys. Rev. Lett. **104**, 252701 (2010).
[10] T. Sikkeland, Ark. Fys. **36**, 539 (1967).
[11] Yu.Ts. Oganessian, Lecture Notes in Physics **33**, 221 (1975).
[12] M. Schädel *et al.*, Phys. Rev. Lett. **41**, 469 (1978).
[13] M. Schädel *et al.*, Phys. Rev. Lett. **48**, 852 (1982).
[14] M. Schädel, *Radiochemische Messungen der Bildungsquerschnitte von Aktinidenisotopen in der Reaktion von*^{238}U*-Ionen mit* ^{238}U, doctoral thesis, Johannes Gutenberg-University Mainz, 1979, and GSI-Report 79–8, Darmstadt (1979).
[15] J.V. Kratz, M. Schädel, H.W. Gäggeler, Phys. Rev. C **88**, 054615 (2013).
[16] M. Schädel *et al.*, Radiochim. Acta **25**, 111 (1978).
[17] R.W. Lougheed *et al.*, Nucl. Instrum. Methods **200**, 71 (1982).
[18] H. Gäggeler *et al.*, Phys. Rev. Lett. **45**, 1824 (1980).
[19] T. Sikkeland, A. Ghiorso, M.J. Nurmia, Phys. Rev. **172**, 1232 (1968).
[20] V.V. Volkov in *Proceedings of the International Conference on Reactions Between Complex Nuclei*, Nashville, 1974. (North-Holland, Amsterdam, 1974) Vol. II, p. 363.
[21] C. Riedel, W. Nörenberg Z. Phys. **A290**, 385 (1979).
[22] J.V. Kratz *et al.*, Phys. Rev. C **33**, 504 (1986).
[23] K.J. Moody *et al.*, Phys. Rev. C **33**, 1315 (1986).
[24] R.B. Welch *et al.*, Phys. Rev. C **35**, 204 (1987).
[25] K.E. Gregorich *et al.*, Phys. Rev. C **35**, 2117 (1987).
[26] D.C. Hoffman *et al.*, Phys. Rev. C **31**, 1736 (1985).
[27] D. Lee *et al.*, Phys Rev. C **27**, 2656 (1983).
[28] V.I. Zagrebaev and W. Greiner, Nucl. Phys. **A944**, 257 (2015).
[29] M. Schädel *et al.*, Phys. Rev. C **33**, 1547 (1986).
[30] J.F. Wild *et al.*, *Transfer Cross Sections from the Reaction of* ^{48}Ca *and* ^{254g}Es, Lawrence Livermore National Laboratory, Nuclear Chemistry Division Annual Report FY83, UCAR 10062-81/1, pp. 144–147.
[31] E.K. Hulet *et al.*, Phys. Rev. Lett. **56**, 313 (1986).
[32] E.K. Hulet *et al.*, Phys. Rev. C **40**, 770 (1989).
[33] M. Schädel *et al.*, Nucl. Instr. Meth. **A264**, 308 (1988), and Radiochim. Acta **48**, 171 (1989).
[34] A. Türler and K.E. Gregorich, Experimental Techniques. In *The Chemistry of Superheavy Elements*, 2nd edn. (eds. M. Schädel, D. Shaughnessy), p. 261 (Springer, Berlin, 2014).
[35] H. Koura *et al.*, *Chart of the Nuclides 2014* (Japan Atomic Energy Agency, Tokai, 2015) and `http://wwwndc.jaea.go.jp/CN14/index.html`.
[36] V.F. Comas *et al.*, Eur. Phys. J. A **49**, 112 (2013).
[37] C. Golabek et al., Eur. Phys. J. A **43**, 251 (2010).
[38] Yu.Ts. Oganessian *et al.*, Atomnaya Energiya **28**, 393 (1970).
[39] M. Götz, S. Götz, J.V. Kratz *et al.*, *Radiochemical study of the kinematics of multinucleon transfer reactions in* $^{48}Ca+^{248}Cm$ *collisions 10% above the Coulomb barrier*, to be published (2016).
[40] V.I. Zagrebaev and W. Greiner, Phys. Rev. C **83**, 044618 (2011).
[41] J. Dvorak *et al.*, Nucl. Instr. Meth. **A652**, 687 (2011), and priv. comm.

EPJ Web of Conferences **131**, 04002 (2016) DOI: 10.1051/epjconf/201613104002
Nobel Symposium NS160 – Chemistry and Physics of Heavy and Superheavy Elements

Entrance channel effects in superheavy element production

Avazbek Nasirov[1,2,a], *Giorgio* Giardina[3,4,5,b], *Giuseppe* Mandaglio[3,4,5,c], and *Akhtam* Muminov[2]

[1] Joint Institute for Nuclear Research, 141980 Dubna, Russia
[2] Institute of Nuclear Physics, 100214 Tashkent, Uzbekistan
[3] Dipartimento di Scienze Matematiche e Informatiche, Scienze Fisiche e Scienze della Terra, University of Messina, 98166 Messina, Italy
[4] INFN Gruppo Collegato di Messina, Messina, Italy
[5] Dipartimento di Scienze Chimiche, Biologiche, Farmaceutiche ed Ambientali, University of Messina, 98166 Messina, Italy

Abstract. The difference between evaporation residue cross sections measured in the cold (X+^{208}Pb, ^{209}Bi) and hot (^{48}Ca+actinides) fusion reactions can be related to the stage of compound nucleus (CN) formation and/or to the stage of its survival against fission. The cold fusion reactions are favorable in synthesis of the superheavy elements (SHE) with charge numbers $Z < 112$ in comparison with the hot fusion reactions due to small excitation energy and large fission barrier of the CN formed in these reactions. The strong decrease of the cross sections of the synthesis of the SHE $Z = 113$ in the cold fusion reactions in comparison with the ones in the hot fusion reactions is the result of the increase of hindrance to the CN formation in the cold fusion reactions. The origin of the intrinsic fusion barrier, $B^*_{\rm fus}$, causing the strong decrease of the probability $P_{\rm CN}$ in the cold fusion is discussed.

1. Introduction

The behaviour of the experimental cross sections obtained in synthesis of the SHE with charge numbers $Z = 110$–118 can be explained by the strong role of the entrance channel characteristics and fission barrier of the CN in the reaction mechanism. The SHE with $Z = 110$–112 were synthesized in cold fusion reactions with Pb and Bi targets which have large binding energies [1]. But the smallest cross section of the synthesis of the SHE with $Z = 113$ observed in the ^{70}Zn + ^{209}Bi [2] reaction demonstrated the hopelessness of using the cold fusion reactions to synthesize heavier $Z > 113$ elements. The reactions with the projectile ^{48}Ca and actinide targets are favourable in the synthesis of the new SHE Fl, 115, Lv, 117, and 118 [3, 4].

The evaporation residue (ER) cross sections obtained at the synthesis of the SHE $Z = 106$–111 in the cold fusion reactions were sufficiently larger than with the ones observed in the hot fusion reactions (see Fig. 1) since the CN formed in the last reaction has a large

[a] e-mail: nasirov@jinr.ru
[b] e-mail: ggiardina@unime.it
[c] e-mail: gmandaglio@unime.it

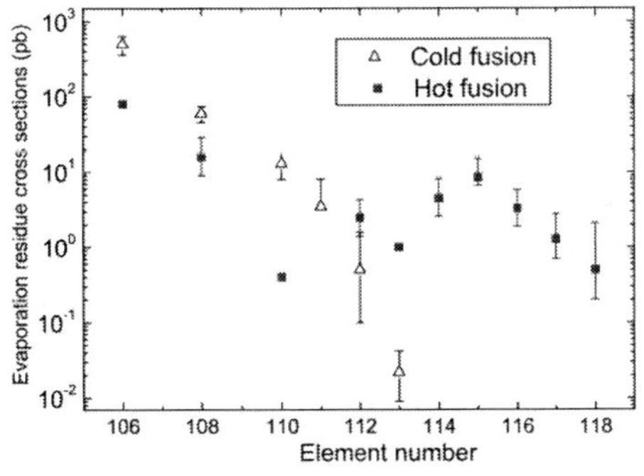

Figure 1. The comparison of the ER cross sections obtained in the cold (open triangles) and hot (solid squares) fusion reactions used to synthesize SHE.

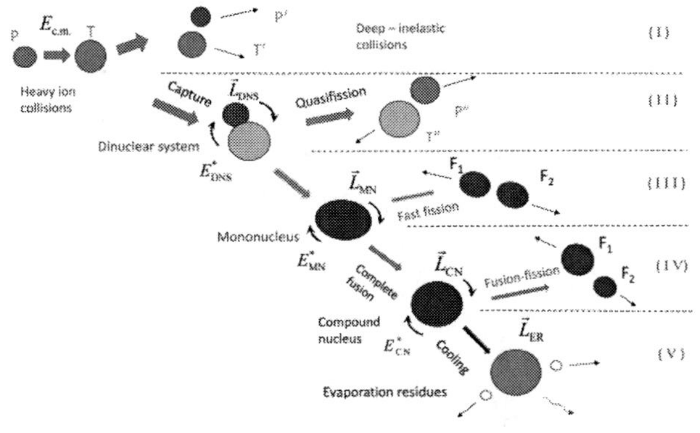

Figure 2. Sketch of reaction channels (steps I to V) leading to the formation of binary products and reducing the probability for the synthesis of superheavy elements (V) in a collision of a projectile ion P with a target nucleus T. P′ and T′ are deep-inelastic collision (incomplete momentum transfer) products, (P″ and T″) quasifission (total momentum transfer) products and (F_1 and F_2) fast fission and fusion-fission products.

excitation energy. Consequently, the survival probability of the very excited CN during de-excitation cascade is small and this fact leads to decreasing the cross section of the synthesis in the hot fusion reactions. But starting the synthesis of Cn ($Z = 112$) the evaporation residues' yield in the experiments by the cold fusion became lower in comparison with the ones observed in the reactions with the ^{48}Ca projectile. This circumstance can be explained by the increase of the hindrance in formation of the CN in the cold fusion reactions caused by the dominance of the quasifission process.

In spite of the successes in the SHE synthesis there is no unambiguous mechanism of the complete fusion of nuclei in the heavy-ion collisions. We will discuss the processes preceding the formation of SHE, and the influence of the entrance channel on the probability of the CN formation.

A theoretical model to calculate the cross section of processes in heavy-ion collisions has been developed on the basis of the realistic concept about reaction mechanism [5, 6]. The sequence of the stages of the heavy-ion collisions leading to the synthesis of SHE (σ_{ER}) shown in Fig. 2 is calculated by the corresponding equations:

$$\sigma_{\mathrm{ER}}(E) = \sum_{\ell=0}^{\ell_d(E)} (2\ell+1)\sigma_{\mathrm{cap}}(E,\ell)P_{\mathrm{CN}}(E,\ell)W_{sur}(E,\ell), \tag{1}$$

where σ_{cap}, P_{CN}, and W_{sur} are the capture cross section, fusion probability, and survival probability, respectively. The methods of calculation used to estimate the ER cross sections are listed here:

1. The partial capture cross sections are found by the solution of the equations of the relative and tangential motions. As a result the values of the orbital angular momentum ($0 < \ell < \ell_d$) contributing to the interaction of nuclei are determined [5, 7].
2. Changes of the initial charge and mass numbers of the interacting nuclei at the capture stage is calculated by the set of transport equations-master equations. The transition coefficients depend on the shell effects in nuclei due to the used microscopic method of their calculation. The method of calculation is presented in the appendix of Ref. [8].
3. The probability P_{CN} of the CN formation from the dinuclear system (DNS) state is calculated statistically by the use of the branching ratio of level densities on the Businaro-Gallone, on the minimum of the potential well and on the maximum of the way to symmetric configuration [9].
4. The ER yield after emission neutrons, γ quanta and charged particles (protons and α particles) from the heated and rotating compound nucleus during de-excitation cascade is determined by the advanced statistical model [10, 11].

2. The advantage of cold fusion reactions in synthesis of SHE $Z < 112$

The CN is formed by the transformation of the DNS in competition with the quasifission. The last process is the decay of the DNS formed at capture of the projectile nucleus by the target nucleus without reaching the CN equilibrium shape. See stage (II) in Fig. 2. The time needed to transform the DNS into a CN is large in the case of collision of the massive nuclei since there is the intrinsic fusion barrier which must be overcome, and the alternative quasifission process has the huge cross section and its angular distribution may be very wide [12, 13]. But the realistic lifetime of the DNS may be not enough for the complete fusion. As a result the quasifission is the dominant process in the stage (II) of the DNS evolution to the complete fusion.

The branching ratio of the competition between complete fusion and quasifission processes is determined by the peculiarities of the DNS which is calculated as a sum of the total interaction potential V of nuclei and reaction energy balance (Q_{gg}):

$$U(Z_1, A_1, Z_2, A_2, \ell; R) = Q_{gg} + V(Z_1, A_1, Z_2, A_2, \ell; R) - V_{rot}^{CN}(\ell), \tag{2}$$

where A_i and Z_i ($i = 1, 2$) are mass and charge numbers of the DNS fragments; R is the internuclear distance between their centres-of-mass; $Q_{\text{gg}} = B_1 + B_2 - B_{\text{CN}}$ is the reaction energy balance; B_1, B_2 and B_{CN} are the binding energies of the interacting nuclei and CN, respectively, which are obtained from the nuclear mass tables in references [14, 15]; $V_{rot}^{CN}(\ell)$ is the CN rotational energy.

The nucleus-nucleus potential $V(A, Z, \ell; R)$ consists of the three parts:

$$\begin{aligned} V(Z_1, A_1, Z_2, A_2, \ell; R) &= V_{\text{Coul}}(Z_1, A_1, Z_2, A_2; R) + V_{\text{nucl}}(Z_1, A_1, Z_2, A_2; R) \\ &\quad + V_{\text{rot}}(Z_1, A_1, Z_2, A_2, \ell; R), \end{aligned} \tag{3}$$

where V_{Coul}, V_{nucl}, and V_{rot} are the Coulomb, nuclear, and rotational potentials, respectively. The methods of calculation of these three parts of the nucleus-nucleus potential are presented in appendix A of Refs. [5, 7].

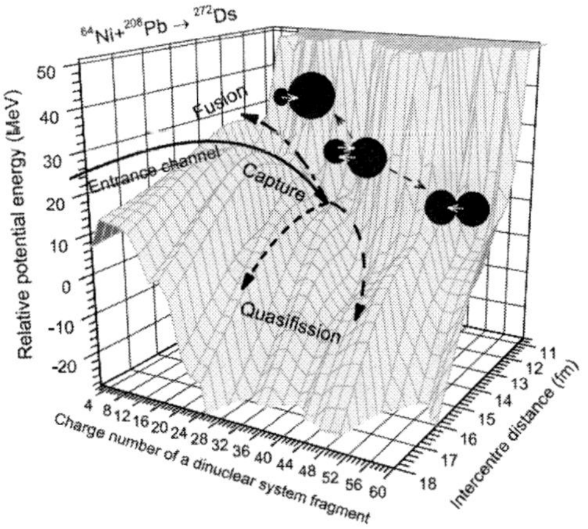

Figure 3. PES calculated for the DNS formed via ^{64}Ni + ^{208}Pb. The arrows show conditionally directions of the DNS evolution to complete fusion and quasifission after capture in the entrance channel.

The difference in the evolution path of the DNS formed in the ^{64}Ni + ^{208}Pb and ^{34}S + ^{244}Pu reactions in the direction of the complete fusion is seen from the comparison of Figs. 3 and 4, respectively, which are presented as a function of the charge number of the DNS fragments and internuclear distance R between their centres-of-masses. The position of the DNS energy path on the potential energy surface (PES) at capture (the end of the solid arrow in Figs. 3 and 4) is important for the transformation of the DNS into the CN. The capture on the PES for the ^{34}S + ^{244}Pu reaction ($Z_p = 16$) is close to the maximum (Businaro-Gallone point) on the way to complete fusion while the initial position of the ^{64}Ni + ^{208}Pb system at capture is placed sufficiently lower ($Z_p = 28$). The difference in the potential energy axis between the minimum of the potential well where the path of the relative kinetic energy trapped ($Z_{\text{capture}} \sim Z_{\text{projectile}}$) and Businaro-Gallone point is used as the intrinsic fusion barrier for the complete fusion B^*_{fus} (see Fig. 5). The appearance of the B^*_{fus} is related to the large binding energy of nucleons in the nuclei with magic proton and neutron numbers. For the heavier and mass symmetric DNS system the depth of this valley is much lower and the value of B^*_{fus} increases causing the hindrance to transformation of the DNS into CN through nucleon transfer.

Therefore, the fusion cross section of the ^{64}Ni + ^{208}Pb reaction is smaller than the one for the ^{34}S + ^{244}Pu reaction but the excitation energy E^*_{CN} of the CN formed in the latter reaction is large ($E^*_{\text{CN}} > 35$ MeV). The large excitation energy and small fission barrier for the isotopes of Ds with the neutron numbers $N = 164$–168 (see Fig. 6) make the survival probability $W_{\text{sur}}(E, \ell)$ of the CN ^{278}Ds formed in the ^{34}S + ^{244}Pu reaction very small. Consequently, the cross section of the synthesis of Ds in the latter reaction is smaller than the one obtained in the cold fusion reaction, ^{64}Ni + ^{208}Pb (see Fig. 1). The values of the fission barrier calculated for the isotopes of Ds and Cn by Kowal *et al.* [16] are presented in Fig. 6 as a function of the neutron number.

The large fission barrier and small excitation energy of the CN ^{272}Ds ($N = 162$) formed in the ^{64}Ni + ^{208}Pb reaction increase very strongly its survival probability. The gain in the surviving stage (V) of the cold fusion reaction can guarantee large cross section of the synthesis in spite of the strong hindrance to complete fusion. Figure 7 obtained from Ref. [17] demonstrates that the fusion excitation function (dashed curve) is much lower than the quasifission (dotted) curve due to the large value B^*_{fus} in the ^{64}Ni + ^{208}Pb reaction. The experimental data on the synthesis of Ds in the ^{64}Ni + ^{208}Pb reaction were reproduced

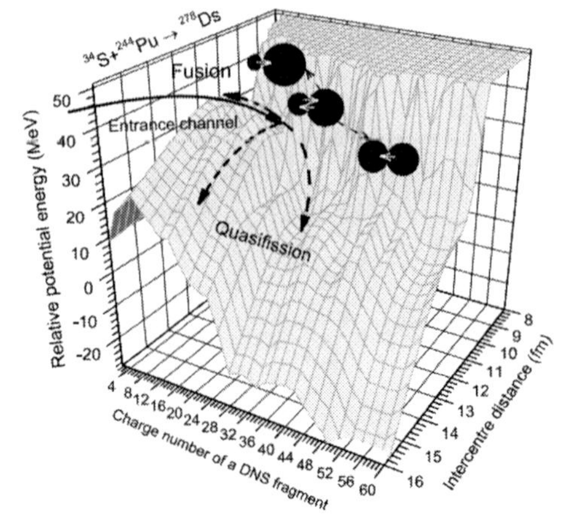

Figure 4. The same as in Fig. 3, but for the reaction $^{34}S + ^{244}Pu$.

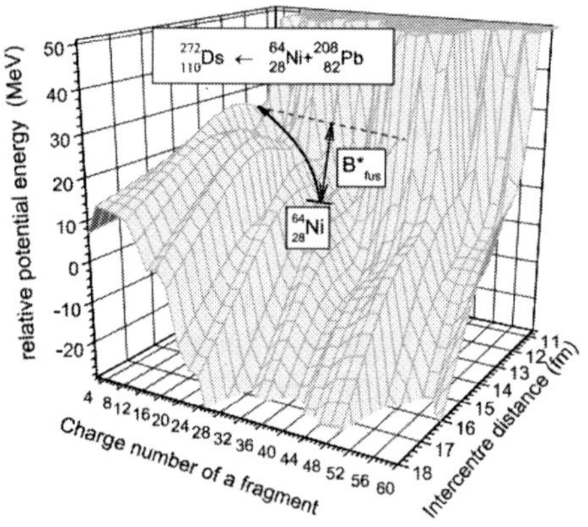

Figure 5. The intrinsic fusion barrier, $B^*_{\rm fus}$, is determined by the difference between the values of the PES corresponding to the entrance channel ($Z = 28$) and its maximum value (similar to the Businaro-Gallone point) in the way to fusion.

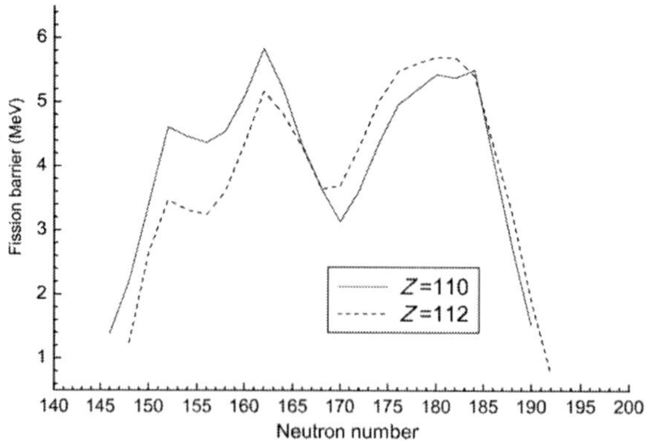

Figure 6. Fission barriers for the isotopes Ds and Cn obtained from Ref. [16] as a function of neutron numbers.

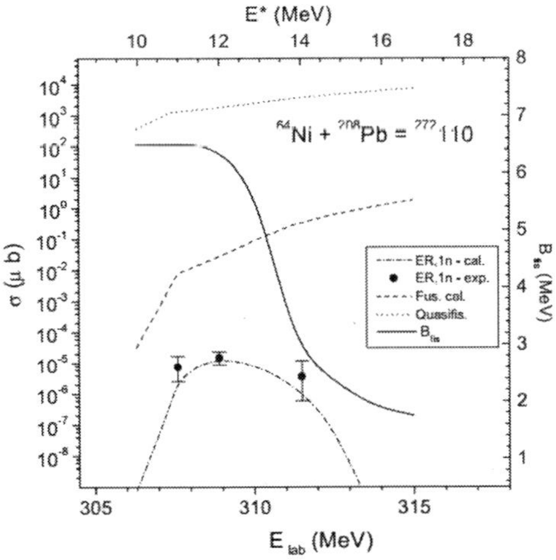

Figure 7. Theoretical quasifission (dotted curve), fusion (dashed curve) and $1n$-evaporation residue (dot-dashed) excitation functions (left axis) for the reaction ^{64}Ni + ^{208}Pb obtained from Ref. [17]. The solid curve shows the fission barrier for the compound nucleus ^{272}Ds (right ordinate axis). The experimental cross sections of the evaporation residue formation in $the 1n$-channel are shown by the solid circles [1].

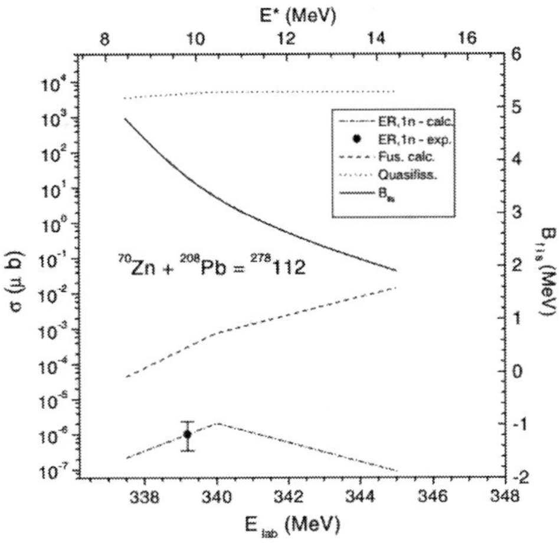

Figure 8. Same as in Fig. 7 but for the ^{64}Ni + ^{208}Pb reaction.

well. Nevertheless the ER cross section in this cold fusion reaction is larger than the one obtained in the hot fusion ^{34}S + ^{244}Pu reaction. It is seen that, the smallness of the ER cross section in the ^{34}S + ^{244}Pu reaction is related to the small fission barrier (3.8 MeV) and large excitation energy of the CN, which is formed. The restriction to the minimum value of E_{CN} is determined by the landscape of potential energy surface (PES) which is high in the entrance channel for the mass asymmetric ^{34}S + ^{244}Pu reaction.

3. The advantage of the hot fusion reactions in synthesis of SHE $Z \geq 112$

The hot fusion reactions with the ^{48}Ca beam are favourable in the synthesis of SHE $Z \geq 112$ (see Fig. 9). The doubly-magic ^{48}Ca is a neutron-rich isotope and its use as a beam increases

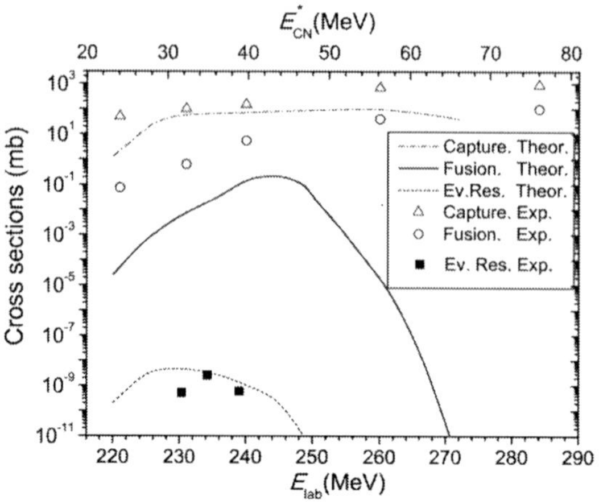

Figure 9. Theoretical capture (dash-dotted curve), fusion (solid curve), and evaporation residue (dashed) excitation functions for the $^{48}Ca+^{238}U$ reaction. The experimental cross sections of the evaporation residue are presented by solid squares [18].

the survival probability of the heated CN. The main effects of the use of ^{48}Ca beam are the large fusion cross section encouraging synthesis of the SHE $Z = 112$–118 and the decrease of the excitation energy E^*_{CN} which is large in the mass asymmetric reactions. The intrinsic fusion barrier is small for the $^{48}Ca+^{238}U$ reaction since the position of the capture point on the PES landscape is placed high as in the case of the $^{34}S+^{244}Pu$ reaction presented in Fig. 4. The fact of the cross sections of the synthesis of the SHE $Z \geq 112$ in the hot reaction become larger than the ones in the cold fusion (see Fig. 1) is the result of the dramatic hindrance to the formation of CN in the cold fusion $^{70}Zn+^{208}Pb$ (dashed curve in Fig. 8) reaction leading to the CN with $Z = 112$. The large value of B^*_{fus} is a ramatic hindrance to complete fusion. The hot fusion reaction is more favourable in comparison with the cold fusion reaction is explained by the effect of the entrance channel: large fusion cross section in the $^{48}Ca+^{238}U$ reaction (see Fig. 9). Therefore, the hot fusion (mass asymmetric) reactions are out of competition in the synthesis of new SHE $Z \geq 112$ since the B^*_{fus} values are relatively small and the condition of the CN formation is favourable. The dependence of the fission barrier [16] on the neutron numbers in the isotopes of the SHE Cn presented in Fig. 6 shows that the fission barrier of the CN ^{286}Cn ($N = 174$) formed in the $^{48}Ca+^{238}U$ reaction is a little larger ($B_f = 5.01$ MeV) than the one ($B_f = 4.29$ MeV) of the CN ^{278}Cn ($N = 166$) formed in the $^{70}Zn+^{208}Pb$ reaction. The large fusion cross sections in the entrance channel and large fission barriers $B_f = 6.34$ and 6.07 MeV [16] for the CN formed in the $^{48}Ca+^{244}Pu$ and ^{248}Cm reactions are favourable in the synthesis of Fl and Lv.

A.K. Nasirov is grateful to the Nobel Foundation for supporting his participation in the Nobel Symposium NS160.

References

[1] S. Hofmann, Rev. Mod. Phys. **72**, 733 (2000).
[2] K. Morita *et al.*, J. Phys. Soc. Jpn. **81**, 103201 (2012).
[3] Yu.Ts. Oganessian *et al.*, Phys. Rev. C **69**, 054607 (2004).
[4] Yu.Ts. Oganessian *et al.*, Phys. Rev. C **74**, 044602 (2006).

[5] A. Nasirov *et al.*, Nucl. Phys. **A759**, 342 (2005).
[6] K. Kim *et al.* Phys. Rev. C **91**, 064608 (2015).
[7] G. Fazio *et al.*, J. Phys. Soc. Jpn. **72**, 2509 (2003).
[8] A.K. Nasirov *et al.*, Phys. Rev. C **84**, 044612 (2011).
[9] A.K. Nasirov *et al.*, Eur. Phys. J. **A34**, 325 (2007).
[10] A. D'Arrigo *et al.*, J. Phys. G **20**, 365 (1994).
[11] R.N. Sagaidak *et al.*, J. Phys. G **24**, 611 (1998).
[12] A.K. Nasirov *et al.*, J. Phys.: Conf. Series **205**, 012018 (2010).
[13] C. Simenel *et al.*, Phys. Lett. **B710**, 607 (2012).
[14] P. Möller, J.R. Nix, W.D. Myers, W.J. Swiatecki, At. Data Nucl. Data Tables **59**, 185 (1995).
[15] G. Audi and A.H. Wapstra, Nucl. Phys. **A595**, 509 (1995).
[16] M. Kowal *et al.* Phys. Rev. C **82**, 014303 (2010).
[17] G. Giardina *et al.* Eur. Phys. J. **A8**, 2056 (2000).
[18] Yu.Ts. Oganessian *et al.* Phys. Rev. C **70**, 064609 (2004).

EPJ Web of Conferences **131**, 04003 (2016) DOI: 10.1051/epjconf/201613104003
Nobel Symposium NS160 – Chemistry and Physics of Heavy and Superheavy Elements

Characterizing the mechanism(s) of heavy element synthesis reactions

Walter Loveland[a]

Department of Chemistry, Oregon State University, Corvallis, OR 97331, USA

Abstract. A review of the current state of our understanding of complete fusion reaction mechanisms is presented, from the perspective of an experimentalist. For complete fusion reactions, the overall uncertainties in predicting heavy element synthesis cross sections are examined in terms of the uncertainties associated with the calculations of capture cross sections, fusion probabilities and survival probabilities.

1. Introduction

Formally, the cross section for producing a heavy evaporation residue, σ_{EVR}, in a fusion reaction can be written as

$$\sigma_{\mathrm{EVR}}(E) = \frac{\pi h^2}{2\mu E}\sum_{\ell=0}^{\infty}(2\ell+1)T(E,\ell)P_{\mathrm{CN}}(E,\ell)W_{\mathrm{sur}}(E,\ell) \qquad (1)$$

where E is the center of mass energy, and T is the probability of the colliding nuclei to overcome the potential barrier in the entrance channel and reach the contact point. (The term "evaporation residue" refers to the product of a fusion reaction followed by the evaporation of a specific number of neutrons.) P_{CN} is the probability that the projectile-target system will evolve from the contact point to the compound nucleus. W_{sur} is the probability that the compound nucleus will decay to produce an evaporation residue rather than fissioning. Conventionally the EVR cross section is separated into three individual reaction stages (capture, fusion, survival) motivated, in part, by the different time scales of the processes. However, one must remember that the W_{sur} term effectively sets the allowed values of the spin. This effect is shown in Fig. 1 where the capture cross sections for several reactions are shown without and with the spin limitation posed by the survival probabilities.

Several successful attempts have been made to describe the cross sections for evaporation residue formation in cold fusion reactions [2–5]. In Fig. 2a, I show some typical examples of post dictions of the formation cross sections for elements 104–113 in cold fusion reactions. The agreement between theory and experiment is impressive because the cross sections extend over six orders of magnitude, i.e., a robust agreement. Because the values of $\sigma_{\mathrm{capture}}$ are well known or generally agreed upon (see below), the values of the product $P_{\mathrm{CN}} \cdot W_{\mathrm{sur}}$ are the same in most of these post dictions. However, as seen in Fig. 2b, the values of P_{CN} differ significantly in these post dictions [2–5], and differ from measurements of P_{CN} [6]. A similar situation occurs in predictions of cross sections for hot fusion reactions. These are clear-cut cases in which a simple agreement between theory and experiment in postdicted cross sections is not sufficient to indicate a real understanding of the phenomena involved.

[a] e-mail: `lovelanw@onid.orst.edu`

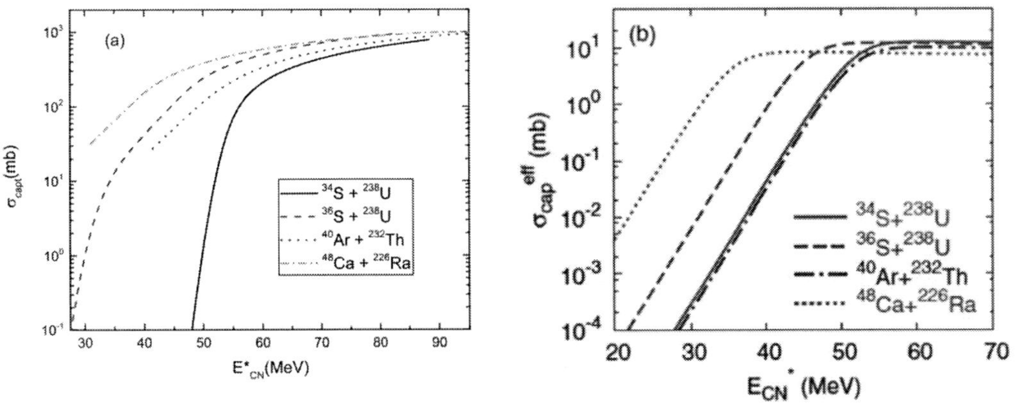

Figure 1. (a) Calculated capture cross sections for some typical reactions. (b) the "spin-limited" capture cross sections for the reactions in (a) [1].

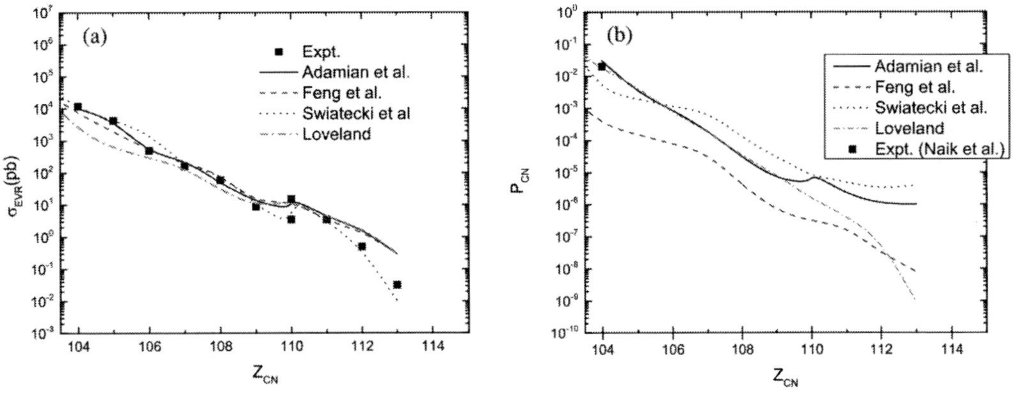

Figure 2. (a) Typical predictions of the formation cross sections of elements 104–113 using cold fusion reactions. (b) P_{CN} values for the predictions in panel (a). The references cited in the legends refer to Adamian [4], Feng [5], Swiatecki [2], and Loveland [3]. The additional reference in panel (b) is to the data of Naik *et al.* [6].

We might ask what the overall uncertainties are in the current phenomenological models for predicting heavy element production cross sections. This is an item of some controversy. Some feel the uncertainties in typical predictions are factors of 2–4 [7] while others estimate these uncertainties to be 1–2 orders of magnitude [8, 9].

2. Capture cross sections

The capture cross section is, in the language of coupled channel calculations, the "barrier crossing" cross section. It is the sum of the quasifission, fast fission, fusion-fission and fusion-evaporation residue cross sections. The barriers involved are the interaction barriers and not the fusion barriers. There are several models for capture cross sections. Each of them has been tested against a number of measurements of capture cross sections for reactions that, mostly, do not lead to the formation of the heaviest nuclei. In general, these models are able to describe the magnitudes of the capture cross sections within 50% and the values of the interaction barriers within 20%. The most robust of these models takes into account the effects of target and projectile orientation/deformation upon the collisions, the couplings

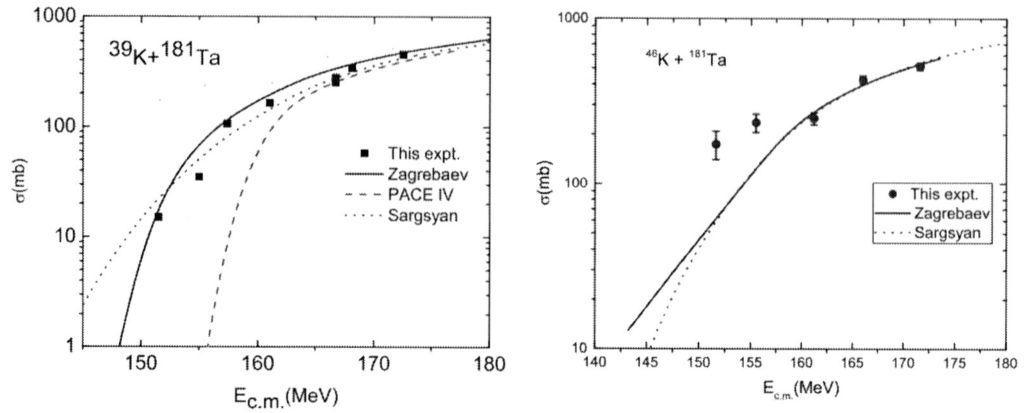

Figure 3. The capture-fission excitation functions for the reaction of ^{39}K (left) and ^{46}K (right) with ^{181}Ta.

associated with inelastic excitations of the target and projectile and the possibility of one or two neutron transfer processes. Loveland [10] has compared calculations of the capture cross sections for reactions that synthesize heavy nuclei with the measured cross sections. Good agreement between the measured and calculated values of the cross sections occurs for all reactions. The ratio of calculated to observed capture cross sections varies from 0.5 to 2. Nominally, given the other uncertainties in estimating σ_{EVR}, this seems generally acceptable. However, from the point of view of an experimentalist, it is not acceptable. The capture cross section is relatively easy to measure and an uncertainty of a factor of 50% may mean having to run an experiment for several months longer to be successful in a synthetic effort.

Future synthetic efforts with heavy nuclei may involve the use of very neutron-rich beams such as radioactive beams. While we understand that such efforts are not likely to produce new superheavy elements due to the low intensities of the radioactive beams [3], there may exist a window of opportunity to make new neutron-rich isotopes of elements 104–108 [11]. Our ability to predict the capture cross sections for the interaction of very neutron-rich projectiles with heavy nuclei is limited [11] and this is especially true near the interaction barrier where the predictions of models of the capture cross sections may differ by orders of magnitude [11]. As part of the effort to study the scientific issues that will be relevant at next generation radioactive beam facilities, such as FRIB, we have started to use the ReA3 facility at the NSCL to study capture processes with fast-releasing beams such as the potassium isotopes. We chose to study the interaction of ^{39,46}K with ^{181}Ta. The first *preliminary* results from that experiment are shown in Fig. 3. The ^{39}K + ^{181}Ta reaction results seem to be well understood within conventional pictures of capture [12, 13] while the neutron-rich ^{46}K results suggest an unusual near barrier fusion enhancement.

3. Survival probabilities, W_{sur}

Formally W_{sur} can be written as

$$W_{\mathrm{sur}}(E_{\mathrm{c.m.}}) = P_{xn}(E^*_{\mathrm{CN}}) \prod_{i=1}^{x} \frac{\Gamma_n(E^*_i)}{\Gamma_n(E^*_i) + \Gamma_f(E^*_i)} \tag{2}$$

where P_{xn} is the probability of emitting x (and only x) neutrons from a nucleus with excitation energy E^*, and Γ_n and Γ_f are the partial widths for decay of the completely fused system by either neutron emission or fission, respectively. For the most part, the formalism for calculating the survival, against fission, of a highly excited nucleus is understood. There are significant uncertainties, however, in the input parameters for these calculations and care must be used in treating some situations. "Kramers effects" and the overall fission barrier height are found [14] to have the biggest effect on the calculated cross sections.

A recent experiment concerning survival probabilities in hot fusion reactions showed the importance of "Kramers effects" [15]. The nucleus ^{274}Hs was formed at an excitation energy of 63 MeV using the ^{26}Mg + ^{248}Cm reaction. ^{274}Hs has several interesting properties. The liquid drop model fission barrier height is zero and there is a subshell at $N = 162$, $Z = 108$. In the formation reaction, P_{CN} is measured [16] to be 1.0. By measuring the angular distribution of the fission associated neutrons, Yanez *et al.* [15] were able to deduce a value of Γ_n/Γ_{total} for the first chance fission of ^{274}Hs ($E^* = 63$ MeV) of 0.89 ± 0.13! A highly excited fragile nucleus with a vanishingly small fission barrier decayed $\sim 90\%$ of the time by emitting a neutron rather than fissioning. Conventional calculations with various values of the fission barrier height were unable to reproduce these results. The answer to this dilemma is to consider the effects of nuclear viscosity to retard fission [17], the so-called Kramers effects. These Kramers effects are the reason that hot fusion reactions are useful in heavy element synthesis, in that the initial high excitation energies of the completely fused nuclei do not result in catastrophic losses of surviving nuclei [18].

With respect to fission barrier heights, most modern models do equally well/poorly in describing fission barrier heights for Th-Cf nuclei. Afanasjev *et al.* [19] found the average deviation between the calculated and known inner barrier heights was $\sim$0.7 MeV amongst various models. Bertsch *et al.* [20] estimate the uncertainties in fission barrier heights are 0.5–1.0 MeV in known nuclei. Kowal *et al.* [21] found for even-even nuclei with $Z = 92-98$ the difference between measured and calculated inner barrier heights was 0.8 MeV. Baran *et al.* [22] found very large, i.e., several MeV, differences between various calculated fission barrier heights for $Z = 112$–120. In summary, fission barrier heights are known within 0.5–1.0 MeV. For super-heavy nuclei, the change of fission barrier height by 1 MeV in each neutron evaporation step can cause an order of magnitude uncertainty in the $4n$-channel. For the $3n$-channel, the uncertainty is about a factor of four [14].

An additional problem is that at the high excitation energies characteristic of hot fusion reactions, the shell effects stabilizing the fission barrier are predicted [23] to be "washed out" with a resulting fission barrier height < 1 MeV for some cases. Furthermore the rate of damping of the shell effects differs from nucleus to nucleus. This point is well illustrated by the calculations in reference [24] of the "effective" fission barrier heights for the ^{48}Ca + ^{249}Bk reaction.

Measurements of fission barrier heights are difficult and the results depend on the models used in the data analysis. Recently Hofmann *et al.* [25] have deduced the shell-correction energies from the systematics of the Q_α values for the heaviest nuclei and used these shell-correction energies to deduce fission barrier heights. The deduced barrier heights for elements 118 and 120 may be larger than expected.

4. Fusion probability, P_{CN}

The fusion probability, P_{CN}, is the least known (experimentally) of the factors affecting complete fusion reactions and perhaps the most difficult to model. The essential task is to measure the relative amounts of fusion-fission and quasifission in a given reaction. Experimentally this is done using mass-angle correlations where it is difficult to measure, with any certainty, the fraction of fusion reactions when that quantity is less than 1%. (For cold fusion reactions, P_{CN} is predicted to take on values of 10^{-2} to 10^{-6} for reactions that make elements

104–113.) The reaction of ^{124}Sn with ^{96}Zr can be used to illustrate the uncertainties in the theory to estimate P_{CN} where [10] various theoretical estimates of P_{CN} range from 0.0002 to 0.56 and where the measured value is 0.05 as well as the data shown in Fig. 2b.

Where we have made progress in understanding P_{CN} is in the excitation energy dependence of P_{CN}. Zagrebaev and Greiner [26] have suggested the following ad hoc functional form for the excitation energy dependence of P_{CN}

$$P_{CN}(E^*, J) = \frac{P^0_{CN}}{1 + \exp[\frac{E^*_B - E^*_{int}(J)}{\Delta}]} \tag{3}$$

where P^0_{CN} is the fissility-dependent "asymptotic" (above barrier) value of P_{CN} at high excitation energies, E^*_B is the excitation energy at the Bass barrier, $E^*_{int}(J)$ is the internal excitation energy $[E_{c.m.} + Q - E_{rot}(J)]$, J is the angular momentum of the compound nucleus, and Δ (an adjustable parameter) is taken to be 4 MeV. This formula describes the extensive data of Knyazheva *et al.* [27] for the $^{48}Ca + {}^{154}Sm$ reaction very well [10]. A generalization of this formula has been used to describe the excitation energy dependence of P_{CN} for the reactions of ^{48}Ca with ^{238}U, ^{244}Pu, and ^{248}Cm [28].

It is also clear that P_{CN} must depend on the entrance channel asymmetry of the reaction. Numerous scaling factors to express this dependence have been proposed and used. An extensive survey of P_{CN} in a large number of fusing systems was made by du Rietz *et al.* [29]. They thought that perhaps some fissility related parameter would be the best way to organize their data on P_{CN} and its dependence of the properties of the entrance channel in the reactions they studied. They found the best fissility-related scaling variable that organized their data was $x_{duRietz} = 0.75x_{eff} + 0.25x_{CN}$. The parameters x_{eff} and x_{CN} are the associated fissilities for the entrance channel and the compound system, respectively. The following equations can be used to calculate these quantities

$$x_{CN} = \frac{(Z^2/A)_{CN}}{(Z^2/A)_{critical}}; \quad (Z^2/A)_{critical} = 50.883 \cdot \left[1. - 1.7826\left(\frac{A-2Z}{A}\right)^2\right];$$

$$x_{eff} = \frac{4Z_1Z_2/[A_1^{1/3}A_2^{1/3}(A_1^{1/3} + A_2^{1/3})]}{(Z^2/A)_{critical}}.$$

In Fig. 4, I show most of the known data on P_{CN} using the du Rietz scaling variable. There is no discernable pattern. Restricting the choice of cases to those in a narrow excitation energy bin improves the situation somewhat but it is clear we are missing something in our semi-empirical systematics.

Some progress has been made in calculating P_{CN} using TDHF calculations. Wakhle *et al.* [30] made a pioneering study of the $^{40}Ca + {}^{238}U$ reaction. The capture cross sections predicted by their TDHF calculations agreed with measured capture cross sections [31] within $\pm 20\%$. In addition they were able to predict the ratio of fusion to capture cross sections, $\sigma_{fus}/\sigma_{capture}$, as 0.09 ± 0.07 at 205.9 MeV and 0.16 ± 0.06 at 225.4 MeV in agreement with reference [31] who measured ratios of 0.06 ± 0.03 and 0.14 ± 0.05, respectively. Whether TDHF calculations will become a predictive tool for heavy element synthesis remains to be seen.

5. Predictions for the production of elements 119 and 120

Loveland [10] has shown that the current predictions for the production cross sections for elements 119 and 120 differ by 1–3 orders of magnitude, reflecting the uncertainties discussed above. For the reaction $^{50}Ti + {}^{249}Bk \rightarrow 119$, the uncertainties in the predicted maximum

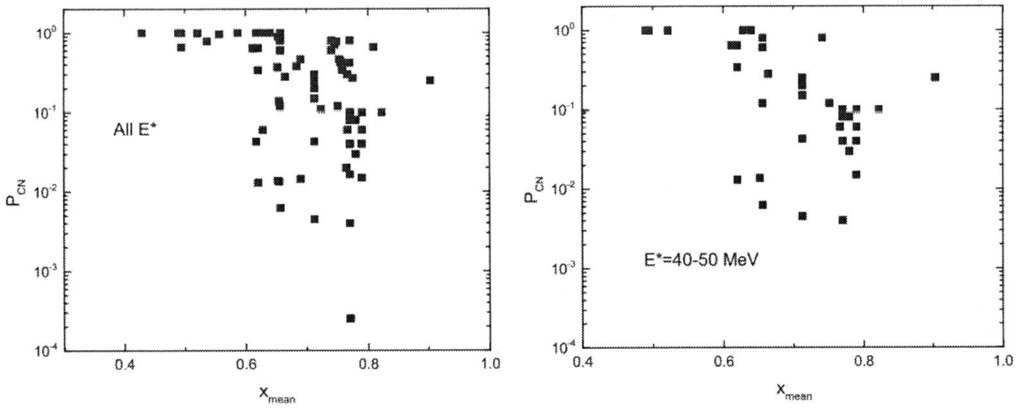

Figure 4. Fissility dependence of P_{CN} with and without excitation energy sorting.

cross sections for the 3*n* and 4*n* channels differ by "only" a factor of 20–40 while larger uncertainties are found in the predictions for the $^{54}Cr + ^{248}Cm \rightarrow$ 120 reaction. The energies of the maxima of the 3*n* and 4*n* excitation functions are uncertain to 3–4 MeV, a troublesome situation for the experimentalist.

6. Conclusions

I conclude that: (a) capture cross sections should be measured for reactions of interest. (b) We need better and more information on fission barrier heights, and their changes with excitation energy for the heaviest nuclei. (c) We need to devise better methods of measuring P_{CN} and more TDHF calculations of P_{CN}. (d) The current uncertainty in calculated values of σ_{EVR} is at least 1–2 orders of magnitude. (e) New opportunities for making neutron-rich actinides with RNBs may exist.

This work was supported, in part, by the Director, Office of Energy Research, Division of Nuclear Physics of the Office of High Energy and Nuclear Physics of the U.S. Department of Energy under Grant DE-SC0014380 and the National Science Foundation under award 1505043.

References

[1] J. Hong, G.G. Adamian, N.V. Antonenko, Phys. Rev. C **92**, 014617 (2015).
[2] W.J. Swiatecki, K. Siwek-Wilczynska, J. Wilczynski, Phys.Rev. C **71**, 014602 (2005).
[3] W. Loveland, Phys. Rev. C **76**, 014612 (2007).
[4] G.G. Adamian, N.V. Antonenko, W. Scheid, Nucl. Phys. **A678**, 24 (2000).
[5] Z.-Q. Feng, G.-M. Jin, J.-Q. Li, W. Scheid, Phys. Rev. C **76**, 044606 (2007).
[6] R.S. Naik *et al.*, Phys. Rev. C **76**, 054604 (2007).
[7] G.G. Adamian, N.V. Antonenko, W. Scheid, Phys. Rev. C **69**, 014607 (2004).
[8] V.I. Zagrebaev, Y. Aritomo, M.G. Itkis, Y.T. Oganessian, M. Ohta, Phys. Rev. C **65**, 014607 (2001).
[9] V.I. Zagrebaev and W. Greiner, Nucl. Phys. **A944**, 257 (2015).
[10] W. Loveland, Eur. Phys. J. **A51**, 120 (2015).
[11] W. Loveland, J. Phys. Conf. Series **420**, 012004 (2013).
[12] `http://nrv.jinr.ru/nrv/`

[13] V.G. Sargsyan (private communication).
[14] H. Lu, D. Boilley, EPJ Web of Conferences **62**, 03002 (2013).
[15] R. Yanez *et al.*, Phys. Rev. Lett. **112**, 152702 (2014).
[16] M.G. Itkis *et al.*, AIP Conf. Proc. **853**, 231 (2006).
[17] H.A. Kramers, Physica **7**, 284 (1940).
[18] A.N. Andreyev *et al.*, Heavy Ion Fusion: Exploring the Variety of Nuclear Properties (Singapore, World Scientific, 1994), 260.
[19] A.V. Afanasjev, H. Abushara, P. Ring, Int. J. Mod. Phys. E **21**, 1250025 (2012).
[20] G. Bertsch, W. Loveland, W. Nazarewicz, P. Talou, J. Phys. G **42**, 1 (2015).
[21] M. Kowal, P. Jachimowicz, A. Sobiczewski, Phys. Rev. C **82**, 014303 (2010).
[22] A. Baran, M. Kowal, P.-G. Reinhard, L.M. Robledo, A. Staszczak, M. Warda, Nucl. Phys. **A944**, 442 (2015).
[23] J.C. Pei, W. Nazarewicz, J.A. Sheikh, A.K. Kerman, Phys. Rev. Lett. **102**, 192501 (2009).
[24] G. Giardina *et al.*, J. Phys. Conf. Series **282**, 012006 (2011).
[25] S. Hofmann *et al.*, Eur. Phys. J. **A52**, 116 (2016).
[26] V. Zagrebaev and W. Greiner, Phys. Rev. C **78**, 034610 (2008).
[27] G.N. Knyazheva *et al.*, Phys. Rev. C **75**, 064602 (2007).
[28] E.M. Kozulin *et al.*, Phys. Rev. C **90** 054608 (2014).
[29] R. du Rietz *et al.*, Phys. Rev. C **88**, 054618 (2013).
[30] A. Wakhle *et al.*, Phys. Rev. Lett. **113**, 182502 (2014).
[31] W.Q. Shen *et al.*, Phys. Rev. C **36**, 115 (1987).

EPJ Web of Conferences **131**, 04004 (2016) DOI: 10.1051/epjconf/201613104004
Nobel Symposium NS160 – Chemistry and Physics of Heavy and Superheavy Elements

Quasifission in heavy and superheavy element formation reactions

D.J. Hinde[1,a], *M.* Dasgupta[1], *D.Y.* Jeung[1], *G.* Mohanto[1,b], *E.* Prasad[1,c], *C.* Simenel[1], *J.* Walshe[1], *A.* Wahkle[1,d], *E.* Williams[1], *I.P.* Carter[1], *K.J.* Cook[1], *Sunil* Kalkal[1], *D.C.* Rafferty[1], *R.* du Rietz[1,e], *E.C.* Simpson[1], *H.M.* David[2], *Ch.E.* Düllmann[2,3,4], and *J.* Khuyagbaatar[2,3]

[1] Department of Nuclear Physics, Research School of Physics and Engineering, Australian National University, Canberra, ACT 2601, Australia
[2] GSI Helmholtzzentrum für Schwerionenforschung, 64291 Darmstadt, Germany
[3] Helmholtz Institute Mainz, 55099 Mainz, Germany
[4] Johannes Gutenberg Universität Mainz, 55099 Mainz, Germany

Abstract. Superheavy elements are created in the laboratory by the fusion of two heavy nuclei. The large Coulomb repulsion that makes superheavy elements decay also makes the fusion process that forms them very unlikely. Instead, after sticking together for a short time, the two nuclei usually come apart, in a process called quasifission. Mass-angle distributions give the most direct information on the characteristics and time scales of quasifission. A systematic study of carefully chosen mass-angle distributions has provided information on the global trends of quasifission. Large deviations from these systematics reveal the major role played by the nuclear structure of the two colliding nuclei in determining the reaction outcome, and thus implicitly in hindering or favouring superheavy element production.

Superheavy elements (SHE) are formed by heavy-ion fusion reactions. Their cross sections are significantly suppressed [1] by the quasifission process [2]. This non-equilibrium process results when the combined di-nuclear system, formed initially when the two nuclear surfaces stick together, subsequently separates into two (fission-like) fragments. This occurs very rapidly, typically $\sim 10^{-20}$ s, well-before a compact compound nucleus is formed [2–4]. The probability of quasifission (P_{QF}) can be very large, thus the complementary probability of compound nucleus formation ($P_{CN} = 1 - P_{QF}$) can be lower than 10^{-3} in unfavourable cases. Understanding the competition between quasifission and fusion is thus very important in predicting the optimal fusion reactions to use to form new elements and isotopes in the superheavy mass region.

A key quantity characterizing the non-equilibrium quasifission process is the "sticking time" between contact of the two nuclear surfaces [5] and breakup (scission). Since the two colliding nuclei always approach each other along the beam axis, and after contact rotate

[a] e-mail: `david.hinde@anu.edu.au`

[b] Current address: BARC, Mumbai, India

[c] Permanent address: Department of Physics, School of Mathematical and Physical Sciences, Central University of Kerala, Kasaragod 671314, India

[d] Current address: National Superconducting Cyclotron Laboratory, Michigan State University, East Lansing, Michigan 48824, USA

[e] Current address: Malmö University, Faculty of Technology and Society, 205 06 Malmö, Sweden

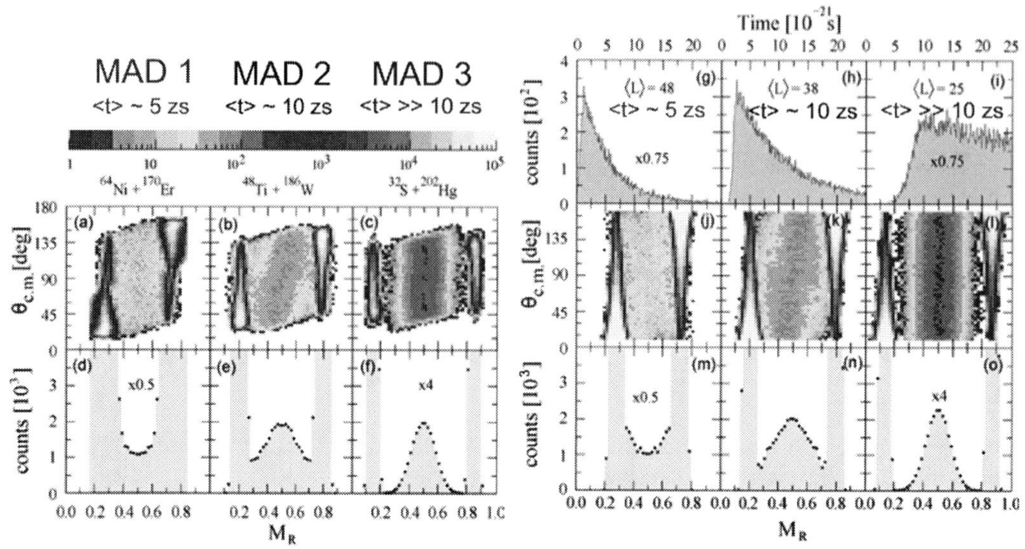

Figure 1. Panels (a)–(c) show the experimental MADs, with (d)–(f) showing the projections onto M_R, for the reactions ^{64}Ni + ^{170}Er, ^{48}Ti + ^{186}W and ^{32}S + ^{202}Hg, which all form ^{234}Cm. The multiplicative factors shown scale the *y*-axis. The simulated MADs for same reactions and beam energies are shown in panels (j),(k),(l), with the corresponding M_R spectra in panels (m),(n),(o), and the sticking time distributions used for the simulations in (g),(h),(i). The calculated mean angular momentum following capture, and the deduced approximate mean sticking times are shown in panels (g),(h),(i) (adapted from Ref. [12]).

with angular velocities that can be calculated, measurement of the rotation angle allows estimation of the sticking time. As the system rotates, mass flow also occurs between the two nuclei. A measurement of both mass and angle, for quasifission events over a wide range of angles, is called a mass-angle distribution (MAD). This gives the most direct information on the dynamics, as long as the system undergoes less than a full rotation (taking $\sim 10^{-20}$ s). This is usually the case for collisions of heavy nuclei, as shown by first measurements at GSI [2, 6], and later results from ANU [3, 4, 7–13]. The kinematic coincidence technique used in the measurements [2, 3, 14] provides direct information on the mass-ratio of the fragments at scission; thus, the data are represented in terms of mass ratio M_R at scission, rather than pre- or post-evaporation masses of the fragments themselves, which cannot be uniquely determined by this method.

Measured MAD and deduced quasifission sticking time distributions are shown in Fig. 1, for reactions forming the compound nucleus ^{234}Cm [3]. The MAD can be categorised by their distinctly different characteristics: minimum mass yield at symmetry associated with short sticking times; a mass-angle correlation with peak yield at symmetry; and no significant mass-angle correlation associated with long sticking times. These are assigned to the categories MAD1, MAD2 and MAD3 respectively [3]. It is clear that there is a strong correlation between the MAD type and the entrance channel charge product. However, for particular cases, we have found that detailed entrance channel properties are extremely important in determining the sticking times and MAD characteristics, including neutron richness [11, 15], and shell structure including static deformation [10] and magic numbers [11]. This multi-variate dependence has resulted in a complex picture.

To clarify and improve our quantitative understanding of the role of shell structures in the dynamics of quasifission, we make an analogy with the liquid drop model approach to nuclear masses, in which localized shell effects can be quantified when the underlying smooth

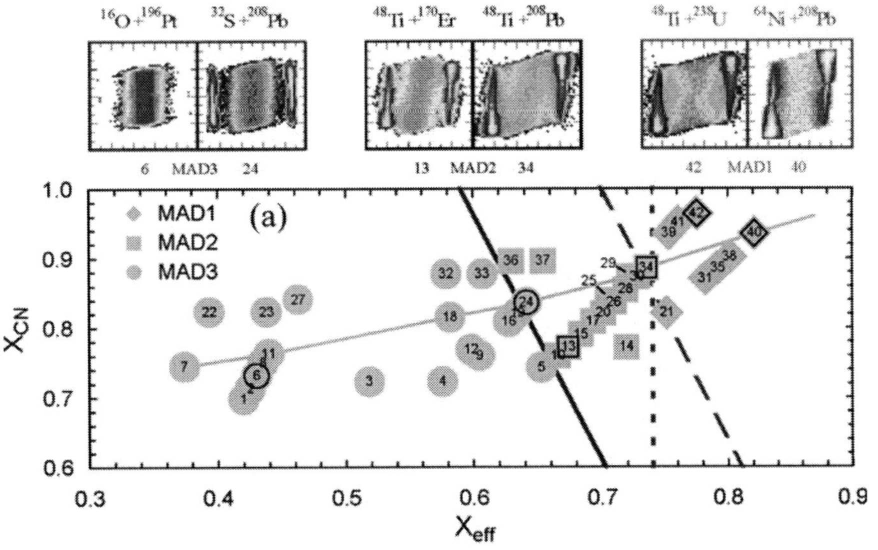

Figure 2. MAD types are plotted as a function of the effective fissility parameter in the entrance channel, x_{eff}, and the compound nucleus fissility parameter, x_{CN} (see text). The numbers inside the symbol indicating the classification of MAD observed refer to the reaction number in Ref. [3]. The diagonal blue line represents the empirical boundary between MAD1 reactions with no mass-angle correlation (left) and MAD2 that have (right). The dashed line indicates the estimated boundary of MAD1 reactions, which show no peak at mass-symmetry in the angle-integrated fission mass distribution. The thin line represents the locus of reactions with ^{208}Pb. Examples of MAD are shown in the panels above, identified by their reaction number.

(liquid drop) trends are well defined. To define the smooth trends in quasifission dynamics, a large number of MAD measurements have been selected, at beam energies somewhat above the capture barrier (typically by ~6%). Here the known effects of deformation alignment [9, 10, 14, 16] and shell structure observed in measurements at below-barrier energies [11, 17] are much reduced [10, 18, 19]. However, the beam energies should not be too far above the capture barriers, otherwise high angular momenta would be introduced in the collisions, which would then not be representative of heavy element formation reactions.

Using these measured MADS, reaction outcomes (defined for simplicity in terms of the MAD types described above) are plotted against the variables that reflect the balance between nuclear and Coulomb forces during the collision, which are expected to determine a smooth dependence of outcomes of reactions forming very heavy elements. Reference [20] proposed that there should be scaling behavior between reactions, based on the schematic "chaotic regime dynamics" model of fusion of heavy nuclei, which couples shape changes of the nuclear system to internal nucleonic degrees of freedom (one-body dissipation).

The experimental MAD types are plotted as a function of two scaling parameters expected to most strongly define the reaction outcome: (i) the effective fissility parameter of the entrance channel, x_{eff}, and (ii) the fissility parameter of the compound nucleus, x_{CN}. The former is most relevant to necked shapes close to the (generally) mass-asymmetric contact configuration in the entrance channel, whilst the latter is applicable to shapes without a constricted neck [21], and where the mass-asymmetry degree of freedom is not constrained.

Figure 2 shows the experimental MAD type as a function of these two variables, with a clear clustering of reactions of each type. The boundary across which a mass-angle correlation becomes significant (between MAD classes 3 and 2) shows a dependence on both variables, with a stronger dependence on the entrance-channel than on the compound nucleus fissility.

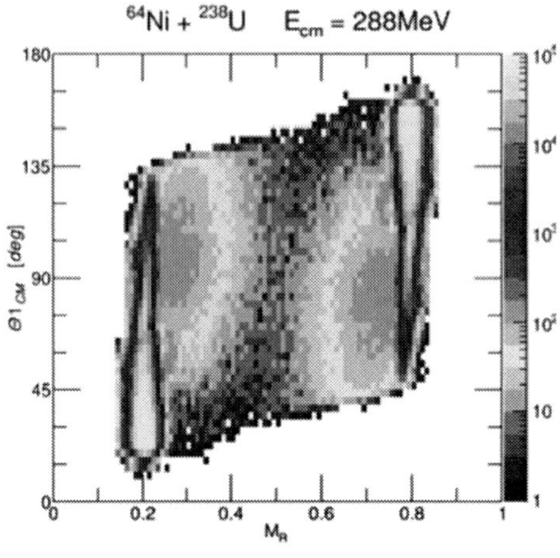

Figure 3. Preliminary mass-angle distribution for the $^{64}Ni + ^{238}U$ reaction, a candidate reaction to form element 120, measured at an above-barrier energy. This shows a strong mass-angle correlation indicating that typically the sticking time for this reaction is short, with limited mass evolution towards mass-symmetry. Recent TDHF calculations for this reaction [25] predict quasifission to be the outcome, with similar characteristics to those measured. It can also be seen that measurement over a wide angular range is necessary even to get a less-than-complete picture of the total mass-ratio spectrum. Mass-symmetric events peak around 90°, but could well be associated with the tail of the quasifission mass distribution. Measurement at angles closer to the beam axis should give greater sensitivity to a fusion-fission component.

The full blue line is our estimate of this boundary based on the current data. Mass-angle distributions for reactions on this line should show similar mass-angle correlations, associated with similar reaction trajectories and timescales. The same should be true for reactions on nearby parallel lines. The equation of this boundary line is $0.75\, x_{eff} + 0.25\, x_{CN} = 0.68$, giving x_{eff} three times the weight of x_{CN}. In terms of the MAD, this defines a mean fissility parameter x_m which seems to be a good predictor of the quasifission characteristics observed at the higher (above-barrier) beam and excitation energies selected here. The thin full line almost orthogonal to the others represents the locus of reactions with the doubly-magic ^{208}Pb. At the above-barrier energies chosen, no significant departure from the systematic behaviour is seen for reactions on this line.

The behaviour is very different at sub-barrier energies. Here it is well-known that collisions with the tips of prolate deformed nuclei are those that result in capture [5], due to the lower capture barrier for tip collisions. Fission angular distributions [14], mass distributions [9, 22, 23] and MAD [10, 13] point to the changing nature (shorter sticking time) of quasifission under these circumstances. Microscopic time-dependent Hartree-Fock (TDHF) calculations of quasifission masses and angles give a good match [13] to experimental results for the reaction $^{40}Ca + ^{238}U$, including the transition from sub-barrier to above-barrier energies (at which all collision orientations contribute). This is also consistent with the absence of evaporation residue yields at sub-barrier energies [24]. For very heavy systems, such as the $^{64}Ni + ^{238}U$ reaction, even at an above-barrier energy, the MAD (shown in Fig. 3) indicates that the majority of capture reactions result in rapid re-separation, with limited mass evolution towards symmetry. Extending the experimental MAD to angles even closer to the beam axis would increase sensitivity to a fusion-fission component (expected close to mass-symmetry), as the quasifission events at this energy are concentrated close to 90°.

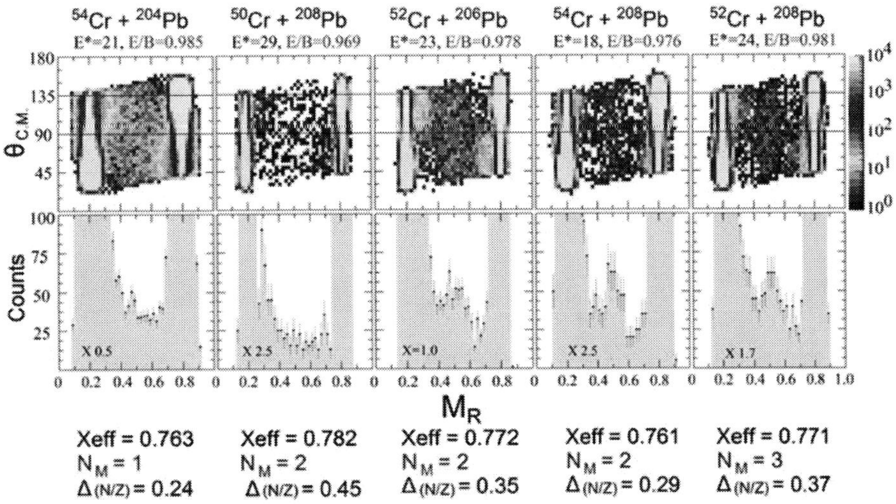

Figure 4. MAD and projected mass-ratio distributions for backward angles from 90° to 135° (as indicated in the MAD plots), for reactions of ^{50,52,54}Cr isotopes with ^{204,206,208}Pb. Sub-barrier beam energies (denoted by E/B), resulted in the low excitation energies E^*. The entrance channel fissility, x_{eff}, the number of entrance-channel magic numbers, and the mismatch in N/Z values are indicated for each reaction (see text). The reaction outcome changes from a minimum in yield at mass-symmetry (left) to a narrow peak at symmetry, having no evidence of a mass-angle correlation.

Sub-barrier reactions with (spherical) closed-shell nuclei show contrasting behaviour to reactions with heavy statically deformed nuclei. It has been shown [11] that increasing the "magicity" in the entrance channel at sub-barrier energies results in narrower mass distributions, correlated with a reduced mass-angle correlation, indicative of longer sticking times. Indeed, for the ^{48}Ca + ^{208}Pb reaction, measured mass distributions [17] appear consistent with a fusion-fission mechanism. These results led to the conclusion that "magicity" plays its strongest role when the N/Z values of the projectile and target nuclei are matched. When this is the case, transfer reactions that destroy entrance-channel magicity (as in the ^{40}Ca + ^{208}Pb reaction [11]) are minimized, preserving the closed shell nature of the collision partners as long as possible during the merging of the two nuclei [11].

Recent ANU results, for the reaction of isotopes of Cr with Pb, are shown in Fig. 4. The left three reactions all form the compound nucleus ^{258}Sg. These measurements were at sub-barrier energies, resulting in the low excitation energies (above the ground-state) as indicated. They support the picture from the Ca + Pb data. The difference between the N/Z values of the target and projectile nuclei is denoted by $\Delta_{N/Z}$ in the figure. The panels are ordered from left to right first by the number of magic numbers in the entrance channel (N_M), and then by $\Delta_{N/Z}$. The left-most reaction has only a single magic number in the entrance channel, and shows a U-shaped mass distribution, consistent with MAD class 1. With two magic numbers, the reactions better matched in N/Z ($\Delta_{N/Z}$ closer to zero) show a peak at mass-symmetry, associated with an angle-independent ridge in the MAD. With three magic numbers, but less favourable $\Delta_{N/Z}$, a similar result is observed. As the beam energy is increased, this mass-symmetric peak becomes a smaller and smaller fraction of the total fission yield within the range $0.3 < M_R < 0.7$. These reactions show a very similar behaviour to the ^{40,44,48}Ca + ^{204,208}Pb reactions; however, the transition from a U-shaped mass distribution to a narrow mass-symmetric component in the mass distribution is in some ways an even more drastic change in reaction outcome. The very sudden change in outcome, depending on neutron number and magicity, will be a severe challenge for models of quasifission to reproduce. And yet it is this level of sensitivity that models must strive to reproduce, to optimise experimental opportunities to create new isotopes of superheavy elements. Investigating the relationship between these changing quasifission characteristics and changes in the superheavy element yield would be of great value.

References

[1] D.J. Hinde, M. Dasgupta, A. Mukherjee, Phys. Rev. Lett. **89**, 282701 (2002).
[2] J. Tōke *et al.*, Nucl. Phys. **A440**, 327 (1985).
[3] R. du Rietz *et al.*, Phys. Rev. C **88**, 054618 (2013).
[4] R. du Rietz *et al.*, Phys. Rev. Lett. **106**, 052701 (2011).
[5] M. Dasgupta, D.J. Hinde, N. Rowley, A.M. Stefanini, Annu. Rev. Nucl. Part. Sci. **48**, 401 (1998).
[6] W.Q. Shen *et al.*, Phys. Rev. C **36**, 115 (1987).
[7] R.G. Thomas *et al.*, Phys. Rev. C **77**, 034610 (2008).
[8] R. Rafiei *et al.*, Phys. Rev. C **77**, 024606 (2008).
[9] D.J. Hinde, R. du Rietz, M. Dasgupta, R.G. Thomas, L.R. Gasques, Phys. Rev. Lett. **101**, 092701 (2008).
[10] D.J. Hinde *et al.*, Phys. Rev. Lett. **100**, 202701 (2008).
[11] C. Simenel *et al.*, Phys. Lett. **B710**, 607 (2012).
[12] E. Williams *et al.*, Phys. Rev. C **88**, 034611 (2013).
[13] A. Wakhle *et al.*, Phys. Rev. Lett. **113**, 182502 (2014).
[14] D.J. Hinde *et al.*, Phys. Rev. C **53**, 1290 (1996).
[15] K. Hammerton *et al.*, Phys. Rev. C **91**, 041602 (2015).
[16] D.J. Hinde *et al.*, Phys. Rev. Lett. **74**, 1295 (1995).
[17] E.V. Prokhorova *et al.*, Nucl. Phys. **A802**, 45 (2008).
[18] K. Nishio *et al.*, Phys. Rev. C **77**, 064607 (2008).
[19] I.M. Itkis *et al.*, Phys. Rev. C **83**, 064613 (2011).
[20] J.P. Blocki, H. Feldmeier, W.J. Swiatecki, Nucl. Phys. **A459**, 145 (1986).
[21] S. Bjørnholm and W.J. Swiatecki, Nucl. Phys. **A391**, 471 (1982).
[22] C.J. Lin *et al.*, Phys. Rev. C **85**, 014611 (2012).
[23] K. Nishio *et al.*, Phys. Rev. C **82**, 044604 (2010).
[24] S. Mitsuoka, H. Ikezoe, K. Nishio, J. Lu, Phys. Rev. C **62**, 054603 (2000).
[25] K. Sekizawa and K. Yabana, Phys. Rev. C **93**, 054616 (2016).

EPJ Web of Conferences **131**, 04005 (2016) DOI: 10.1051/epjconf/201613104005
Nobel Symposium NS160 – Chemistry and Physics of Heavy and Superheavy Elements

Fusion-fission probabilities, cross sections, and structure notes of superheavy nuclei

Michał Kowal[1,a], *Tomasz* Cap[1], *Piotr* Jachimowicz[2], *Janusz* Skalski[1], *Krystyna* Siwek-Wilczyńska[3], and *Janusz* Wilczyński[1,b]

[1] National Centre for Nuclear Research, Hoża 69, 00-681 Warsaw, Poland
[2] Institute of Physics, University of Zielona Góra, Szafrana 4a, 65516 Zielona Góra, Poland
[3] Institute of Experimental Physics, Faculty of Physics, University of Warsaw, Warsaw, Poland

Abstract. Fusion – fission probabilities in the synthesis of heaviest elements are discussed in the context of the latest experimental reports. Cross sections for superheavy nuclei are evaluated using the "Fusion by Diffusion" (FBD) model. Predictive power of this approach is shown for experimentally known Lv and Og isotopes and predictions given for $Z = 119$, 120. Ground state and saddle point properties as masses, shell corrections, pairing energies, and deformations necessary for cross-section estimations are calculated systematically within the multidimensional microscopic-macroscopic method based on the deformed Woods-Saxon single-particle potential. In the frame of the FBD approach predictions for production of elements heavier than $Z = 118$ are not too optimistic. For this reason, and because of high instability of superheavy nuclei, we comment on some structure effects, connected with the K-isomerism phenomenon which could lead to a significant increase in the stability of these systems.

1. Introduction

The uncharted region of the Z–N plane can answer many questions of fundamental importance for science. One of them is: "What can be the largest possible atomic number Z of an atomic nucleus?". The current answer was gained thanks to the tremendous advance in experimental studies achieved in recent years [1–5] via fusion-evaporation reactions, once the element $Z = 118$ has been synthesized [6, 7]. The question whether we can go still further waits for an answer as attempts to go beyond $Z = 118$ hit two obstacles: (*i*) a difficulty or impossibility of making targets from Es and heavier actinides, and (*ii*) reactions with heavier projectiles like ^{50}Ti, ^{54}Cr, ^{58}Fe, and ^{64}Ni, did not produce any ERs up to now. The cross sections for the synthesis of heaviest elements are of the order of picobarns or even femtobarns. It happens because of a very strong hindrance occurring in the fusion process itself. The fusion is a highly dissipative process in which the fusing system, staring from a touching configuration must passage over the saddle point into the configuration of the compound nucleus.

The main issue therefore boils down to the problem of motion in the potential surface with a local minimum (representing the meta-stable state) and the barrier protecting this minimum which one should overcome to fuse two nuclei. Taking into account the influence

[a] e-mail: `michal.kowal@ncbj.gov.pl`
[b] Deceased

of the environment requires addition of the damping force and the stochastic fluctuations. In the case of so called "fusion-by-diffusion" model (FBD) [8], the stochastic process of shape fluctuation leads to the overcoming of the saddle point. The fluctuating force, which is de facto caused by shape fluctuations, is assumed as a white noise to obtain the solution of the Smoluchowski diffusion equation in the deformation space along the fission valley. Thus, basically we touch the famous Kramers problem [9]; the relaxation from the meta-stable state with the assumption of the white noise and with damping proportional to the velocity. Induced or spontaneous relaxation from the meta-stable state via fluctuating potential barriers occurs in many branches of physics: phase transitions, bi-stability in quantum optic, spontaneous emission from excited states, electron transport in semiconductors, kinetics of chemical reactions, etc. [10].

The fundamental assumption which allows to investigate the mechanism of the formation of superheavy compound nuclei is the Bohr hypothesis, which states that the synthesis of a compound system can be treated as a Markow-type process (a stochastic process without any memory effect). This implies that the exit channel is completely independent of the entrance channel and both are independent of the intermediate stage of the reaction leading to the compound nucleus. The Bohr hypothesis can be justified mainly due to the different time scales of the particular stages. According to this hypothesis the total probability for the synthesis of a new superheavy element can be factorized into three independent ingredients:

$$P_{tot} = P_{cap} \times P_{fus} \times P_{sur}, \tag{1}$$

where P_{cap} stands for the probability of overcoming the Coulomb barrier, called the "capture process", P_{fus} is the formation probability that the nucleus, starting from the touching configuration, will finish up with a compound nuclear shape, and P_{sur} is the probability that the compound nucleus will survive against fission. The cross sections for the synthesis of superheavy nuclei are dramatically small because the fusion probability P_{fus} is hindered (in some reactions even by several orders of magnitude) due to the fact that the saddle configuration of the heaviest compound nuclei is much more compact than the configuration of two colliding nuclei at sticking. Survival probabilities, P_{sur}, are in turn extremely sensitive to the height of the fission barrier, especially in case of hot fusion reactions, because at each step of the deexcitation cascade the competition between neutron emission and fission strongly depends on the difference of energy thresholds for these two decay modes.

Low probabilities of for superheavy nuclei production and their high instability makes it worthwhile to search for long-lived exotic configurations. Obvious candidates are high-K isomers or ground states with a high K quantum number.

2. Model fusion-by-diffusion with macroscopic-microscopic input

The FBD model [8] in its extended version [11] serves to calculate cross sections for the synthesis of superheavy nuclei. The extension consists in including the angular momentum dependence of all probabilities of successive stages in Eq. (1). In other words, for each angular momentum (ℓ) the partial evaporation residue cross section $\sigma_{ER}(\ell)$ for production of a given final nucleus in its ground state is factorized as the product of the partial capture cross sections:

$$\sigma_{ER}(\ell) = \lambda\!\!\!^{-2}\pi \sum_{l=0}^{\infty}(2l+1)P_{cap}(\ell)P_{fus}(\ell)P_{sur}(\ell), \tag{2}$$

where $\lambda\!\!\!^{-2} = \hbar^2/2\mu E_{c.m.}$ is the wavelength (μ is the reduced mass of the colliding system and $E_{c.m.}$ is the energy in the center of mass).

It is assumed in the FBD model that after the contact of the two nuclei, a neck between them grows rapidly at an approximately fixed mass asymmetry and constant length of the system. This "neck zip" is expected to carry the system towards the bottom of the asymmetric fission valley. This is the "injection point" from where the system starts its climb uphill over the saddle in the process of thermal fluctuations in the shape degrees of freedom. Theoretical justification of the above picture of fast zipping the neck was given in Ref. [12], where the later stage of the stochastic climb uphill was described by solving the two-dimensional Langevin equation. Theoretical location of an effective injection point can be deduced from this model [12]. Also in a modified fusion-by-diffusion model [13] the location of the injection point was estimated theoretically.

Competition between neutron emission and fission can be resolved in the spirit of the transition state theory with the help of the well known formulas. All details regarding the calculations of the survival probability, P_{sur}, can be found in Ref. [11]. In case of calculating multiple evaporation (xn) channels a simplified algorithm [14], avoiding the necessity of using the Monte Carlo method, was applied. The most sensitive point here is the knowledge of the fission barrier height. The most important in fission-barrier calculations is its reliability which, to large extent, is based on the model used for calculating energy of a nucleus as a function of deformation. Energy maps are necessary to appreciate fission barriers. In our case multi-dimensional energy landscapes are calculated within the Microscopic-Macroscopic (MM) model based on the deformed Woods-Saxon potential [15]. The Strutinski shell and pairing correction [16] is taken for the microscopic part. For the macroscopic part we use the Yukawa plus exponential model [17] with parameters specified in [18]. Thus, all parameter values are kept exactly the same as in all recent applications of the model to heavy and superheavy nuclei. Mononuclear shapes can be parameterized via spherical harmonics, $Y_{lm}(\vartheta, \varphi)$, by the following equation of the nuclear surface:

$$R(\vartheta, \varphi) = c(\{\beta\}) R_0 \left(1 + \sum_{\lambda=1}^{\infty} \sum_{\mu=-\lambda}^{+\lambda} \beta_{\lambda\mu} Y_{\lambda\mu}\right), \tag{3}$$

where $c(\{\beta\})$ is the volume-fixing factor and R_0 is the radius of a spherical nucleus. This parameterization has its limitations; certainly, it is not suitable for too elongated shapes. However, for moderately deformed saddle points in superheavy nuclei it excellently reproduces all shapes generated by other parametrizations, e.g. by Ref. [19], as we checked in numerous tests. For nuclear ground states it is possible to confine the analysis to axially-symmetric shapes, with the expansion truncated at β_{80}.

All details of calculations of ground-state and saddle-point properties as masses, shell corrections, pairing energies, and deformations can be found in Ref. [20, 21]. However, in cited papers all tables are limited to even–even nuclei. This is why the fission barrier heights for other nuclei have been calculated separately by adding the quasiparticle energy, $E_q = \sqrt{(\varepsilon_q - \lambda)^2 + \Delta^2}$, where ε_q is the energy of the odd nucleon state q, λ is the Fermi energy and Δ is the pairing-gap energy. Calculations have been performed without blocking.

It should be emphasized that in the frame of this method the largest ground-state shell effect ($\sim$9 MeV) is observed for the semi-magic nucleus ^{270}Hs ($Z = 108$, $N = 162$). The second minimum ($\sim$7 MeV) of the shell correction is located around the nucleus $Z = 114$ and $N = 184$. When superposed with the weakly deformation-dependent macroscopic part, this component is largely responsible for the emergence of global minima in superheavy nuclei. Concerning the fission barriers, one obtains three areas with clearly raised barriers: around $N \approx 152$, $N = 162$, and $N \approx 180$. There is also a region of low barriers around $N = 170$.

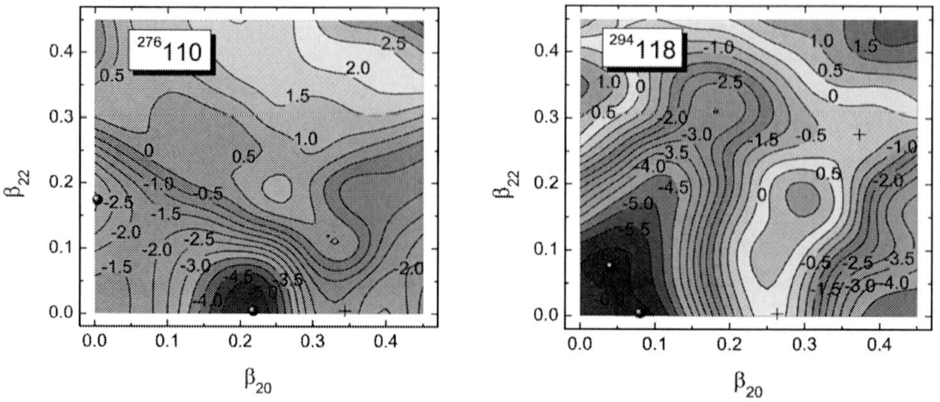

Figure 1. Energy surfaces for two nuclei known experimentally. See text for details.

Finally, let us emphasize that our calculated first and second fission barriers in actinides are in a very good agreement with the experimental/empirical data: the root mean square deviation is 0.7 MeV [22] for the second, and 0.5 MeV [20] for the first fission barriers. In our opinion, such agreement justifies using our model for predictions in heavier nuclei.

3. Predictions for the synthesis cross sections of Z = 114–120

Since the inclusion of triaxiality is absolutely necessary in the superheavy region, we start with showing in Fig. 1 two landscapes in the (β_{20}, β_{22}) plane of experimentally known nuclei: $^{276}110$ and $^{294}118$. The potential energy surface calculated in the multi-dimensional space, is projected in Fig. 1 on the (β_{20}, β_{22}) plane. This means that it is shown as a function of β_{20}, β_{22}, but at each given (β_{20}, β_{22}) point it is minimized in the remaining degrees of freedom. Here the minimization has been performed over β_{30}, β_{40}, β_{42}, β_{44}, β_{50}, β_{60}, β_{70}, and β_{80}. The energy on the map is normalized in such a way that its macroscopic part is put equal to zero at the spherical shape of a nucleus. In $^{276}110$ one can see two minima: prolate - a prolate ground state and a prolate superdeformed one, see left panel of Fig. 1. The map shows also the axial fission saddle. The nucleus $Z = 118$, $N = 176$ is nearly spherical as can be seen in right panel of Fig. 1. There is also a secondary oblate superdeformed minimum. Competition between triaxial and axial saddles is clearly visible.

In the FBD model there is one fit parameter, namely s_{inj}, which is defined as the excess of the total length of the combined system over the length of the initial system (at the touching configuration) when the neck-zip process brings the system to the asymmetric fission valley. Based on the nuclear masses and fission barriers deduced from the energy maps (as shown above for $^{276}110$ and $^{294}118$) one can adjust this parameter to the experimental synthesis cross section at the maximum of a given (xn) excitation function. Details of this fit procedure with the specification of reaction data used can found in Ref. [23]. The compilation of the so-deduced s_{inj} values is displayed in Fig. 2 as a function of the kinetic energy excess, $E_{c.m.}-B_0$, above the Coulomb barrier, B_0, in the case of two sets of theoretical input data: based on the deformed Woods-Saxon model [20, 21] described here (left panel) and assuming the fission-barrier heights [24] and the ground-state masses [25] of P. Möller *et al.* (right panel).

Based on what we known from the dynamics of nucleus-nucleus collisions one can expect that the injection distance will increase with the decreasing energy: $E_{c.m.}-B_0$. It is exactly the case if we are using our data. Such a very good correlation between the s_{inj} values and

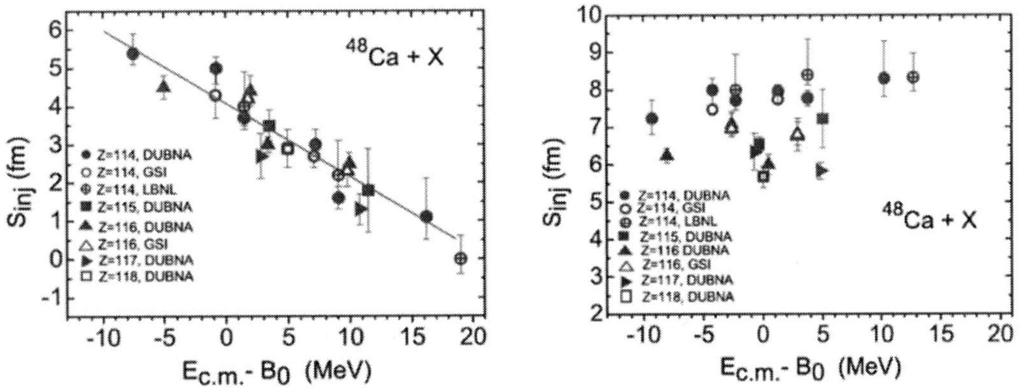

Figure 2. Systematics of the injection-point distance, s_{inj}, as a function of the kinetic energy excess determined using our MM approach [20, 21] on the left, and with P. Möller's compilation of nuclear data [24, 25] on the right.

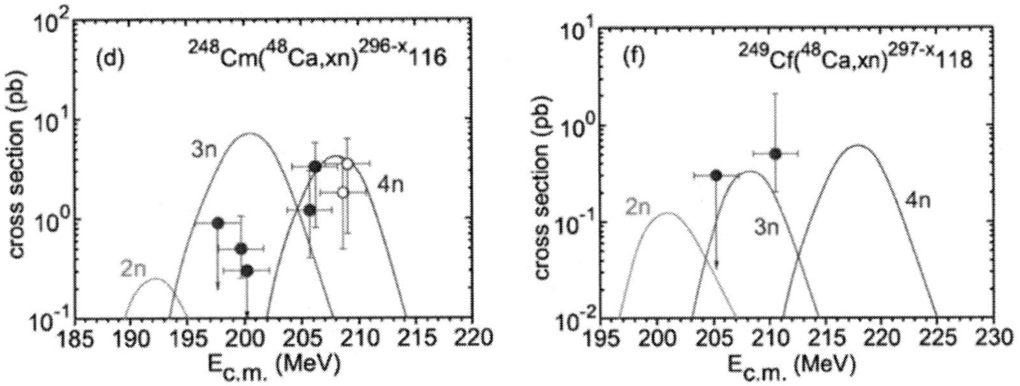

Figure 3. Energy dependence of the cross section for synthesis of superheavy nuclei in hot fusion reactions. Full circles represent data for $3n$, $4n$, and $5n$ reaction channels obtained in Dubna experiments for elements $Z = 116$ and $Z = 118$. Open circles represent data taken at GSI.

the corresponding energies $E_{c.m.}-B_0$ can be viewed as an argument in favour of our MM results because such a striking correlation would be very unlikely if the theoretical barrier heights were inconsistent with experimental values. In contrast, the right panel of this figure demonstrates the evident inconsistency of the set of s_{inj} values obtained for the barriers of Ref. [24] in which the individual points seem to be almost randomly scattered and do not show any correlation with energy.

In Ref. [23] we calculated the excitation functions for known superheavy elements starting from Flerovium up to Oganesson, using results of our MM model. In Fig. 3 we show two examples, namely ^{292}Lv, ^{293}Lv, ^{294}Lv and ^{293}Og, ^{294}Og, ^{295}Og. Taking into account that the fission barriers could not be calculated more precisely than up to 0.5 MeV, the obtained agreement with experimental data can be considered satisfactory.

Then we want to give predictions for currently unknown superheavy elements $Z = 119$ and $Z = 120$. Particulary interesting is the nucleus $^{302}120$, as two unsuccessful attempts to produce it have already taken place at GSI, providing a one-event cross-section limit of $0.58(^{134}_{48})$ pb [26] or 90 fb in [27], and in Dubna [28], providing the limit of 400 fb. The cross-section estimates [29] do not support a possibility of an easy production of this superheavy isotope in the laboratory. It is worth noting that with the barrier of the order of 10 MeV,

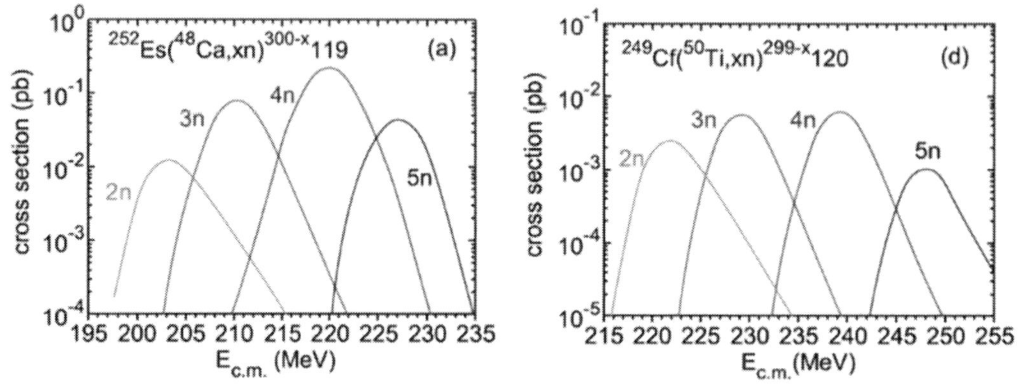

Figure 4. Synthesis cross sections of yet undiscovered superheavy nuclei of $Z = 119$ and $Z = 120$ predicted by using the fusion-by-diffusion model.

as obtained in the frame of the self-consistent theory [30], producing superheavy $Z = 120$ nuclei should not pose any difficulties. We chose these two cases of fusion reactions for which our predictions are the most optimistic. For $Z = 120$, it turned out to be the reaction ^{249}Cf(^{50}Ti, xn)$^{299-x}$120. The corresponding excitation function is shown on the right hand side in Fig. 4. The largest cross section of about 6 fb is expected for the $3n$ and $4n$ evaporation channels. One very important remark is required here: carrying out this reaction requires a change of the projectile from ^{48}Ca used successfully up to now to ^{50}Ti with which no hot superheavy-element synthesis was completed successfully. Generally, prospects for the synthesis of element $Z = 120$ are considerably worse than those for $Z = 119$. To obtain $Z = 119$, our calculations prefer the ^{252}Es(^{48}Ca, xn)$^{300-x}$119 reaction, for which a synthesis cross section of about 0.2 pb in the $4n$ channel at $E_{c.m.} \simeq 220$ MeV is expected. This case is shown in the left panel in Fig. 4.

Closing this section we want to emphasize that the existing theoretical evaluations of the fission barriers differ significantly. Even the results of the two models based on the microscopic-macroscopic approach differ dramatically for some nuclei. Our calculations indicate, in contrast to the self-consistent mean-field studies, that fission barriers, which are still quite substantial for some $Z = 118$ nuclei, become lower than 5.5 MeV for $Z = 126$. Quite recently we noted a dramatic divergence in calculated fission barriers [32]. A brief review of recent progress in theoretical studies on fission barriers and fission half-lives of even–even superheavy nuclei can be found in Ref. [31].

4. K-isomerism and possible stability enhancement

A recent theoretical study [33] of barriers within both the MM Woods-Saxon and Skyrme SLy6 Hartree-Fock plus BCS models do not predict much chance to produce even heavier ($Z > 126$) nuclei. Moreover, it seems that two current methods of making superheavy elements in the laboratory, cold and hot fusion reactions, seem to reach their limits. On the other hand, not all superheavy isotopes with $Z < 118$ have been produced yet. Therefore, while pondering upon possible new reactions leading towards the island of stability, it may be worthwhile to search for long-lived exotic superheavy configurations. Obvious candidates are high-K isomers, for which increased stability is expected due to some specific hindrance mechanisms. In our recent work [34] we went even further: from systematic calculations it turned out that some nuclei can have a *ground state* with a high-K quantum number.

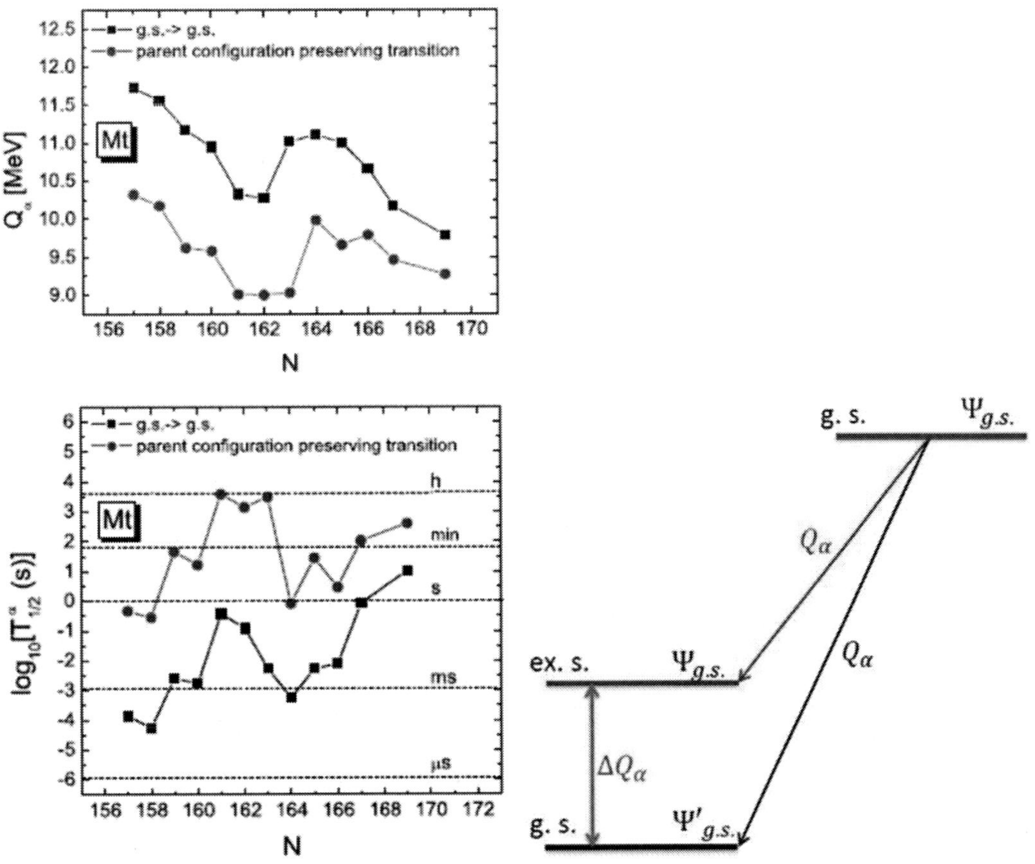

Figure 5. Q_α values calculated with blocking following from the WS model and the increase in α– half-life induced by ΔQ_α [36], assuming a complete decay hindrance to different configurations (left panel). Schematic representation of appropriate transitions (right panel).

The structure of odd-odd nuclei is typically more complicated than that of odd-A systems. If we disregard collective vibrations, the ground-state configuration is a result of coupling the unpaired neutron and proton to a total angular momentum. The energy ordering of coupled configurations is usually attributed to a residual neutron-proton interaction. This is why the investigation of odd-odd nuclei requires a significant expansion of the used method. This was done in Ref. [35]. An α-decay hindrance of a high-K configuration is expected when the same configuration in the daughter has a sizable excitation. If configuration-changing transitions had been strictly forbidden, the hindrance would have been determined by this excitation ΔQ_α which is schematically shown in Fig. 5 (right panel). Also in Fig. 5 (left panel), a decrease in energy release, ΔQ_α, and $\log_{10} T_\alpha$ calculated according to [36], for ground-state to ground-state and configuration-preserving transitions are shown for various Meitnerium isotopes. The latter half-life would correspond to an absolute hindrance of configuration-changing decays. For our favourite case, ^{272}Mt, such hindrance reaches six orders of magnitude. This takes us from the lifetime of milliseconds to hours. Let us stress that the whole argument is based on both the presence of deformed semi-magic shells at $Z = 108$ and $N = 162$ and the position of high-Ω intruder orbitals just above that shell.

5. Summary

It may be that the heaviest element that can be created and identified will simply be limited by our synthesis techniques. Predictions based on the FBD model are rather pessimistic, giving the cross sections at the level of femtobarn. Synthesis of elements $Z > 118$ will require a projectile heavier than ^{48}Ca (as fermium, $Z = 100$, cannot be used as a target, while einsteinium, $Z = 99$, is extremely difficult), and such reactions have thus so far proven fruitless. Might it be possible, therefore, to obtain superheavy isotopes with a structure that slows down fission and α decay? Of particular interest would be half-lives of several seconds or even minutes, since this would make it possible to analyze the chemical properties of individual radioactive atoms. The α-decay which dominates in the extreme superheavy region is delayed by differences in structure of parent and daughter configurations, resulting either from their very different deformations or a different angular momentum coupling of unpaired nucleons.

M.K. and J.S. were co-financed by the National Science Centre under Contract No. UMO-2013/08/M/ST2/00257 (LEA COPIGAL). One of the authors (P.J.) was cofinanced by Ministry of Science and Higher Education: "Iuventus Plus" grant No. IP2014 016073.

References

[1] Yu.Ts. Oganessian and V.K. Utyonkov., Nucl. Phys. **A944**, 62 (2015).
[2] V.K. Utyonkov *et al.*, Phys. Rev. C **92**, 034609 (2015).
[3] G. Münzenberg, Nucl. Phys. **A944**, 5(2015).
[4] S. Hofmann *et al.*, Eur. Phys. J. **A48** 62, (2012).
[5] J. Khuyagbaatar *et al.*, Phys. Rev. Lett. **112**, 172501 (2014).
[6] Yu.Ts. Oganessian *et al.*, Phys. Rev. C **87**, 057302 (2013).
[7] Yu.Ts. Oganessian *et al.*, Phys. Rev. C **87**, 054621, (2013).
[8] W.J. Świątecki, K. Siwek-Wilczyńska, J. Wilczyński, Phys. Rev. C **71**, 014602 (2005).
[9] H.A. Kramers, Physica **7**, 284 (1940).
[10] H. Haken, *Synergetics* (Springer, Berlin, 1978).
[11] T. Cap, K. Siwek-Wilczyńska, J. Wilczyński, Phys. Rev. C **83**, 054602 (2011).
[12] D. Boilley *et al.*, Phys. Rev. **C 84**, 054608 (2011).
[13] Zu-Hua Liu and Jing-Dong Bao, Phys. Rev. C **81**, 044606 (2010).
[14] T. Cap *et al.*, Acta Phys. Pol. **B43**, 297 (2012).
[15] S. Ćwiok *et al.*, Comp. Phys. Comm. **46**, 379 (1987).
[16] V.M. Strutinski, Sov. J. Nucl. Phys.**3**, 449 (1966); Nucl. Phys. **A95**, 420 (1967).
[17] H.J. Krappe, J.R. Nix, A.J. Sierk, Phys. Rev. C **20**, 992 (1979).
[18] I. Muntian, Z. Patyk, A. Sobiczewski, Acta Phys. Pol. **B32**, 691 (2001).
[19] P. Möller and A. Iwamoto, Phys. Rev. C **61**, 047602 (2000).
[20] M. Kowal, P. Jachimowicz and A. Sobiczewski, Phys. Rev. C **82**, 014303 (2010).
[21] M. Kowal, P. Jachimowicz, J. Skalski, `arXiv:1203.5013`, (2012).
[22] P. Jachimowicz, M. Kowal, J. Skalski, Phys. Rev. C **85**, 034305 (2012).
[23] K. Siwek-Wilczyńska *et al.*, Phys. Rev. C **86**, 014611 (2012).
[24] P. Möller *et al.*, Phys. Rev. C **79**, 064304 (2009).
[25] P. Möller *et al.*, At. Data Nucl. Data Tables **59**, 185 (1995).
[26] S. Hofmann *et al.*, Eur. Phys. J. **A52**, 180 (2016).

[27] S. Hofmann, Russ. Chem. Rev. **78**, 1123 (2009).
[28] Yu.Ts. Oganessian *et al.*, Phys. Rev. C **79**, 024603 (2009).
[29] K. Siwek-Wilczyńska, T. Cap, J. Wilczyński, Int. J. Mod. Phys. **E19**, 500 (2010).
[30] A. Staszczak, A. Baran, W. Nazarewicz, Phys. Rev. C **87** (2013) 024320.
[31] A. Baran *et al.*, Nucl. Phys. **A944**, 442 (2015).
[32] W. Brodzinski *et al.*, JPS Conf. Proc. **6**, 020054 (2015).
[33] W. Brodzinski and J. Skalski, Phys. Rev. C **88**, 044307 (2013).
[34] P. Jachimowicz, M. Kowal, J. Skalski, Phys. Rev. C **92**, 044306 (2015).
[35] P. Jachimowicz, M. Kowal, J. Skalski, Phys. Rev. C **89**, 024304 (2014).
[36] G. Royer, Nucl. Phys. **A848**, 279 (2010).

EPJ Web of Conferences **131**, 05001 (2016) DOI: 10.1051/epjconf/201613105001
Nobel Symposium NS160 – Chemistry and Physics of Heavy and Superheavy Elements

First ionization potential of the heaviest actinide lawrencium, element 103

Tetsuya K. Sato[1,a], *Masato* Asai[1], *Anastasia* Borschevsky[2], *Thierry* Stora[3], *Nozomi* Sato[1], *Yusuke* Kaneya[1,4], *Kazuaki* Tsukada[1], *Christoph E.* Düllmann[5,6,7], *Klaus* Eberhardt[5,7], *Ephraim* Eliav[8], *Shinichi* Ichikawa[9], *Uzi* Kaldor[8], *Jens V.* Kratz[7], *Sunao* Miyashita[10], *Yuichiro* Nagame[1], *Kazuhiro* Ooe[11], *Akihiko* Osa[1], *Dennis* Renisch[7], *Jörg* Runke[6], *Matthias* Schädel[1], *Petra* Thörle-Pospiech[5,7], *Atsushi* Toyoshima[1], and *Norbert* Trautmann[7]

[1] Japan Atomic Energy Agency, Tokai, Ibaraki 319-1195, Japan
[2] The Van Swinderen Institute for Particle Physics and Gravity, University of Groningen, 9700 AB Groningen, The Netherlands
[3] ISOLDE, CERN, CH-1211 Geneva 23, Switzerland
[4] Graduate School of Science and Engineering, Ibaraki University, Mito, Ibaraki 310-8512, Japan
[5] Helmholtz-Institut Mainz, 55099 Mainz, Germany
[6] GSI Helmholtzzentrum für Schwerionenforschung, 64291 Darmstadt, Germany
[7] Institut für Kernchemie, Johannes Gutenberg-Universität Mainz, 55099 Mainz, Germany
[8] School of Chemistry, Tel Aviv University, 69978 Tel Aviv, Israel
[9] Nishina Center for Accelerator-Based Science, RIKEN, 2-1 Hirosawa, Wako, Saitama 351-0198, Japan
[10] Graduate School of Science, Hiroshima University, Kagamiyama, Higashi-Hiroshima 739-8526, Japan
[11] Institute of Science and Technology, Niigata University, Niigata 910-2181, Japan

Abstract. The first ionization potential (IP_1) of element 103, lawrencium (Lr), has been successfully determined for the first time by using a newly developed method based on a surface ionization process. The measured IP_1 value is $4.963^{0.08}_{0.07}$ eV. This value is the smallest among those of actinide elements and is in excellent agreement with the value of 4.963(15) eV predicted by state-of-the-art relativistic calculations also performed in this work. Our results strongly support that the Lr atom has an electronic configuration of $[Rn]7s^2 5f^{14} 7p^1_{1/2}$, which is influenced by strong relativistic effects. The present work provides a reliable benchmark for theoretical calculations and also opens the way for studies on atomic properties of heavy elements with atomic number $Z > 100$. Moreover, the present achievement has triggered a controversy on the position of lutetium (Lu) and Lr in the Periodic Table of Elements.

1. Introduction

The chemical properties of an element are primarily governed by the configuration of electrons in its valence shell. The relativistic effects on the electronic structure of heavy elements in the seventh row of the Periodic Table become so strong that in some cases even their ground-state configurations may differ from those expected from lighter elements in the same group [1].

[a] Corresponding author: sato.tetsuya@jaea.go.jp

According to the actinide concept established by G.T. Seaborg [2], element 103, lawrencium (Lr), is placed at the tail end of the actinide series of the Periodic Table as the heaviest actinide element. By analogy with Lu, which is the lanthanide homologue of Lr and has an electronic structure of $[Xe]4f^{14}6s^25d^1$, the electronic configuration of Lr would be expected to be $[Rn]5f^{14}7s^26d^1$. Theoretical relativistic calculations, however, have predicted that the ground-state configuration of Lr is $[Rn]5f^{14}7s^27p^1_{1/2}$ since the $7p_{1/2}$ orbital would be stabilized below the 6*d* orbital in Lr by strong relativistic effects [3]. Since this prediction came out about 40 years ago, Lr has attracted interest of theoretical and experimental chemists.

Different electronic configuration yields different chemical behaviour. If the ground state configuration of Lr has a $7p_{1/2}$ valence electron, this would have implications of high volatility of Lr, as Eichler pointed out as a result of a semi-empirical consideration [4]. This prompted Jost and his colleagues to investigate the volatility of atomic Lr using a gas chromatographic technique [5]. In their experiment, no evidence for Lr as a volatile element with a $7p_{1/2}$ configuration was found.

In practice, chemical investigation of heavy elements such as Lr poses some difficulties. Elements heavier than fermium (Fm, $Z = 100$) must be produced at accelerators using reactions of heavy ions with heavy target materials. Moreover, both half-lives and cross sections of the isotopes of the still heavier elements are rapidly decreasing. Thus, they are usually available in quantities of a few atoms only at a time. Consequently, beginning with about the end of the actinides, properties of the elements must be studied on an atom-at-a-time scale. So far, it has been experimentally shown that Lr exhibits a stable +3 oxidation state in solution [6], and its ionic radius was evaluated as 88.6 ± 0.3 pm [7]. Based on the empirically developed "actinide concept", and in agreement with theoretical calculations, in today's Periodic Table, the actinide element series terminates with Lr. However, it has not yet been experimentally confirmed that Lr has appropriate properties as the last member of the actinide series.

2. The first ionization potential of Lr

Experimental determination of the ground-state electronic structure of Lr could demonstrate the magnitude of the influence of relativistic effects and provide new insights into the chemical properties of the heaviest elements.

The first ionization potential (IP_1) is a fundamental physical and chemical property of an element, which is qualitatively defined as the amount of energy required to remove the valence electron of an atom to form a positive ion. Experimentally determined IP_1, therefore, would give information on the outermost electronic orbital of the atom. In addition, IP_1 is one of the few physical quantities which can be estimated directly by theoretical calculations. A precise and accurate experimental determination of IP_1 can therefore benchmark electronic structure calculations.

IP_1 values of weighable amounts of nuclear-reactor-produced heavy elements up to einsteinium (Es, $Z = 99$) have been measured by resonance ionization mass spectrometry (RIMS) [8]. RIMS was also applied to an investigation of Fm with a sample of 2.7×10^{10} atoms of ^{255}Fm (half-life $T_{1/2} = 20.1$ h). In that experiment, the atomic level structure, but not the IP_1, was determined [9]. Recently, resonance ionization laser ion source (RILIS) studies optimized for short-lived nuclides made it possible to determine IP_1 of astatine (At, $Z = 85$) using ^{199}At ($T_{1/2} = 7.2$ s) produced in the proton-induced spallation reaction of uranium [10]. IP_1 values of heavy elements with $Z \geq 100$, however, could not be determined experimentally so far, as the produced amounts of the atoms were too low for the above techniques.

A sufficiently long-lived and detectable isotope for chemical experiments with Lr is ^{256}Lr ($T_{1/2} = 27$ s). It is produced at a rate of one atom per several seconds in the fusion-evaporation reaction of a ^{249}Cf target with a ^{11}B beam [11]. A new and highly efficient experimental method was thus required to determine an IP_1 value of Lr with such short-lived Lr atoms and in such low amounts.

In order to determine the IP_1 value of Lr, we have developed a novel method based on the surface ionization technique. The surface ionization is an ionization process which takes place on a solid surface kept at high temperature [12, 13]. Based on the Saha-Langmuir equation, an analytical model describes the surface ionization process in a hollow tube (cavity)-type ion-source. The ionization efficiency, I_{eff}, of an element in a small hot cavity can be expressed as:

$$I_{\mathrm{eff}} = \frac{N \exp\left(\frac{\phi - \mathrm{IP}_1^*}{kT}\right)}{1 + N \exp\left(\frac{\phi - \mathrm{IP}_1^*}{kT}\right)}, \tag{1}$$

where ϕ is the work function which is material-dependent, k the Bolzmann constant, T the temperature of the ionizing surface, and N a parameter that depends on the effective number of atom-surface interactions in the cavity. IP_1^*, the effective IP_1, is directly related to IP_1 as:

$$\mathrm{IP}^* = \mathrm{IP}_1 - kT\,\ln\left(\frac{Q_1}{Q_0}\right), \tag{2}$$

where Q_i and Q_o are the partition functions of electrons in an atom and an ion at a given temperature, respectively, which can be calculated using excitation energies and statistical weights of their ground and excited states. In the present study, tantalum (Ta) was chosen as a cavity material. The IP_1 value of the element of interest is deduced from Eq. (1) where the parameter N is determined from a set of experimental ionization efficiencies (I_{eff}) of known IP_1 of some elements [14].

A schematic experimental set-up used in the measurement is shown in Fig. 1. To investigate surface ionization behaviour of short-lived Lr and other isotopes, we employed an ISOL (Isotope Separator On-Line) system which has a surface ion-source coupled to an aerosol gas-jet transport system at the tandem accelerator facility of Japan Atomic Energy Agency (JAEA). The ion-source has been newly developed for this purpose. The set-up consists of a target recoil chamber, a gas-jet transport system [15], a surface ion source installed in the ISOL system [16], and α/γ spectrometry systems. For an efficient α-particle measurement, we applied a rotating catcher wheel apparatus MANON (Measurement system of Alpha particle and spontaneous fission events ON-line) [17]. Nuclear reaction products produced in the target recoil chamber were attached on CdI_2 aerosol flowing into the chamber with 1.4 L/min He carrier gas. The radioactivity-laden aerosol particles were transported continuously into the ion-source via a Teflon capillary. Surface-ionized nuclear reaction products were extracted from the ion source and accelerated for mass separation. After the mass separation, ions of the isotope of interest are collected and detected to determine the ionization efficiency. For ^{256}Lr, I_{eff} values of $(33 \pm 4)\%$ and $(36 \pm 7)\%$ were measured at 2700 K and 2800 K, respectively [14].

To determine the value of the parameter N in Eq. (1) at each temperature, short-lived lanthanide and alkali isotopes ^{142,143}Eu, ^{143}Sm, ^{148}Tb, ^{153,154}Ho, ^{157}Er, ^{162}Tm, ^{165}Yb, ^{168}Lu and ^{80}Rb produced in nuclear reactions of ^{11}B beams with mixed lanthanide targets and a Ge target were provided to the experimental set-up and their ionization efficiencies were measured. Figure 2 shows the I_{eff} values at 2800 K as a function of IP_1*. The IP_1* value for each element was calculated with Eq. (2) using energies and statistical weights of low-lying excited states in the ion and the atom of each element provided by the National Institute of Standards and Technology (NIST) atomic database [18]. The determined N

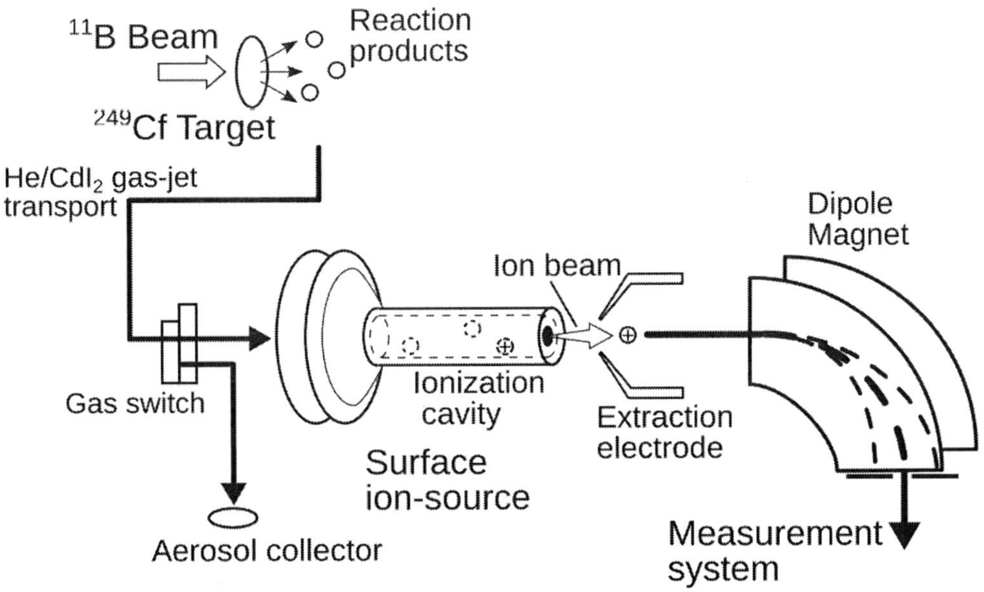

Figure 1. Schematic diagram of the experimental set-up for a measurement of the first ionization potential of Lr. The set-up was constructed based on the JAEA-ISOL system.

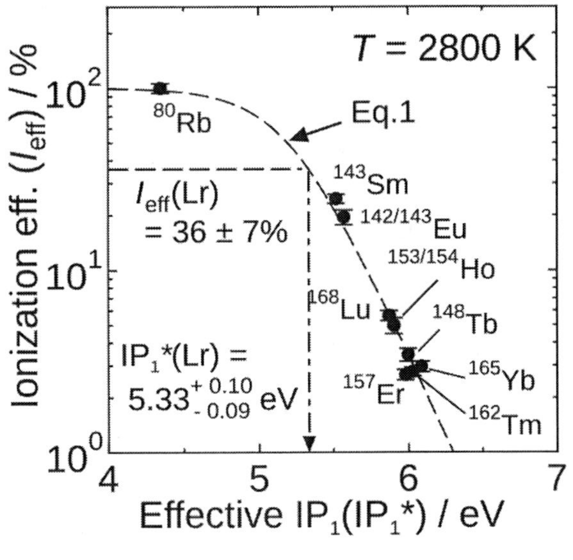

Figure 2. The ionization efficiency (I_{eff}) of various short-lived isotopes as a function of the effective IP_1 (IP_1^*) at 2800 K. The dashed curve is obtained by fitting the experimental data using Eq. (1). The position of the measured Ieff value of Lr, (36 ± 4)%, is also shown. From the fitted Eq. (1) with $N = 50 \pm 3$, IP_1^* of Lr is calculated to be $5.33^{+0.10}_{-0.09}$ eV. This corresponds to an IP_1 value of $4.97^{+0.13}_{-0.11}$ eV [14].

values at $T = 2700$ K and 2800 K were 43 ± 3 and 50 ± 3, respectively. Using the N values, the Lr IP_1* values at $T = 2700$ K and 2800 K were calculated to be $5.29^{+0.08}_{-0.07}$ eV and $5.33^{+0.10}_{-0.09}$ eV respectively. Since the excitation energies of Lr and Lr^+, which are required to calculate IP_1 using Eq. (2) are unavailable, we employed theoretical values proposed in [19]. Then, IP_1 values of $4.95^{+0.10}_{-0.08}$ and $4.97^{+0.13}_{-0.11}$ were determined at $T = 2700$ K and 2800 K, respectively, for Lr. Based on these results, our experimentally determined value for IP_1 of Lr is $4.96^{+0.08}_{-0.07}$ eV.

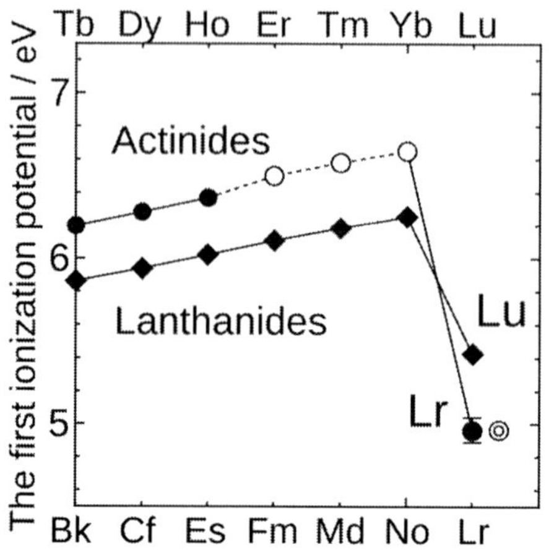

Figure 3. Variation of the experimental first ionization potentials of actinides(•) and lanthanides(♦). Open circles indicate estimated values for the heavy actinides [20], while the double circle shows the theoretically calculated value [14].

A state-of-the-art relativistic calculation of the IP_1 of Lr was also performed. The calculation was conducted using the relativistic coupled cluster approach with single, double, and perturbative triple excitations (DC CCSD(T)), and corrected for the Breit contribution and Lamb shift. The calculated IP_1 value is 4.963(15) eV. Our experimental result on the first ionization potential of $4.96^{+0.08}_{-0.07}$ is in excellent agreement with the theoretical value calculated in this work [14].

3. Where is the appropriate place of Lr in the Periodic Table?

The measured IP_1 value of Lr, 4.96 eV, is shown together with the theoretical values and experimental and/or estimated values [20–22] of lanthanide and actinide elements in Fig. 3. It is known that the IP_1 value increases monotonically from terbium to ytterbium and decreased dramatically at lutetium in the lanthanide series. As shown in Fig. 3, IP_1 of Lr is distinctly lower than that of Lu, 5.425871(12) eV [21] and the lowest among the actinides. This validates the position of Lr as the last actinide element (f-block element) in the Periodic Table, and quantitatively reflects and confirms the theoretically predicted situation of closed $5f^{14}$ and $7s^2$ shells with an additional weakly-bound electron in the valence orbital. The IP_1 of Lr is surprisingly low, with the valence electron even weaker bound than in sodium (5.1291 eV) [22]. The metallicity of Lr can be thus compared with an alkali metal.

On the other hand, our experimental result strongly supports that the outermost electronic orbital of Lr is the $7p_{1/2}$, due to the influence of the strong relativistic effects. This proposed electronic configuration of Lr, $[Rn]5f^{14}7s^27p^1_{1/2}$, suggests it would also not be out of the place in the p-block. A transition element is defined as "an element whose atom has an incomplete d sub-shell, or which can give rise to cations with an incomplete d sub-shell" by the International Union of Pure and Applied Chemistry (IUPAC) [23]. According to this definition, Lr would not belong to the transition elements.

It was also suggested that Lu and Lr should be placed below scandium and yttrium in the d-block, based on its ground-state valence-electron configuration and self-consistent group trends for various chemical and physical properties [24].

Since the introduction of the "actinide concept" as the most dramatic modern revision of the Periodic Table in the 1940s, the element with atomic number 103, lawrencium, played a crucial role as the last element in the actinide series. This special position has placed this element into the focus of discussion on the influence of relativistic effects on electronic structure and motivated the determination of its atomic properties. Now, our study not only provides new information, both experimental and theoretical, but also poses a new question: "what are the appropriate positions of Lu and Lr in the Periodic Table?" This issue was already on the agenda of 48th IUPAC Council Meeting held in Korea in 2015 [25]. An IUPAC task group has been formed in order to make a recommendation regarding the membership of group 3 of the Periodic Table: Project No. 2015-039-2-200 [26].

We thank the JAEA tandem accelerator crew for supplying intense and stable beams for the experiments. The ^{249}Cf was made available by H. Nitsche (University of California, Berkeley); it was produced in the form of ^{249}Bk through the former Transplutonium Element Production Program at Oak Ridge National Laboratory (ORNL) under the auspices of the Director, Office of Science, Office of Basic Energy Sciences, Chemical Sciences, Geosciences, and Biosciences Division of the US Department of Energy. Financial support by the Helmholtz-Institut Mainz is acknowledged. This work has been partly supported by the Grant-in-Aid for Scientific Research (C) no. 26390119 of the Ministry of Education, Science, Sports and Culture (MEXT).

References

[1] P. Pyykkö, Chem. Rev. **88**, 563 (1988).
[2] G.T. Seaborg, Science **104**, 379 (1946).
[3] J.-P. Desclaux and B. Fricke, J. Phys. **41**, 943 (1980).
[4] B. Eichler and S. Hübener, Inorg. Chim. Acta, **146**, 261 (1988).
[5] D.T. Jost*et al.*, Inorg. Chim. Acta, **146**, 255 (1988).
[6] R.J. Silva *et al.*, Inorg. Nucl. Chem. Lett. **6**, 733 (1970).
[7] D.C. Hoffmann *et al.*, J. Radioanal. Nucl. Chem. **124**, 135 (1988).
[8] J. Peterson *et al.*, J. Alloy. Comp. **271–273**, 876 (1998).
[9] M. Sewtz *et al.*, Phys. Rev. Lett. **90**, 163002 (2003).
[10] S. Rothe *et al.*, Nature Comm. **4**, 1835 (2013).
[11] N. Sato *et al.*, Radiochim. Acta **102**, 211 (2014).
[12] É.Y. Zandberg and N. Ionov, Sov. Phys. Usp. **2**, 255 (1959).
[13] R. Kirchner, Nucl. Instr. Meth. **A292**, 203 (1990).
[14] T.K. Sato *et al.*, Nature **520**, 209 (2015).
[15] T.K. Sato *et al.*, J. Radioanal. Nucl. Chem. **303**, 1253 (2015).
[16] T.K. Sato *et al.*, Rev. Sci. Instrum. **84**, 023304(5) (2013).
[17] Y. Nagame *et al.*, J. Nucl. Radiochem. Sci. **3**, 85 (2002).
[18] A. Kramida *et al.*, NIST Atomic Spectra Database version 5.1 (2013); `http://physics.nist.gov/asd` (accessed July 2014).
[19] A. Borschevsky *et al.*, Eur. Phys. J. **D45**, 115 (2007).
[20] J. Sugar, J. Chem. Phys. **60**, 4103 (1974).
[21] H. Maeda *et al.*, J. Phys. B **22**, L511 (1989).
[22] Handbook of Chemistry and Physics, 65th ed., eds. R.C. Weast, M.J. Astle, W.H. Beyer (CRC, 1985).
[23] IUPAC Gold book `http://goldbook.iupac.org/`
[24] W.B. Jensen, Found Chem. **17**, 23–31 (2015).
[25] 48th IUPAC Council Meeting Final Agenda Book.
[26] E. Scherri, Chem. International, March-April 22–23 (2016).

EPJ Web of Conferences **131**, 05002 (2016) DOI: 10.1051/epjconf/201613105002
Nobel Symposium NS160 – Chemistry and Physics of Heavy and Superheavy Elements

On the way to unveiling the atomic structure of superheavy elements

Mustapha Laatiaoui[1,2,a]

[1] Helmholtz-Institut Mainz, Staudinger Weg 18, 55128 Mainz, Germany
[2] GSI Helmholtzzentrum für Schwerionenforschung GmbH, Planckstr. 1, 64291 Darmstadt, Germany

Abstract. Optical spectroscopy of the transfermium elements (atomic number $Z > 100$) is nowadays one of the most fascinating and simultaneously challenging tasks in atomic physics. On the one hand, key atomic and even nuclear ground-state properties may be obtained by studying the spectral lines of these heaviest elements. On the other hand, these elements have to be produced "online" by heavy-ion induced fusion-evaporation reactions yielding rates on the order of a few atoms per second at most, which renders their optical spectroscopy extremely difficult. Only recently, a first foray of laser spectroscopy into this heaviest element region was reported. Several atomic transitions in the element nobelium ($Z = 102$) were observed and characterized, using an ultra-sensitive and highly efficient resonance ionization technique. The findings confirm the predictions and additionally provide a benchmark for theoretical modelling. The work represents an important stepping stone towards experimental studies of the atomic structure of superheavy elements.

1. Introduction

The experimental investigation of the atomic spectra of elements beyond fermium lies at the frontier of contemporary atomic physics research [1]. The observation of atomic transitions in these very heavy elements, for instance, would provide a stringent test of modern atomic theories and models addressing relativistic and quantum electrodynamic (QED) effects [2, 3]. In addition, through hyperfine structure and isotope shift measurements, nuclear properties may be obtained in a comprehensive and nuclear model-independent way [4]. These properties include the spin, the magnetic dipole, and the electric quadrupole moments, and the changes in the mean-square charge radii.

However, despite the strong efforts undertaken and the large optical spectroscopy research programs initiated in the last century, optical spectroscopy was deemed to reach a barrier of applicability at the element einsteinium [5] and, since more than a decade now, at fermium [6]. The transfermium elements remained out of reach for such an approach until recently, as neither primordial isotopes of these elements exist nor can macroscopic samples be synthesized. These radioactive elements have to be produced by heavy-ion induced fusion-evaporation reactions yielding low rates, which inevitably calls for ultra-efficient and fast spectroscopy techniques of extreme sensitivity. Moreover, the lack of any existing tabulated spectral lines for these elements represents a big hurdle for any laser

[a] e-mail: `m.laatiaoui@gsi.de`

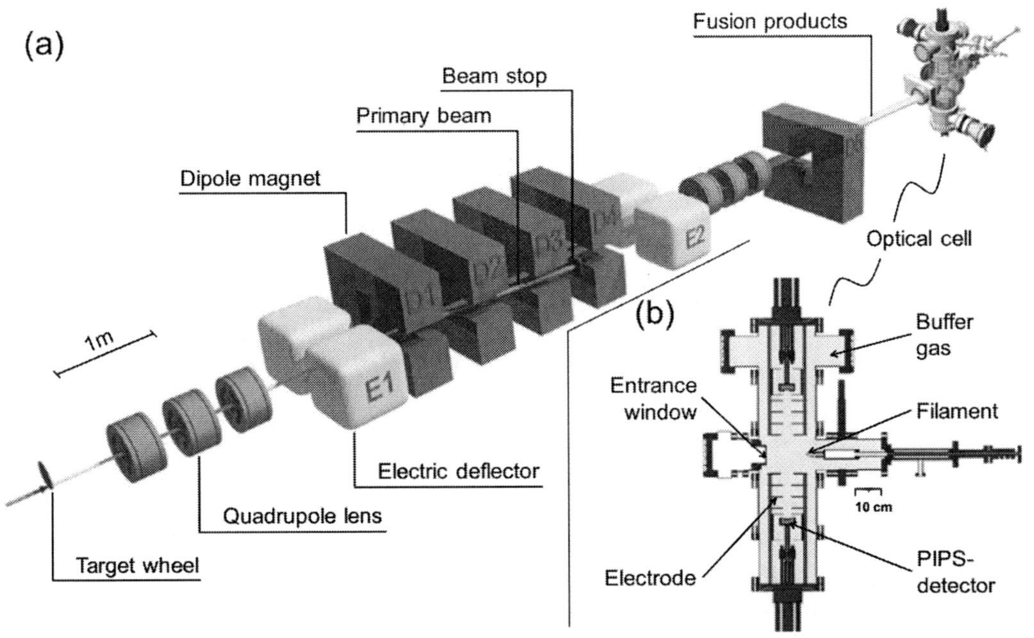

Figure 1. The experimental setup. (a) Schematic overview of the velocity filter SHIP [11] with a buffer-gas stopping cell coupled to it. A thin entrance Mylar foil separates the buffer gas volume inside the cell from the high vacuum chamber of the SHIP beamline. (b) Interior of the cell, see text for details.

spectroscopy experiment, making an extensive search for atomic transitions within a spectral range proposed by state-of-the-art atomic model calculations necessary.

Laser resonance ionization spectroscopy of the element nobelium (No) has long been anticipated at the GSI in Darmstadt [7]. To this end, the so-called RAdiation Detected Resonance Ionization Spectroscopy (RADRIS) technique was employed [8, 9]. In August 2015, atomic transitions for this element were observed for the first time, revealing atomic structure information at production rates of a few atoms per second at most [10]. Further, an upper limit value for the first ionization potential (IP) has been obtained. The investigations were focused on the radionuclide ^{254}No, but were extended to even ^{252}No with the lowest production rate among the investigated isotopes.

2. Experimental details

Figure 1 shows the experimental setup used for laser spectroscopy of nobelium. It consists of a buffer-gas stopping cell, which was installed in the focal plane of the Separator for Heavy Ion reaction Products (SHIP) [11]. The cell exhibits an entrance window, a catcher filament, and two identical extraction electrode systems with a corresponding α-particle detector (Passivated Implanted Planar Silicon semiconductor detector, PIPS). The radionuclide ^{254}No with a half-life of $T_{1/2} = 51.2$ s was chosen for the first laser spectroscopy experiments. This was produced at SHIP in the reaction ^{208}Pb(^{48}Ca,2n)^{254}No with a cross section of about 2 μb [12], cf. Table 1. The rate of product ions delivered to the cell was regularly checked by a retractable position-sensitive 16-strip silicon detector placed in front of the entrance window.

The RADRIS measurements were organized in user-defined measurement cycles comprising a beam-on and a beam-off period to allow a temporal disentanglement of the preparation of the sample atoms from the actual spectroscopy investigations. During the

Table 1. Nobelium isotopes investigated in the RADRIS experiments. The isotopes were produced in the $2n$ channel of the nuclear reactions ^{48}Ca+ ^{208,206}Pb using lead sulfide (PbS) targets. The reaction cross sections are taken from reference [12]. The applied measurement cycles comprising beam-on and beam-off periods were optimized upon the isotope half-life [19].

Isotope	Half-life	^{48}Ca-beam energy	Lead isotope	PbS-Target thickness	Cross-section	Beam-on/-off cycle	RADRIS efficiency
^{254}No	51.2 s	217 MeV	^{208}Pb	470 μg/cm^2	2 μb	25 s/5 s	(6.4 ± 1)%
^{252}No	2.4 s	217 MeV	^{206}Pb	460 μg/cm^2	0.5 μb	3 s/3 s	(3.3 ± 1)%

beam-on period, the fusion products emerging from the target were separated from the primary beam by SHIP before entering through a 3 μm-thick entrance foil (Mylar) into the cell filled with 95 mbar high-purity argon gas. After thermalization, the fusion products remaining in a positive charged state were accumulated on a catcher filament, a 125 μm thin tantalum wire, where they were neutralized. In a subsequent step, the fusion-product beam was turned off. Then, the accumulated atoms were evaporated from the filament, by heating this briefly to about 1350 K. Nobelium atoms were then ionized in a two-step photoionization process [13] by pulsed lasers, see inset in Fig. 2a. In a first step ($\overline{\nu}_1$) during this process, a valence electron is excited from the $^1S_0$ ground state into a $^1P_1$ excited state. In a second step ($\overline{\nu}_2$), it is excited further into the continuum beyond the ionization potential such that $\overline{\nu}_1 + \overline{\nu}_1 \geq$ IP leading to the ionization of the atom. Once the ionization occurred, the laser-created ions were promptly guided by suitable electric fields to the PIPS detector where they were unambiguously identified via their unique radioactive decay fingerprint.

Up to four pulsed dye lasers operated at a pumping wavelength of 248 nm and one Nd:YAG pumped Optical Parametric Oscillator (OPO) operated in a frequency mixing mode were simultaneously used to search for the $^1P_1$ excited state in nobelium. The laser wavelengths were monitored using a HighFinesse-Ångstrom wavelength meter WS/7-UVU that was calibrated to an internal neon lamp. The second step was provided by excimer laser light at 351–355 nm wavelength with high power so that an efficient excitation to the continuum was possible. The reader is referred to references [10, 14] for more details on the method and the laser systems used in these experiments.

3. Results

Based on the latest multi-configuration Dirac-Fock (MCDF) [15, 16] and relativistic coupled cluster (RCC) [17, 18] calculations a spectral range between 28887 and 33191 cm^{-1} was chosen for the search for the nobelium $^1S_0$-$^1P_1$ ground-state transition, cf. Fig. 2.

The range up to 30530 cm^{-1} was scanned more than twice in ≈ 0.9 cm^{-1} steps by the dye lasers. The OPO system scanned multiple times in 3 cm^{-1} steps the range between 30000 cm^{-1} and 33191 cm^{-1}. The events registered during the scans were analyzed and potential candidates for the location of the resonance were identified. In the following experiments, these candidates were checked at smaller scan steps and higher statistics. In total, more than 6000 laser scan steps were conducted before discovering the resonance located at 29961.5 cm^{-1}. Figure 2a shows a typical scan across the resonance, which is marked by an increase in the registered ^{254}No α-decay count rate on the PIPS detector. From the saturation characteristics for this first-step excitation, an Einstein coefficient for spontaneous emission A_{ki} of $(4.2^{+2.6}_{-2.8}) \cdot 10^8$ /s was experimentally determined, see Fig. 2b. This is in good agreement with the A_{ki}-predictions [15–17], corroborating the observation of the $^1S_0$-$^1P_1$ ground-state transition. All the predicted values for the transition energy are in agreement with the measured one, which is far more precise, and thus provides a powerful benchmark. Noticeable, however, is the striking agreement of the RCC calculated value with

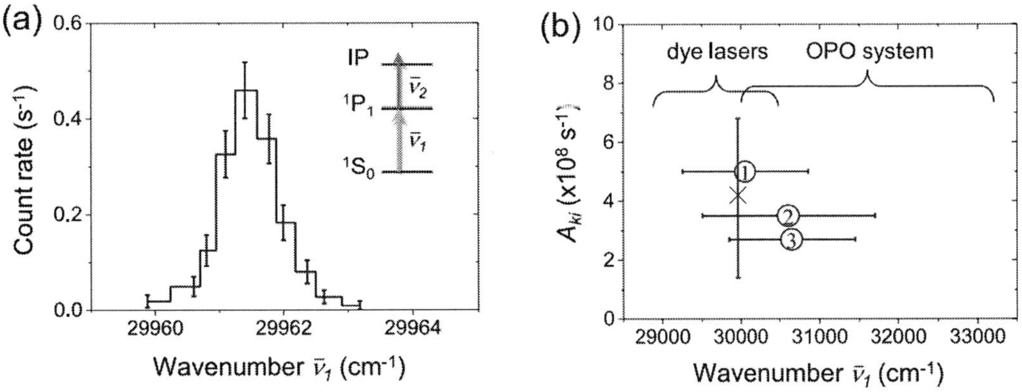

Figure 2. The first-step excitation. (a) The observed first-step resonance in nobelium in terms of registered ^{254}No α-decay count rates normalized to $6.2 \cdot 10^{12}$ ^{48}Ca projectiles per second. Inset: a simplified ionization scheme. (b) Overview on the measured ($\times$) and predicted ($\circ$) first-step excitation energy and the Einstein coefficient for spontaneous emission A_{ki} for the $^1P_1$ atomic state in nobelium. The calculated values are taken from RCC- [17] (1) and MCDF calculations [16] (2) and [15] (3), respectively, with an error estimate for (2) as reported in reference [10].

the measurement. This supports that a rather conservative error estimate for the transition energy was given in reference [17]. Also worthy of mentioning is the general overestimation of the energy by all theoretical models. At least in the case of the MCDF calculations, one may attribute this to the treatment of correlation effects that positively contribute to the orbital energy.

In total, less than 10000 atoms delivered to the gas cell were sufficient to measure the shown resonance shown in Fig. 2a. The overall efficiency of the setup is determined by the ratio of the decay count rate measured with the PIPS detector at maximum of a resonance to the implantation rate of ions delivered to the cell and measured with the 16-strip silicon detector, both normalized to the primary beam intensity. This amounted to 6.4% in the case of ^{254}No and was achieved over rather long measurement times. By conditioning the filament surface, i.e. by heating the tantalum wire under vacuum conditions to about 1800 K, the overall efficiency was boosted for a short time by nearly a factor of two. This behaviour may be attributed to an increased retention of impurities by a clean tantalum filament surface, hence, preventing unwanted associative desorption processes. Using ^{206}Pb-targets, it was possible to extend the measurements to ^{252}No at a production cross section of only 0.5 μb (cf. Table 1). It turned out that the overall efficiency strongly depends on the isotope's half-life as a result of the applied measurement cycles, which in turn were optimized with respect to the half-life. The smaller the ratio of the beam-on to the beam-off period, the smaller the efficiency. Nonetheless, the latter amounted to 3.3% for ^{252}No ($T_{1/2} = 2.4$ s) proving the high efficiency of the applied spectroscopy technique.

4. Conclusions

Recently, the element nobelium on atom-at-a-time quantities was successfully probed by lasers, for the first time since its official discovery in 1958. At present this is the heaviest element, for which laser spectroscopic investigations were performed. The long-awaited results on the level search on nobelium confirmed a predicted strong ground-state atomic transition, the $^1S_0$-$^1P_1$ transition.

From the accuracy viewpoint, the error estimates from the MCDF calculations were realistic for this transition. In comparison, RCC calculations delivered a predicted value of a striking agreement with the measured one. Given a sufficient treatment of the Coulomb correlations, the latter approach may give insights on the strength of QED effects, which could have been underestimated in nobelium so far.

In addition, the measurements revealed a high efficiency of the RADRIS setup for the different nobelium isotopes, paving the way for isotope shift and hyperfine structure studies in the region of deformed nuclei around the neutron shell closure $N = 152$ and $Z = 102$. Moreover, the heaviest actinide element, lawrencium, which is produced at lower rates than the investigated nuclides and exhibits a different atomic structure, is now within reach of RADRIS experiments.

The results thus provide valuable anchor points for theoretical models and opens up new perspectives for future precision measurements of atomic and nuclear properties of the elusive transfermium elements using laser spectroscopy.

References

[1] H. Backe *et al.*, Nucl. Phys. **A944**, 492 (2015).
[2] P. Schwerdtfeger *et al.*, Nucl. Phys. **A944**, 551 (2015).
[3] E. Eliav, S. Fritzsche, U. Kaldor, Nucl. Phys. **A944**, 518 (2015).
[4] P. Campbell, I.D. Moore, M.R. Pearson, Prog. Part. Nucl. Phys. **86**, 127 (2016).
[5] E.F. Worden *et al.*, in *The Chemistry of the Actinide and Trans-actinide Elements*, eds. L.R. Morss, M. Edelstein, J. Fuger, vol. 3, chap. 16.4 (Springer, 2008).
[6] M. Sewtz *et al.*, Phys. Rev. Lett. **90**, 163002 (2003).
[7] H. Backe *et al.*, Eur. Phys. J. **D45**, 99 (2007).
[8] W. Lauth *et al.*, Phys. Rev. Lett. **68**, 1675 (1992).
[9] M. Laatiaoui *et al.*, Eur. Phys. J. **D68**, 71 (2014).
[10] M. Laatiaoui *et al.*, Nature (2016), in press. `http://dx.doi.org/10.1038/nature19345`.
[11] S. Hofmann and G. Münzenberg, Rev. Mod. Phys. **72**, 733 (2000).
[12] Yu.Ts. Oganessian *et al.*, Phys. Rev. C **64**, 054606 (2001).
[13] V.S. Letokhov, *Laser Photoionization Spectroscopy* (Academic Press, 1987).
[14] F. Lautenschläger *et al.*, Nucl. Instr. Meth. **B383**, 115 (2016).
[15] P. Indelicato *et al.*, Eur. Phys. J. **D45**, 155 (2007).
[16] Y. Liu, R. Hutton, Y. Zou, Phys. Rev. A **76**, 062503 (2007).
[17] A. Borschevsky *et al.*, Phys. Rev. A **75**, 042514 (2007).
[18] V.A. Dzuba, M.S. Safronova, U.I. Safronova, Phys. Rev. A **90**, 012504 (2014).
[19] M. Laatiaoui *et al.*, Hyperfine Int. **227**, 69 (2014).

EPJ Web of Conferences **131**, 05003 (2016) DOI: 10.1051/epjconf/201613105003
Nobel Symposium NS160 – Chemistry and Physics of Heavy and Superheavy Elements

Mass measurements and ion-manipulation techniques applied to the heaviest elements

Michael Block[1,2,3,a]

[1] GSI Helmholtzzentrum für Schwerionenforschung, 64291 Darmstadt, Germany
[2] Helmholtz Institute Mainz, 55099 Mainz, Germany
[3] Johannes Gutenberg-Universität, 55099 Mainz, Germany

Abstract. High-precision mass measurements of radionuclides with state-of-the-art mass spectrometry allows us to obtain accurate binding energies. These reflect changes in the nuclear structure. Two-nucleon separation energies, for example, are sensitive indicators of shell closures and the onset of deformation. In addition, masses provide a benchmark for theoretical nuclear models and support their improvement with respect to a better predictive power. In the region of the heaviest elements this enables mapping the strength of shell effects and their extension as recently demonstrated for $N = 152$. Accurate masses can furthermore provide anchor points to fix decay chains in the mass surface. Moreover, advanced ion-manipulation techniques developed in the context of mass spectrometry pave the way for novel types of experiments such as trap-assisted decay spectroscopy of nuclear state-selected samples. This opens new perspectives for a mass number assignment of decay chains originating from yet unknown superheavy nuclides.

1. Introduction

Ion-trap based techniques allow us to manipulate ions at low energy in a well-controlled manner to prepare pure samples of radioactive ions for precision measurements. One of the most prominent applications for radioactive nuclides is Penning trap mass spectrometry (PTMS) [1, 2]. In this approach, the mass of an ion is obtained via a measurement of the cyclotron frequency $\nu_c = qB/m$ at which the ion orbits in the confining electromagnetic fields of a Penning trap. The typical relative precision that is nowadays obtained for the masses of radionuclides is on the order of 10^{-8} for short-lived nuclides [3]. For long-lived and stable particles even two to three orders of magnitude lower uncertainties have been reached [4, 5]. The ingredients for mass measurements with highest precision are the selective removal of unwanted nuclides and the minimization of systematic effects well below the anticipated uncertainty. The most common systematic effects in PTMS include the spatial and temporal homogeneity of the confining fields and the ion-ion Coulomb interaction. The systematic effects in PTMS and their impact on the mass uncertainty is well understood and has been discussed in the literature in depth [6, 7]. In general, the measurements in a one-ion-at-a-time procedure and the application of appropriate ion cooling methods reduces the amplitudes of the ion motion such that the ions only probe the most homogeneous region of the trapping fields. Confinement in an ultra-high vacuum environment furthermore reduces

[a] e-mail: m.block@gsi.de

collisions with rest gas atoms. Using appropriate calibrating ions, moreover, absolute mass measurements are feasible using carbon cluster ions as mass reference, for example.

In the context of nuclear physics, and in particular for investigation of the nuclear properties of the heaviest elements, two main features of mass measurements are of interest. On the one hand side the nuclear structure evolution can be tracked by masses and mass differences to identify, for example, shell closures and the onset of deformation. Moreover, accurate direct mass measurements provide anchor points in the mass surface to pin down decay chains originating from superheavy nuclides as demonstrated for example for nobelium isotopes [8]. On the other hand, the high mass resolving power provided by Penning trap techniques is suitable for the separation of unwanted by-products to prepare pure radioactive samples for subsequent experiments such as nuclear decay spectroscopy. These so-called trap-assisted decay spectroscopy experiments have already been performed in lighter nuclides [9] demonstrating that even nuclear state-selected samples can be delivered to a decay spectroscopy setup for certain nuclides [10]. Trap-assisted decay spectroscopy of isotope-selected samples opens new perspectives for an A/Q identification of superheavy nuclides, an application of particular interest for nuclides originating from yet unknown elements.

2. Mass measurements with Penning traps

PTMS has been a well-established mature field for many years [2]. However, the extension to the heaviest nuclides was only achieved less than ten years ago with pioneering measurement with SHIPTRAP at GSI Darmstadt [11]. The major challenges that had to be mastered are the rather low yield and the high energy of the nuclides of interest following their production in nuclear fusion reactions. The present status of PTMS of the heaviest elements has been summarized in recent review articles [12, 13]. Here, the essence of the method with respect to its application to the heaviest elements will be briefly summarized in the following.

The radionuclides of interest are produced in complete fusion-evaporation reactions. Following the separation from the primary beam in an electromagnetic recoil separator, the reaction products are provided with energies in the order of some ten MeV for experiments. So-called buffer gas stopping cells are utilized, also in the case of SHIPTRAP [14], to convert these energetic ions of interest into a low-energy beam suitable for ion trapping. These buffer gas stopping cells, sometimes also referred to as ion catchers, slow down the reaction products in an inert buffer gas atmosphere, usually helium, at pressures of about 50–100 mbar in some ten centimeters. Depending on the element and its ionization potential the major fraction of the slowed ions remain singly- or doubly-charged. These can be extracted within tens of milliseconds from the buffer gas by a combination of electric fields and gas flow through an extraction nozzle. First-generation stopping cells have reached overall efficiencies of about 5–10% [14, 15]. Recently, first cryogenic devices have been introduced boosting the efficiency to a level of about 30–50% [16, 17].

The extracted ions are then transported, cooled, and bunched by radio-frequency quadrupole ion guides. The formed low-energy bunched ion beam is injected into a Penning trap system where the mass of the ion is measured. Pioneering experiments on nobelium and lawrencium nuclides around the deformed neutron shell closure at $N = 152$ have been performed with the Penning trap system SHIPTRAP [11, 18]. It consists of two cylindrical Penning traps that are installed in the bore of a superconducting solenoid featuring two homogeneous centers. The first trap is dedicated to ion preparation by a mass-selective buffer gas cooling technique that allows removing unwanted by-products that may accompany the ions of interest. The mass resolving power achieved in this step is on the order of 100 000, sufficient to separate nuclear isobars. This is, however, not an issue for the heaviest elements where the ions of interest are typically produced in xn-evaporation channels of complete fusion reactions. In the second trap mass measurements can be performed using different techniques. The most exotic nuclides such as ^{256}Lr [18] have been measured by the so-called time-of-flight ion-cyclotron-resonance (ToF-ICR) method as described in reference [19]. This method is rather robust and universal and can be applied to radionuclides with half-lives exceeding about 10 ms. The achievable precision scales inversely proportional with the excitation (and hence measurement) time. The main drawback of this method is the need to scan the excitation frequency around the expected cyclotron frequency and the corresponding number of 30–50 ions that needs to be accumulated in order to determine a mass value. In the above mentioned case of ^{256}Lr a single measurement took about four days, which is challenging as many parameters have to be kept constant to achieve a high precision. This measurement time can be further reduced with a number of upgrades some of which are addressed in Sect. 4. Ultimately single-ion mass measurement techniques have to be employed to foray into the territory of the heaviest elements.

Mass spectrometers such as Penning traps with their high mass resolving power are also useful isobar separators. In many cases even isomeric states can be separated. This opens up interesting perspectives for novel types of experiments. Trap-assisted decay spectroscopy is a powerful combination of Penning traps or other mass separators with decay spectroscopy setups. In the region of heavy nuclides that predominantly decay by α decay efficient and compact α-γ detection setups are suitable. The feasibility of this approach has been demonstrated combing TASISpec [20] with SHIPTRAP [21], for example. Purified and potentially state-selected tailored samples were implanted into the setup where the decay is measured. Similar experiments have already been performed for β-decay studies, for example in Jyväskylä [9] and specifically for measurements of superallowed β decays. The feasibility of α-γ spectroscopy was demonstrated for Rn and Ra isotopes in proof-of-principle experiments connecting TASISpec and SHIPTRAP. Due to the purity of the samples this approach is particularly useful for measuring branching ratios of weak decay branches accurately. This has been shown for example for ^{213}Ra [21].

Another opportunity for applying this technique is the mass identification of members of decay chains originating from yet unknown superheavy nuclides. To this end only a moderate mass resolving power on the order of 300 suffices to distinguish different isotopes. Such a mass resolving power can already be realized with radio frequency mass filters, for example, but of course also with Penning traps and time-of-flight mass spectrometers. Since the slowing down, cooling and bunching of the ions of interest takes place on a time scale on the order of 10–100 milliseconds, this method can be applied to relatively short-lived nuclides. Alternative ideas for mass-selected decay spectroscopy for the A/Q identification of superheavy nuclides are pursued within the FIONA project at Berkeley [22]. There an electromagnetic separator, a Wien filter with unbalanced electric and magnetic field will be utilized. Mass spectrometric A/Q identification techniques with the high resolution achieved in ion traps also provide an option for the detection of long-lived nuclides and nuclides that do not exhibit α decay or spontaneous fission, if a competitive overall efficiency is achieved.

3. Mass measurements and nuclear structure

A key question with respect to the nuclear structure evolution in the heaviest elements is related to the character of shell effects. While in lighter nuclei, for example in the tin isotopic chain, well localized and strong shell closures are observed, in the heaviest nuclides a different behaviour is predicted, see for example [23]. Different nuclear models obtain rather extended regions of enhanced shell stabilization in contrast to well-localized magic numbers.

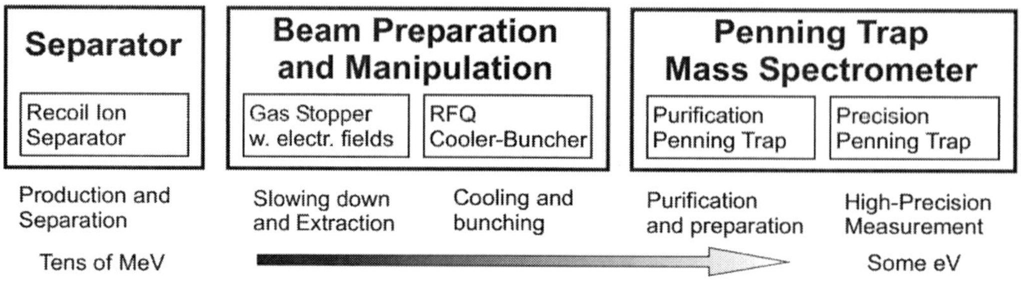

Figure 1. Schematic layout of a Penning trap mass spectrometer suitable for the heaviest elements.

In addition, the emergence of deformed shell closures in the regions of neutron numbers $N = 152$ and $N = 162$ have been predicted and experimentally observed, originally based on experimental Q_α values. The present experimental situation is illustrated by the two-neutron separation energies for the elements from $Z = 98$–100 as displayed in Fig. 1. The values have been calculated based on the experimental data compiled in the most recent Atomic Mass Evaluation (AME 2013) [24]. Despite the large uncertainties in the more exotic nuclides indications of shell closures at $N = 152$ and $N = 162$ are clearly visible.

The AME 2013 considered the results of direct mass measurements performed with SHIPTRAP for the accurate mapping of the $N = 152$ shell effects in nobelium and lawrencium [11, 18]. This deformed shell closure was found to be about $\delta_{2n} = 1.2$ MeV. The $N = 152$ shell is most pronounced at $Z = 100$ with a shell gap of about $\delta_{2n} = 1.5$ MeV. The evolution for different proton numbers is yet not fully established due to a lack of experimental data with sufficient precision. In principle, direct mass measurements are an ideal tool for such studies. However, the low yield of the nuclides of interest for such measurements makes this endeavor tedious. In lighter nuclides between uranium and fermium the availability of macroscopic amounts for certain long-lived nuclides facilitates off-line measurements of some relevant nuclides. First measurements have been performed for the masses of Am, Cm, and Cf nuclides with TRIGA-TRAP [25] and will be extended in the near future. A local mass evaluation in the spirit of the the Atomic Mass Evaluation (AME) furthermore indicated that already a few accurate direct mass measurements can impact the masses of many more nuclides in the region via numerous established decay links.

Thus, a promising strategy for the future is to combine mass spectrometry and α-γ decay spectroscopy: direct mass measurements provide the masses of selected anchor points with high accuracy, which is in particular relevant for odd-odd and odd-A nuclides. In these nuclides the strongest α-decay branch usually populates excited states. Then, the accurate determination of the mass values from a Q_α value require a detailed knowledge of the decay scheme. For even-even nuclides the situation is rather straightforward as the strongest α transitions typically connects the ground states in the mother and daughter nuclides. The different data can be fed into the network of the AME where the masses of all nuclides are evaluated based on known data and their links. The impact of even a few accurate data has been demonstrated for example for the masses of the nobelium isotopes measured at SHIPTRAP [8] and for masses of actinides measured at TRIGA-TRAP [25].

While the experimental data around $N = 152$ is already rather comprehensive the situation for $N = 162$ is worse as indicated by the large uncertainties of many masses in Fig. 2. In addition, the masses of certain nuclides are only known from extrapolations. Nonetheless, the shell effects at $N = 162$ are clearly visible. For many of these nuclides direct mass measurements are presently out of reach due to their low yield and rather short half-lives for some neutron-deficient nuclides. However, even for detailed α-γ spectroscopy

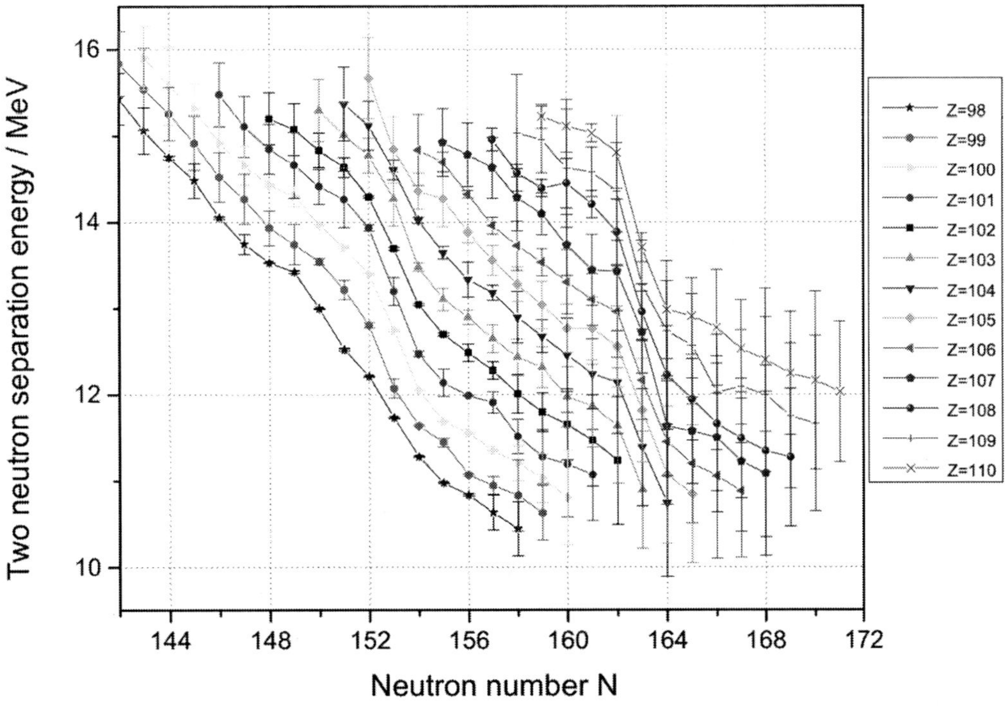

Figure 2. Two neutron separation energies for the elements from $Z = 98$–100. The data are calculated based on the masses from reference [24].

the low cross sections, for example a few picobarn for the Ds isotopes $^{268-270}$Ds, are also challenging.

4. Future perspectives for mass measurements of heavy elements

At present a mass measurement program for nuclides in the region above fermium is ongoing using the SHIPTRAP setup at the GSI in Darmstadt. The lowest yield handled to date was on the order of one particle per minute (delivered to the gas cell) in the case of ^{256}Lr corresponding to a cross section of about sixty nanobarn [18]. Recent upgrades to the SHIPTRAP setup have been implemented to boost its performance further in order to extend the reach to nuclides with much lower yield. First, a new cryogenic stopping cell has been installed. It features a larger stopping volume and a beam injection along the symmetry / extraction axis, which is in contrast to the previous gas cell where the beam was injected almost perpendicular. In addition, the operation at temperatures of about 40 K results in a higher cleanliness since potential impurities freeze out. In off-line measurements using a radioactive recoil ion source a extraction efficiency of up to 70% was determined [17]. For the integration of the cryogenic gas cell, the entire SHIPTRAP setup was recently relocated within the newly arranged experimental area behind the SHIP separator at the GSI. First on-line commissioning experiments with ^{254}No resulted in an increased overall efficiency by at least a factor of five [26]. Further improvements are anticipated.

The second upgrade concerns the measurement technique employed to obtain the cyclotron frequency. Recently, the novel phase-imaging ion-cyclotron-resonance method technique has been developed at SHIPTRAP [27]. In this method the ion motion is imaged utilizing a spatially resolving ion detector, typically a microchannel plate detector with delay line anode. Choosing an appropriate ion optics with distortion free magnification allows to measure the phase of the (radial) ion motion from which the corresponding frequency can be obtained. The systematic uncertainties relevant to this method have been studied in depth [28]. Several mass measurements have shown that the novel method provides a forty times higher mass resolving power and a gain in precision by a factor of about five compared to the conventional scheme [27]. The new method has been applied to mass difference measurements of stable and long-lived nuclides, for example ^{163}Ho and ^{48}Ca, reaching uncertainties of tens of electronvolts [29, 30]. For the application to the heaviest elements in the context of nuclear structure studies uncertainties of a few keV suffice. Then, to reach the same uncertainties the method is about twenty-five times faster than the conventional method. This is beneficial for the access to shorter-lived nuclides. Furthermore, the high resolving power opens up new perspectives to study isomeric states with very low excitation energy, for example in ^{255}Lr.

A key development to extend the reach to heavier nuclides with decreasing yield is the realization of single-ion mass measurement techniques. A technique that has already been developed for electrons, protons, and lighter stable ions, often highly charged, is the so-called Fourier-transform ion-cyclotron-resonance technique (FT-ICR) [31]. The method relies on a non-destructive image current detection. Trapped ions induce image charges into the trap electrodes due to their motion. Such weak currents on the order of 100 femtoamperes can be amplified by resonant tank circuits with high quality factor. For radioactive ions developments are presently underway, for example at TRIGA-TRAP [32], SHIPTRAP [33], and at the NSCL within the SIPT project [34].

In the future also time-of-flight mass spectrometers (ToF-MS) [35] are planned to be used for experiments on the heaviest nuclides. In recent years so-called multi-reflection (MR) ToF-MS have been developed. In these devices cooled short ion bunches are reflected between electrostatic mirrors. Thus, a compact setup can be maintained while providing high mass resolving powers up to about 200 000 for millisecond flight times [36–38]. In contrast to conventional ToF-MS where the flight path limits the achievable precision to typically a level of 10^{-6} these devices can reach relative mass uncertainties of about 10^{-7} or even below [39]. MR-ToF-MS devices are thus suitable for mass measurements with a precision usually required to study global nuclear structure trends. These devices require low-energy bunched beams as in case of PTMS. Hence the beam preparation stages are usually the same involving a buffer gas stopping cell and an RFQ buncher. At RIKEN an MR-ToF-MS setup has been installed behind the new gas-filled recoil separator GARIS II [36]. First on-line measurements have been performed in neutron-deficient Fr isotopes [40].

All the different improvements will allow us to extend the reach of mass spectrometry to more exotic nuclides in the future to make a stronger push into the region of the heaviest elements.

References

[1] K. Blaum, Phys. Rep. **425**, 1 (2006).
[2] H.J. Kluge, Int. J. Mass Spectrom. **349**, 26 (2013).
[3] G. Bollen *et al.*, Phys. Rev. Lett. **96**, 152501 (2006).
[4] R. Van Dyck Jr. *et al.*, Phys. Rev. Lett. **92**, 220802 (2004).
[5] R.S. Van Dyck *et al.*, Int. J. Mass Spectrom. **251**, 231 (2006).
[6] L.S. Brown and G. Gabrielse, Rev. Mod. Phys. **58**, 233 (1986).
[7] G. Bollen *et al.*, J. Appl. Phys. **68**, 4355 (1990).

[8] M. Dworschak *et al.*, Phys. Rev. C **81**, 064312 (2010).
[9] J. Rissanen *et al.*, Phys. Rev. C **83**, 011301 (2011).
[10] L.-L. Andersson *et al.*, GSI Scientific Report 2010, GSI Report 2011-1, p. 213.
[11] M. Block *et al.*, Nature **463**, 785 (2010).
[12] M. Block, Int. J. Mass Spectrom. **349**, 94 (2013).
[13] M. Block, Nucl. Phys. **A944**, 471 (2015).
[14] J. Neumayr *et al.*, Nucl. Instr. Meth. **B244**, 489 (2006).
[15] G. Savard, J. Phys.: Conf. Ser. **312**, 052004 (2011).
[16] S. Eliseev *et al.*, Nucl. Instr. Meth. **B258**, 479 (2007).
[17] C. Droese *et al.*, Nucl. Instr. Meth. **B338**, 126 (2014).
[18] E.M. Ramirez *et al.*, Science **337**, 1207 (2012).
[19] M. König *et al.*, Int. J. Mass Spectrom. **142**, 95 (1995).
[20] L.L. Andersson *et al.*, Nucl. Instr. Meth. **A622**, 164 (2010).
[21] D. Rudolph *et al.*, GSI Scientific Report 2008, GSI Report 2009-1, p. 177.
[22] J.M. Gates, contribution to this issue (2016).
[23] M. Bender, W. Nazarewicz, P.G. Reinhard, Phys. Lett. **B515**, 42 (2001).
[24] N. Wang and M. Liu, J. Phys.: Conf. Ser. **420**, 012057 (2013).
[25] M. Eibach *et al.*, Phys. Rev. C **89**, 064318 (2014).
[26] O. Kaleja *et al.*, GSI Scientific Report, XXXX.
[27] S. Eliseev *et al.*, Phys. Rev. Lett. **110**, 082501 (2013).
[28] S. Eliseev *et al.*, Appl. Phys. B **114**, 107 (2014).
[29] S. Eliseev *et al.*, Phys. Rev. Lett. **115**, 062501 (2015).
[30] F. Köhler *et al.*, Nature Communications **7**,10246 (2016).
[31] A.G. Marshall, C.L. Hendrickson, G.S. Jackson, Mass Spectrom. Rev. **17**, 1 (1998).
[32] J. Ketelaer *et al.*, Eur. J. Phys. A **42**, 311 (2009).
[33] C. Weber *et al.*, Eur. Phys. J. **A25**, 65 (2005).
[34] R. Ringle, S. Schwarz, G. Bollen, Int. J. Mass Spectrom. **349**, 87 (2013).
[35] H. Wollnik, Int. J. Mass Spectrom. **349**, 38 (2013).
[36] Y. Ito *et al.*, Nucl. Instr. Meth. **B317**, 544 (2013).
[37] W.R. Plaß, T. Dickel, C. Scheidenberger, Int. J. Mass Spectrom. **349**, 134 (2013).
[38] R. Wolf *et al.*, Nucl. Instr. Meth. **A686**, 82 (2012).
[39] P. Schury *et al.*, Nucl. Instr. Meth. **B335**, 39 (2014).
[40] P. Schury *et al.*, Nucl. Instr. Meth. **B376**, 425 (2016).

EPJ Web of Conferences **131**, 05004 (2016) DOI: 10.1051/epjconf/201613105004
Nobel Symposium NS160 – Chemistry and Physics of Heavy and Superheavy Elements

Nuclear spectroscopy with GEANT4

The superheavy challenge

Luis G. Sarmiento[a]

Department of Physics, Lund University, 22100 Lund, Sweden

Abstract. The simulation toolkit GEANT4 was originally developed at CERN for high-energy physics. Over the years it has been established as a *swiss army knife* not only in particle physics but it has seen an accelerated expansion towards nuclear physics and more recently to medical imaging and γ- and ion- therapy to mention but a handful of new applications.

The validity of GEANT4 is vast and large across many particles, ions, materials, and physical processes with typically various different models to choose from. Unfortunately, atomic nuclei with atomic number $Z > 100$ are not properly supported. This is likely due to the rather novelty of the field, its comparably small user base, and scarce evaluated experimental data.

To circumvent this situation different workarounds have been used over the years. In this work the simulation toolkit GEANT4 will be introduced with its different components and the effort to bring the software to the heavy and superheavy region will be described.

1. Introduction

During the past decade, ^{48}Ca-induced fusion-evaporation reactions on radioactive actinide targets have been the driving force behind the observations of several correlated α-decay chains, which all terminate by spontaneous fission. These decay chains are interpreted to originate from the production of atomic nuclei with proton numbers $Z = 113$–118 [1]. However, neither their mass, A, nor their atomic number, Z, have been measured directly thus far.

Monte Carlo simulations are already an established method to assess various experimental aspects in nuclear and particle physics among others disciplines. Simulation packages of the passage – and interaction – of particles through matter such as the GEANT4 simulation toolkit [2] have been used systematically to benchmark and characterize complete experimental set-ups.

GEANT4 has its origins in high-energy particle physics. The initial scope quickly widened when it became apparent that such a tool would also benefit the nuclear, accelerator, space, and medical physics community. Lack of theoretical understanding, scarce experimental data, and a relative small community have hindered the expansion of the toolkit towards simulation of atomic nuclei with atomic number $Z > 100$. Recently GEANT4 has been used to verify the coherence and consistency of decay schemes of heavy and superheavy nuclei [3–6] showing that the – arguably new – field of heavy and superheavy nuclei can benefit greatly from native support in the simulation toolkit.

[a] e-mail: `Luis.Sarmiento@nuclear.lu.se`

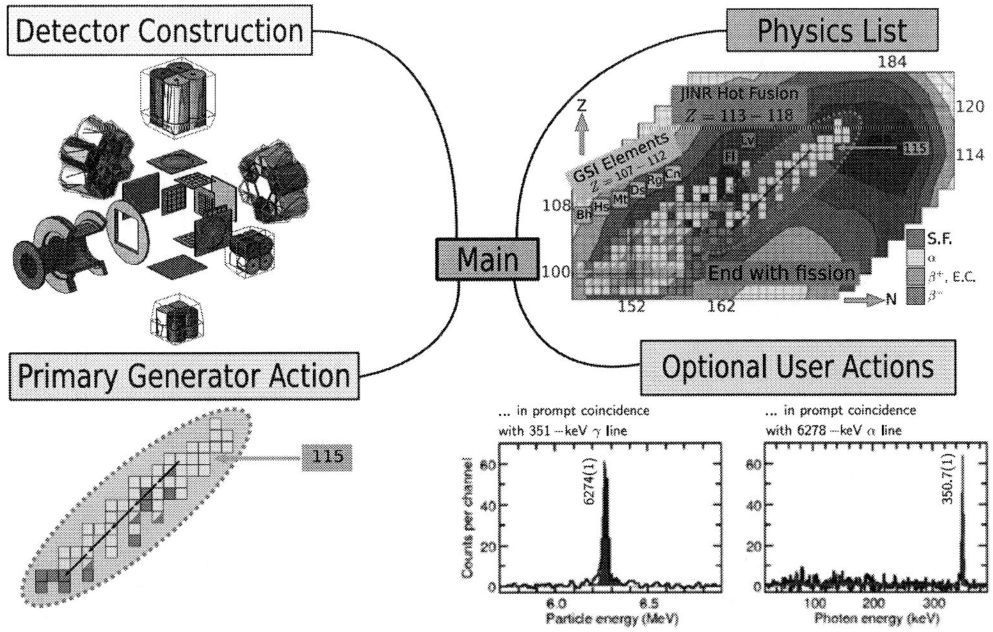

Figure 1. Schematic presentation of the different components of a Monte Carlo simulation using GEANT4. The "laboratory" (geometry, materials, positions) is defined in the *Detector Constructor*. The relevant particles and physical processes to be included in the simulation should go in the *Physics List*. In the *Primary Generator Action* the properties and definitions of the initial particles coming for the "source" to be simulated are included. On top of these three mandatory components the user has the possibility to use some *Optional User Actions*. See text for details.

2. GEANT4 overview

Any GEANT4 simulation requires some basic "ingredients". Figure 1 cartoons the ingredients in particular for the simulation of the TASISPEC experimental set-up [7]. First, we describe the physical space, for instance a laboratory containing a detector set-up, in the *Detector Constructor*. There the geometry, materials, and positions of the different components are defined. Second, since the simulation toolkit covers a wide range of energies and particles, the relevant ones to be used for a specific simulation are included in the *Physics List*; in this example we include the ones in the realm of a typical nuclear physics experiment (α, β^+, or β^- particles, γ rays, ions, electromagnetic interactions, radioactive decay, etc.) and we leave out the high-energy ($\sim$ GeV) and very-low energy ($\sim$ meV) specific physics processes and particles. Third, in the *Primary Generator Action* the properties and definitions of the initial particles to be simulated are included, for instance a radioactive source or an ion beam, in our case a radioactive beam of $^{288}115$ atoms. Last, on top of these three mandatory components the user has the possibility to use some *Optional User Actions* to have more fine control of the the simulation. Simulation control parameters, gating procedures, and user-defined simulation outputs are examples of possible *actions*.

In particular for TASISPEC, the geometry was included in the most realistic, yet practical manner (see upper left of Fig. 1). The remaining *ingredients* of the simulation have been carefully tailored and benchmarked in such a way that it was even possible to use experimental data from calibration sources to determine the experimental thicknesses of the dead layers of the silicon detectors for more than 2000 pixels on a pixel-by-pixel basis [8], and to implement these in the final GEANT4-simulated geometry.

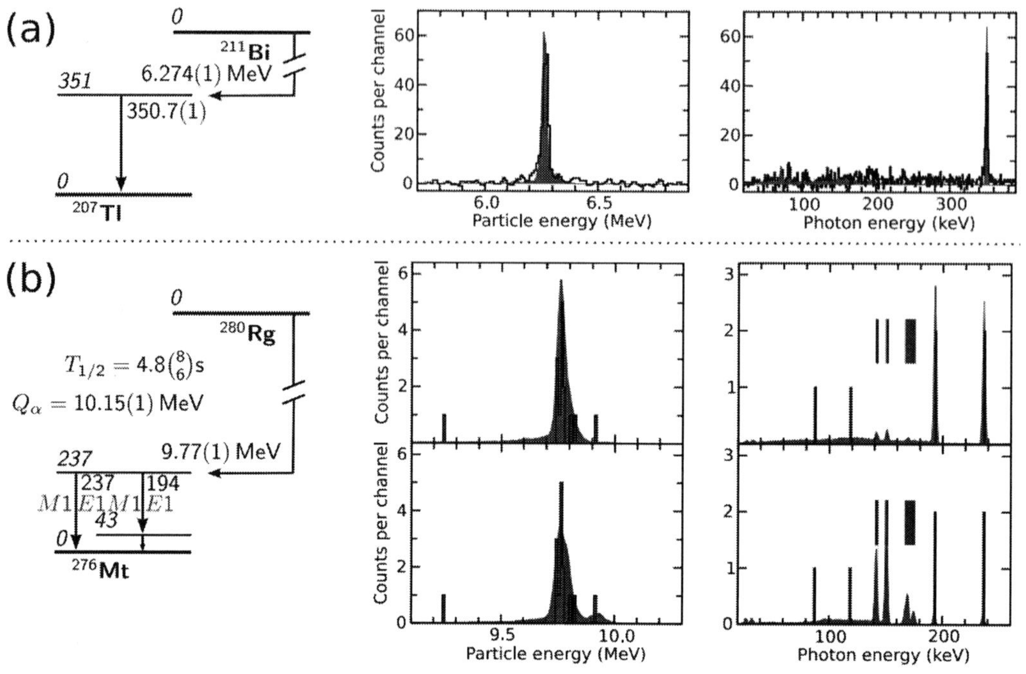

Figure 2. Experimental (black) and simulated (green and red) particle- and photon-energy spectra for the (a) $^{211}Bi \rightarrow {}^{207}Tl$ and (b) $^{280}Rg \rightarrow {}^{276}Mt$ decays. The simulations used 10^5 decays following the schemes shown on the left hand side. The level scheme shown in (a) contains only statistical uncertainties of the peak fit. Level energies and γ-ray energies are in keV. In blue, the K X-ray energies of Mt are indicated in the photon spectra. See text for details.

After the simulated experimental set-up has been benchmarked against calibration sources it is desirable to test its performance under "in-beam" conditions and compare it to the experimental results in the most direct possible way. By using the *Optional User Actions* it is possible to accommodate the output of the simulation into experimental-like events that can be analyzed in the exact same fashion as the experimental data. The deposition of ^{211}Bi ions into TASISpec was investigated; the deposition profile, event rate, and energy distribution of the ions were set to match the experimental conditions after the TASCA separator [9, 10] during the 2012 experiment. The results of the comparison between the simulation and the experiment can be found in the lower-right part of Fig. 1 and they are featured later in Fig. 2.

Once the performance of the simulation has been benchmarked under different conditions the next step is to use the – at this stage – well-established simulation in a self-consistent way to shed light in the interpretation of newly available experimental data in a process dubbed Geant4-aided nuclear spectroscopy.

For the Geant4-aided nuclear spectroscopy we iterate over user-defined nuclear decay schemes which are *fed* into the simulation in a self-consistent procedure. The definition of new decay schemes (or override of the existing ones) is readily possible using User Interface Commands provided by the toolkit itself. The commands allow the user to manipulate the simulation physics input in two ways: (*i*) the radioactive decay (decay type, branching ratios, etc.), and (*ii*) the subsequent nuclear de-excitation (γ-ray transitions, emission of conversion electrons, etc.). The interplay between those two inputs provides valuable information in the analysis of the complex, and oftentimes scarce, experimental data.

3. GEANT4 and the decay of heavy and superheavy atomic nuclei

Ideally, to simulate the implantation and decay of heavy and superheavy atomic nuclei, for example element 115, one should apply the same "recipe" as for the in-beam simulation of ^{211}Bi decays. This means we have already included the nuclear physics processes we are interested in; in particular electromagnetic interactions, radioactive decays, and atomic relaxation. As for the *Primary Generator Action*, i.e. an initial nucleus of element 115, it should be a matter of creating the corresponding particle and its electromagnetic decay properties of the full chain, and include them in the simulation using the existing tools for it. Unfortunately, until rather recently the GEANT4 source code had a hard-coded upper limit in the atomic number of $Z = 100$.

While – now – it is supported to use $Z > 100$, i.e. element 115, as a *particle* – in the toolkit sense of it – and its decay properties (radioactive decay and photon de-excitation) via user-interface commands, elements with $Z > 100$ are most certainly far from being fully supported. There are over 100 GEANT4 classes in the source code that are silently forced to $Z \leq 100$ or even lower atomic numbers, for instance the electronic stopping power for materials which is valid only for $Z \leq 91$ [11].

Of paramount importance in the investigation of heavy and superheavy atomic nuclei is the possibility to investigate their atomic relaxation processes as highly converted γ transitions are responsible for summing effects [6] in the α-decay energies and perhaps more importantly in the emission of characteristic X rays [3]. However, that is one of the aspects which is more difficult to extend in the toolkit due to the complicated many physical processes and databases intertwined for the radiative (fluorescence photons), and non-radiative (Auger and Coster-Kronig electrons) atomic relaxation.

As a workaround it is possible to hard-code the atomic relaxation process as a whole as part of the *Primary Generator Action* [5] for a stationary nucleus, or it is possible to *disguise* heavy nuclei as lighter ones by modifying the atomic properties such as electron binding energies, occupancy, X-ray and Auger energies, etc., of the heaviest odd number element to be taken as the heaviest odd element available in GEANT4, i.e. to convert $Z = 115$ to $Z = 99$ and $Z = 113$ to $Z = 97$, in their atomic properties and study their decay on a step-by-step basis as this method allows most dynamic aspects of the ions to be simulated [12]. Other approaches have ignored the atomic relaxation but allowed hard-coded decay paths covering only γ rays and conversion electrons using the electron binding energies of the heaviest available atom in GEANT4, Fm ($Z = 100$) [13].

The electron binding energies for elements up to $Z = 130$ are readily available in the literature [14]. Fluorescence and Auger energies and their corresponding transitions probabilities can be obtained using a linear extrapolation of the data files for low energy electromagnetic processes for elements $Z = 95$ to 100, which are available in GEANT4 [15]. Proper stopping power tables, bulk density values, and many other quantities associated with a *properly defined* element – or material – are still to be provided.

4. Current state of affairs

As it was previously stated it is necessary to benchmark the simulation performance before GEANT4-aided nuclear spectroscopy can be performed. Figure 2a shows the experimental (black) and simulated (green) particle and photon spectra of the transfer-reaction background isotope, ^{211}Bi, and its α-decay daughter, ^{207}Tl. It is remarkable how well the simulation reproduces the experimental data. The simulation comprised 10^5 decays following the decay scheme shown on the left hand side. The experimentally measured energy values, $E_\alpha = 6274(1)\,\text{keV}$ and $E_\gamma = 350.7(1)\,\text{keV}$, differ slightly from $E_\alpha = 6278.2(7)\,\text{keV}$ and $E_\gamma = 351.07(5)\,\text{keV}$ found in the literature [16], but for the experimental numbers also systematic uncertainties of $E_\alpha \lesssim 10\,\text{keV}$ and $E_\gamma \lesssim 1\,\text{keV}$ need to be considered. The experimental and simulated data was treated in an identical fashion to improve the validity and consistency of the physical assessment of the interpretation of results.

The particle energy spectrum as seen by the simulated set-up was normalized to the number of α particles detected experimentally. The α-photon coincidence efficiency of the simulation is notably very well under control, because the yield of the 351-keV line in the photon energy spectrum is nicely reproduced using a normalization factor identical to the one of the particle spectrum [12].

Once the simulation has been proven to work it is possible to use it to interpret newly available experimental data. This is demonstrated in the study of the α decay of ^{280}Rg. In this case it was necessary to *disguise* the superheavy nuclei as described in the previous section. Figure 2b shows the result of the comparison between the experimental data and the simulation for two cases, namely varying the electromagnetic character of the simulated 194-keV and 237-keV γ-ray transitions: either $E1$ (green) or $M1$ (red). Just like for the ^{211}Bi α decay, the photon spectra were normalized to the number of α particles detected experimentally. It is possible to assign the 194-keV and 237-keV transitions as $E1$ transitions in the decay scheme of Fig. 2b. If the γ-ray transitions had $M1$ character instead, the simulation would suggest the detection of K_α X rays, because $M1$ transitions with energies just above the K-binding energy are highly converted. This initial GEANT4-aided assignment was later independently confirmed [4]. The $E1$ character of those transitions implies very interesting nuclear structure physics [3, 17–19].

5. Future native extension of GEANT4 towards superheavy elements

Currently GEANT4 uses by default the Evaluated Atomic Data Library EADL [20] and optionally the compilation from Bearden [21] for the electron binding energies. Unfortunately those datasets cover only the range $Z \leq 100$. There are many different compilations of binding energies of atomic properties in the literature like the ones from Carlson [22], Sevier [23], Firestone [24], and Lide [25].

In its current – development – form the GEANT4 uses the compilation of electron binding energies which constitute the backbone to the BrIcc code for evaluating theoretical conversion coefficients [26, 27]. The purpose of this choice is to be consistent with the nuclear deexcitation processes simulated in the discrete so called *photon evaporation* where the BrIcc code is used.

The new dataset of electron binding energies includes not only new – and more – values but also brings slightly different electron occupancies of certain orbitals for a number of elements. In order to make the changes backwards-compatible the code need to be tested and the current development effort is being made towards the implementation of a user-interface command that switches to the new databases `fluor_SHE` and `auger_SHE`[1] as well as expanding the applicability of the related data reading routine and the associated sizes of information container arrays throughout the relevant parts across the entire GEANT4 source code.

---

[1] Working names.

6. Closing remarks

It is shown that it is possible to tune the "ingredients" of a GEANT4 simulation to provide a self-consistent cross check of the interpretation of the sometimes very scarce experimental data. Even though quantities such as implantation depth or recoil energy of the nuclei after the α decay are not *strictly* correct they do not necessarily compromise the validity of the spectroscopic results as long as the relevant atomic decay information is properly accounted for. The use of a high-resolution spectroscopic set-up in conjunction with comprehensive Monte Carlo simulations proves to be an effective probe to study the nuclear structure properties of heavy and superheavy man-made elements.

This work is supported by the Swedish Research Council (VR 2013-4271) and the Knut and Alice Wallenberg Foundation (KAW 2015.0021). The author would like to acknowledge the GEANT4 collaboration, in particular the *electromagnetic* group and the other members of the *hadronic* group for their support and valuable discussions.

References

[1] Yu.Ts. Oganessian, Rep. Prog. Phys. **78**, 036301 (2015).
[2] S. Agostinelli *et al.*, Nucl. Instr. Meth. **A506**, 250 (2003).
[3] D. Rudolph *et al.*, Phys. Rev. Lett. **111**, 112502 (2013).
[4] J.M. Gates *et al.*, Phys. Rev. C **92**, 021301(R) (2015).
[5] A. Lopez-Martens *et al.*, Nucl. Phys. **A852**, 15 (2011).
[6] Ch. Theisen *et al.*, Nucl. Instr. Meth. **A589**, 230 (2008).
[7] L.G. Sarmiento *et al.*, Nucl. Instr. Meth. **A667**, 26 (2012).
[8] U. Forsberg *et al.*, EPJ Web of Conferences **66**, 02036 (2014).
[9] M. Schädel, Eur. Phys. J. **D45**, 67 (2007).
[10] J.M. Gates *et al.*, Phys. Rev. C **83**, 054618 (2011).
[11] L.G. Sarmiento, 20th `Geant4` collaboration meeting, Fermilab, USA (2015).
[12] L.G. Sarmiento *et al.*, PoS (**X LASNPA**) 057 (2014).
[13] D.M. Cox *et al.*, Eur. Phys. J. **A51**, 1 (2015).
[14] T.A. Carlson and C.W. Nestor Jr., Atom. Data Nucl. Data Tables **19**, 153 (1977).
[15] `https://geant4.web.cern.ch/geant4/support/download.shtml`.
[16] F.G. Kondev and S. Lalkovski, Nucl. Data Sheets **112**, 707 (2011).
[17] Yue Shi *et al.*, Phys. Rev. C **90**, 014308 (2014).
[18] D. Rudolph *et al.*, Acta Phys. Pol. **B45**, 263 (2014).
[19] D. Rudolph *et al.*, AIP Conf. Proc. **1681**, 030015 (2015).
[20] S.T. Perkins *et al.*, *Tables and Graphs of Atomic Subshell and Relaxation Data Derived from the LLNL Evaluated Atomic Data Library (EADL), $Z = 1$–100, UCRL-50400* **30** (1997).
[21] J.A. Bearden *et al.*, Rev. Mod. Phys. **39**, 125 (1967).
[22] T.A. Carlson, *Photoelectron and Auger spectroscopy* (Plenum, New York, 1975).
[23] K.D. Sevier, Atom. Data Nucl. Data Tables **24**, 323 (1979).
[24] R.B. Firestone *et al.*, *Table of Isotopes, 8th ed.* (John Wiley & Sons, New York, 1996).
[25] D.R. Lide, ed., *CRC Handbook of Chemistry and Physics, 95th ed.* (Boca Raton, 2015).
[26] T. Kibédi *et al.*, Nucl. Instr. Meth. **A589**, 202 (2008).
[27] T. Kibédi *et al.*, Atom. Data Nucl. Data Tables **98**, 313 (2012).

EPJ Web of Conferences **131**, 05005 (2016) DOI: 10.1051/epjconf/201613105005
Nobel Symposium NS160 – Chemistry and Physics of Heavy and Superheavy Elements

ORNL actinide materials and a new detection system for superheavy nuclei

Krzysztof P. Rykaczewski[1,a], *James B.* Roberto[1], *Nathan T.* Brewer[1,2], and *Vladimir K.* Utyonkov[3]

[1] Oak Ridge National Laboratory, Physics Division, Oak Ridge, TN 37831, USA
[2] University of Tennessee, Department of Physics and Astronomy, Knoxville, TN 37996, USA
[3] Joint Institute for Nuclear Research, RU-141980 Dubna, Russian Federation

Abstract. The actinide resources and production capabilities at Oak Ridge National Laboratory (ORNL) are reviewed, including potential electromagnetic separation of rare radioactive materials. The first experiments at the Dubna Gas Filled Recoil Separator (DGFRS) with a new digital detection system developed at ORNL and University of Tennessee Knoxville (UTK) are presented. These studies used ^{240}Pu material provided by ORNL and mixed-Cf targets made at ORNL. The proposal to use an enriched ^{251}Cf target and a large dose of ^{58}Fe beam to reach the $N = 184$ shell closure and to observe new elements with $Z = 124$, 122 and 120 is discussed.

1. ORNL actinide resources

Actinide materials originating from the ORNL's High Flux Isotope Reactor (HFIR) and Radiochemical Engineering Development Center (REDC) have been used in experiments identifying nine superheavy elements, from $Z = 104$ to 106 and from $Z = 113$ to 118. The relatively high power of HFIR combined with its flux trap design provide an intense steady-state neutron flux of $2.4*10^{15}$ neutrons/s/cm^2. This very high neutron flux is essential for actinide production. The adjacent REDC operates several hot cells capable of handling highly radioactive materials from HFIR. REDC provides unique capabilities for the efficient processing and separation of actinide materials including $Z = 98$ ^{252}Cf with a half-life of 2.6 years. ^{252}Cf is produced in HFIR in hundred milligram quantities from Am and Cm feedstock material, mostly recoverable after production and separation cycles lasting from several months to two years. $Z = 97$ ^{249}Bk can be obtained as a bi-product of ^{252}Cf production campaigns [1]. Currently, approximately 10 mg of ^{249}Bk can be separated after a typical ^{252}Cf campaign. Increasing the production of ^{249}Bk requires a dedicated irradiation in HFIR using a suppressed thermal neutron flux tailored for ^{249}Bk production [2]. Such an approach can increase ^{249}Bk production by factors of two or more but is more costly since it is not compatible with standard ^{252}Cf production campaigns. Actinide materials valuable for research on superheavy nuclei include highly enriched ^{237}Np, ^{242}Pu, ^{241,243}Am, ^{244}Cm, and ^{248}Cm. Up to gram quantities of these isotopes are currently in actinide inventories at ORNL, with the exception of ^{248}Cm where the current inventory is 80 mg. REDC is currently recovering about 10 mg/year of highly enriched ^{248}Cm from old ^{252}Cf sources. In addition, about 15 mg of mixed-Cf material, containing longer-lived ^{249}Cf (50%), ^{250}Cf (15%), and

[a] Corresponding author: `rykaczewskik@ornl.gov`

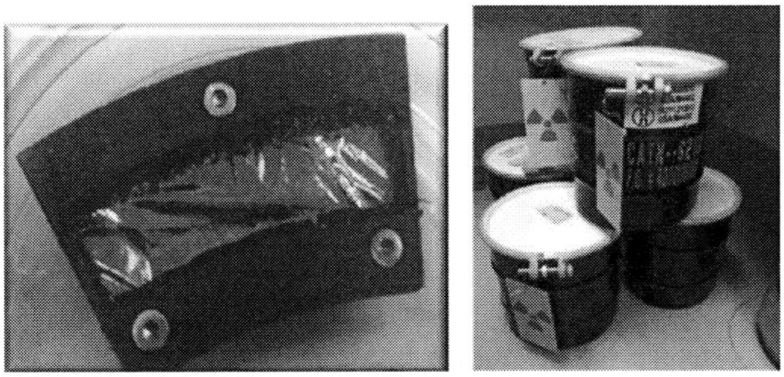

Figure 1. Left panel: One of the mixed-Cf target sectors electrodeposited on Ti foils at ORNL's REDC. Right panel: Twelve target sectors with required shielding arriving at JINR Dubna early 2015.

^{251}Cf (35%), has recently been recovered from these old sources. In 2015, twelve target sectors containing mixed-Cf, with a typical thickness of about 0.33 mg/cm^2 and 3-cm^2 area, were electrodeposited at REDC on Ti-foils for a planned experiment at JINR to search for heavier isotopes of element 118 (see Fig. 1). The total content of ^{252}Cf in all sectors was below two micrograms. However, with a high percentage of ^{250}Cf (half-life $T_{1/2} = 13$ years), the neutron emission from the target material is still quite high, at the level of $2*10^7$ neutrons per second into full solid angle.

The isotope ^{251}Cf ($T_{1/2} = 898$ years) is the most valuable component of the mixed-Cf target enabling the possible synthesis of new $Z = 118$ isotopes, two neutrons closer to the predicted $N = 184$ shell closure. Technically, it would be possible to mass separate tens of milligrams of ^{251}Cf suitable for a target wheel containing nearly 100% enriched ^{251}Cf. This would require an efficient electromagnetic separator dedicated to radioactive materials. A lower current prototype separator is currently operating at ORNL to separate stable isotopes. A similar separator optimized for actinide materials would be more than adequate for the needs of the superheavy element research community.

2. Experiments with the ORNL/UTK digital detection system at the DGFRS

A new detection system for superheavy nuclei has been designed, built and commissioned at ORNL. The concept follows the experience gained during discovery research on proton and α emitters at the Recoil Mass Separator at ORNL's Holifield Radioactive Ion Beam Facility. The new detector system uses a very large Si-wafer, with 128 mm by 48 mm active area, processed to work as a Double-sided Silicon Strip Detector (DSSD) by Micron Semiconductor (BB-17 type, 300 microns thick). This detector is installed at the final focus of the DGFRS, to detect implanted nuclei and observe related α and fission decays. It has a matching size silicon single veto detector and six single detectors of 120 mm by 60 mm area acting as an escape-catching silicon-detector box. This system has replaced the position sensitive silicon detectors used earlier at the DGFRS for studies of superheavy nuclei. All silicon detectors and two Dubna recoil transmission detectors Multi-Wire Proportional Chambers (MWPCs) are operated using Pixie 16 digital signal processing electronics manufactured by XIA [3], in parallel to the local analogue data acquisition system. The digital data acquisition system was developed at the UTK's Digital Pulse Processing Laboratory, by R. Grzywacz and collaborators [4], and at ORNL, with contributions by K. Miernik (Oak Ridge/Warsaw) and N. Brewer (Oak Ridge). All single pulses are analysed on board the Pixie-16 module, while on-board recognized pile-up signals occurring within ten microseconds are recorded as digital

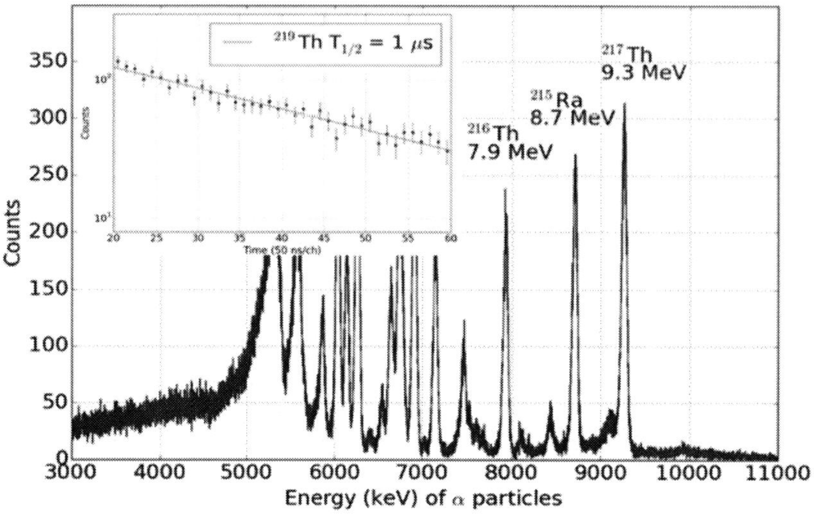

Figure 2. Energy spectrum of α particles emitted from the fusion-evaporation products of the ^{48}Ca+^{nat}Yb reaction and measured using the new digital detection system at the DGFRS. The inset shows the decay pattern of α particles emitted from a 1-μs activity of ^{219}Th derived from collected traces of the DSSD preamplifier signals.

pulse images, with 10-ns time resolution, for initial on-line and further off-line analysis. An example of the capability of the new digital detection system is illustrated in Fig. 2.

The reaction between a $Z = 70$ natural ytterbium target and ^{48}Ca ions was used to calibrate the α-energy scale of the silicon detectors. Short-lived $Z = 90$ thorium activities were produced and implanted in the DSSD, and their decay was recorded. This includes the direct observation of the implantation and decay of ^{219}Th activity having a 1-μs half-life. The ^{219}Th decay curve was obtained through the analysis of recoil-decay signal traces. For the first time, the recoil implantation followed by a decay of such a short-lived activity was directly identified at the DGFRS.

2.1 Fusion-evaporation reactions between a ^{48}Ca beam and $^{239,240}Pu$ targets

The first experiment aiming at new superheavy nuclei and using the new digital detection system was performed at the DGFRS with a ^{48}Ca beam irradiating ^{239}Pu and ^{240}Pu targets (the latter material provided from ORNL). The spectroscopy results include the observation of three decay chains of ^{285}Fl including the previously not detected full energy of the first α particle and evidence for the new even-even isotope ^{284}Fl undergoing fast spontaneous fission [5]. The low cross sections detected in the experiments with $^{239,240}Pu$ targets compared to the larger values observed with $^{242,244}Pu$ targets indicate that the west shore of the Island of Stability has been identified, as illustrated in Fig. 3.

2.2 Irradiation of a mixed-Cf target with a ^{48}Ca beam

The irradiation of the ORNL mixed-Cf target with a ^{48}Ca beam started at the DGFRS at the beginning of October 2015. After nine days of bombardment at about 0.5 particle-μA, a decay chain of the isotope $^{294}118$ was observed through both analogue and digital data acquisitions [6]. The observed decay properties were close to the previously identified four chains of this only known isotope of the heaviest known atomic element. Two α decays followed by a

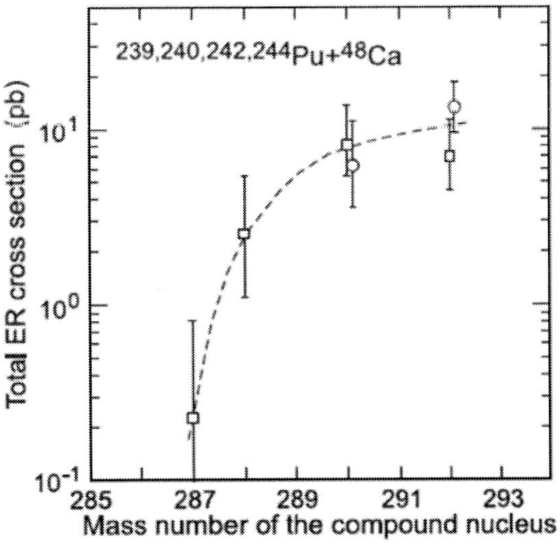

Figure 3. The cross section pattern observed for the $Z = 114$ Fl isotopes produced using a ^{48}Ca beam and four Pu targets, with mass A ranging from 239 to 244 (figure courtesy of Physical Review C). The dramatic drop of cross-section values was interpreted as being caused by the reduction of fission barriers with decreasing neutron number of the Fl compound nucleus [5].

spontaneous fission of ^{286}Fl were detected. The experiment continued with slowly increasing beam current, up to 0.7 particle-μA. However, frequent checks of the target performance, through the observation of α particles emitted from the target, began to show signs of target thickness increase, i.e., of a build-up of an extra layer of material at the target surface. Optical inspection of the targets indeed revealed a layer of material covering the electrodeposited mixed-Cf material. We are in the process of returning the mixed-Cf target to ORNL, for assessing the problem and refurbishing the target sectors.

3. Towards the predicted neutron shell closure at $N = 184$ and new elements

The properties of α decays for nuclei produced in the reactions between radioactive actinide targets and a ^{48}Ca beam point clearly to the large enhancement of stability with increasing neutron number. See, e.g., reference [7]. The largest neutron number reached so far is $N = 177$, in ^{293}Lv and ^{294}117 nuclei. While the spectroscopic evidence is in agreement with a predicted shell closure at $N = 184$, the experiment is far from this neutron number. However, use of heavier and more intense beams and the heaviest actinide targets might bring us closer to $N = 184$. For example, the reaction between a ^{58}Fe beam and a ^{251}Cf target could create the compound nucleus ^{309}124 in an excited state. The isotope ^{309}124 has $N = 185$ neutrons, i.e., one neutron added to the predicted spherical magic core. At known parts of the Segre chart such "core plus one particle" nuclei have a relatively small binding energy for this valence nucleon. One can expect that the evaporation of the 185th neutron is highly probable. The resulting one-neutron evaporation product, the isotope ^{308}124 with $N = 184$ neutrons might have stability against particle evaporation and radioactive decay enhanced by its single- or even doubly-magic character. The decay chain starting from the ^{308}124 activity would reach the known ^{292}Lv isotope after four α decays. The subsequent two α particles of known properties will populate ^{284}Cn which undergoes spontaneous fission with about 90-ms half-life. It means there are several known spectroscopic signatures allowing us to identify the decay of the new nucleus $^{308}124$. The observation of the one-neutron evaporation channel in the ^{58}Fe+^{251}Cf reaction would constitute a revival of the "cold fusion" approach in the search for superheavy nuclei. Of course, the qualitative analysis presented here assumes that $N = 184$ is a good magic number for a new element with high atomic number, $Z = 124$. The HFB calculations summarized recently [8] support the magic character of $N = 184$ at $Z = 124$ and predict a very large fission barrier in this region of nuclei, of the order of 10 MeV. However, microscopic-macroscopic calculations [8] point to rather small fission barriers for the nuclei discussed, closer to 3 MeV, indicating that the magic character of $N = 184$ does not hold for such superheavy elements. As one can expect, the cross-section estimates vary dramatically with fission barrier parameter values. The very preliminary and rough estimates by Krystyna Wilczyńska indicate that the production yield for the xn-evaporation channels might be within experimental reach with the fission barrier parameter even slightly below 5 MeV [9].

However, the experimental attempt to identify the nucleus $^{308}124$ and following α decay products $^{304}122$ and $^{300}120$, i.e., three new elements including a $N = 184$ isotone, will be much better justified if an enriched ^{251}Cf target will be available. Therefore, having the actinide separation capability together with a new generation of accelerator laboratories can elevate superheavy element science from our current capabilities, for instance $^{294}118$, toward further discoveries including several new elements and the magic $^{308}124$ nucleus.

This research was supported by the U.S. Department of Energy (DOE) Office of Science, Office of Nuclear Physics, Isotope Development and Production for Research and Applications Program, under contract DE-AC05-00OR22725 with UT-Battelle, LLC. We are grateful to the ORNL Radiochemical Engineering Development Center and High Flux Isotope Reactor, a DOE Office of Science, Basic Energy Sciences User Facility, for their support in the production and chemical separation of the actinide materials. We also thank our many collaborators at the Flerov Laboratory of Nuclear Reactions (JINR, Dubna, Russia), Lawrence Livermore National Laboratory, Vanderbilt University, and the University of Tennessee–Knoxville, without whom this research would not have been possible.

References

[1] J.B. Roberto *et al.*, Nucl. Phys. **A944**, 99 (2015).
[2] R. Boll *et al.*, Super Heavy Element Target Material Needs and Production at ORNL, unpublished ORNL report, April 2016.
[3] `http://xia.com`
[4] R.K. Grzywacz *et al.*, Nucl. Instr. Meth. **B261**, 1103 (2007).
[5] V. K. Utyonkov *et al.*, Phys. Rev. C **92**, 034609 (2015).
[6] N.T. Brewer *et al.*, to be published.
[7] V. K. Utyonkov *et al.*, ibidem.
[8] A. Baran *et al.*, Nucl. Phys. **A944**, 442 (2015).
[9] K. Wilczyńska, private communication (2016).

EPJ Web of Conferences **131**, 06001 (2016) DOI: 10.1051/epjconf/201613106001
Nobel Symposium NS160 – Chemistry and Physics of Heavy and Superheavy Elements

The discovery of elements 107 to 112

Sigurd Hofmann[1,2,a]

[1] GSI Helmholtzzentrum für Schwerionenforschung, 64291 Darmstadt, Germany
[2] Institut für Physik, Goethe-Universität Frankfurt, 60438 Frankfurt, Germany

Abstract. In the 1960s, theoretical concepts prepared the path to nuclear matter with proton and neutron numbers far beyond the nuclei known at that time. The new laboratory GSI was founded for research on reactions with heavy ions, in particular those for production of the predicted super-heavy nuclei. In this contribution it is presented how the interaction between experiment and theory resulted in a continuous improvement of the experimental set-ups on the one hand, and of the knowledge of the processes during the nuclear reaction and of the properties of the produced nuclei on the other hand. In the course of this work six new elements from 107 to 112 were produced and identified. An overview of the present status of experimental results and a comparison with theoretical interpretations is given.

1. Expectations of research on super-heavy nuclei in the 1960s

The 1960s were a trailblazing decade for research on super-heavy nuclei (SHN). Maria Goeppert Mayer, in her Nobel Lecture in 1963, presented an extended shell model for the nucleus reaching up to proton number $Z = 184$ [1]. Using the right interaction between spin and angular momentum of the nucleons, the known magic numbers appeared as shell gaps up to $Z = 82$ for the protons and $N = 126$ for the neutrons. New magic numbers were predicted at $Z = 126$ and 184, the latter value as a new magic number also for the neutrons.

In the middle of the 1960s, the concept of the macroscopic-microscopic (MM) model for calculating binding energies of nuclei was invented by V.M. Strutinsky [2]. In this model a microscopic shell-correction energy (SCE) which considers the shell structure of the nuclei, is added to the macroscopic part derived from the liquid drop model. Positive or negative SCE values decrease or increase the macroscopic part of the binding energy, respectively.

Using this method a number of the measured phenomena could be naturally explained. In particular, it became possible to calculate the binding energy of a heavy fissioning nucleus as function of deformation and thus to determine the fission barrier. Partial fission half-lives were calculated as a quantum mechanical tunneling process through the so determined fission barrier. The calculations revealed the existence of so called 'islands of stability' located in the vicinity of shell closures for both protons and neutrons far beyond the known nuclei. In addition, instead of the earlier magic number 126 for the protons, a new number 114 was calculated [3–13].

An important result which could be explained quite naturally applying the Strutinsky method for calculation of the structure of the fission barrier were the fission isomers

[a] e-mail: S.Hofmann@gsi.de

discovered by S.M. Polikanov *et al.* in 1962 [14]. They gain their stability from a second minimum in the fission barrier at large deformation.

Great excitement among physicists and chemists created a prediction by T. Sikkeland in 1966. Based on high fission barriers of up to 12.1 MeV for the doubly magic nucleus $Z = 126$ and $N = 184$, probability ratios for neutron emission versus fission (Γ_n/Γ_f) of up to 100 were calculated for compound nuclei (CN) of isotopes of element 126 produced in fusion reactions of heavy beam and target nuclei [15]. Such high survival probabilities resulted in high cross-sections for synthesis of SHN of up to 100 mb. The heaviest nuclei investigated at that time were produced only on a level of 1 nb. These were isotopes of element 105 studied in fusion reactions of light beams with targets of ^{243}Am in Dubna [16] and ^{249}Cf in Berkeley [17].

On that background, the Gesellschaft für Schwerionenforschung (GSI) was founded in 1969 for research on interactions of heavy ions and nuclei with matter with the linear accelerator UNILAC as the central device [18]. One focus of the research program was the synthesis of SHN and the study of their radioactive decay modes.

2. The velocity filter SHIP and early experiments

The main instrument planned for study of SHN in fusion reactions was the velocity filter SHIP. The design was suggested by the research group of H. Ewald and G. Münzenberg from the Justus-Liebig-Universität in Gießen [19]. A schematic three dimensional view of SHIP is shown in Fig. 1. The greatest progress was expected from the short separation time of the fusion products determined by the flight time of the ions through the separator of 1–2 μs.

The velocity filter was the first major instrument ready when first beams were delivered from the UNILAC at the beginning of 1976. Already in May of that year first spectra from our study of the test reaction ^{40}Ar + ^{144}Sm were presented by P. Armbruster at the conference on Nuclei far from Stability in Cargèse [20]. At high background suppression, spectra were measured from the recoiled evaporation residues (ER), from the α decay of those detector implanted neutron-deficient isotopes of elements from Pt to Hg, and from K–X-rays emitted from nuclei after electron capture (EC).

These most successful test experiments led immediately to a search for SHN following Sikkeland's prescription. The reaction studied was ^{136}Xe + ^{170}Er → ^{306}120* in 1976 [21] and ^{65}Cu + ^{238}U → ^{303}121* [22] in 1977. Although the beam times were short, only one and two days were allocated, respectively, cross-section limits of about 1 nb were reached. The promised island of SHN was not found, and technical improvements became necessary.

3. Technical improvements and new results on synthesis and decay

Relatively simple detector devices had been used in the early experiments. These were a 25 mm diameter multi-foil carbon secondary electron detector for a rough Z determination described in [23], a commercial 450 mm^2 Si detector, and a single crystal Ge detector. Measurement of the distribution of reactions products in the focal plane of SHIP revealed that larger area detectors are needed for fully exploiting the separation properties of SHIP.

A first version of position sensitive Si detectors was developed in 1978 and used for experiments. Combining seven single detectors a total area of 87 mm × 27 mm was covered in the focal plane. Two single foil secondary electron TOF detectors of 60 mm × 30 mm [24] and a Ge detector completed the set-up. The new detector system was tested with the study of neutron-deficient α emitters along $N = 82$. A number of new isotopes and α emitting yrast

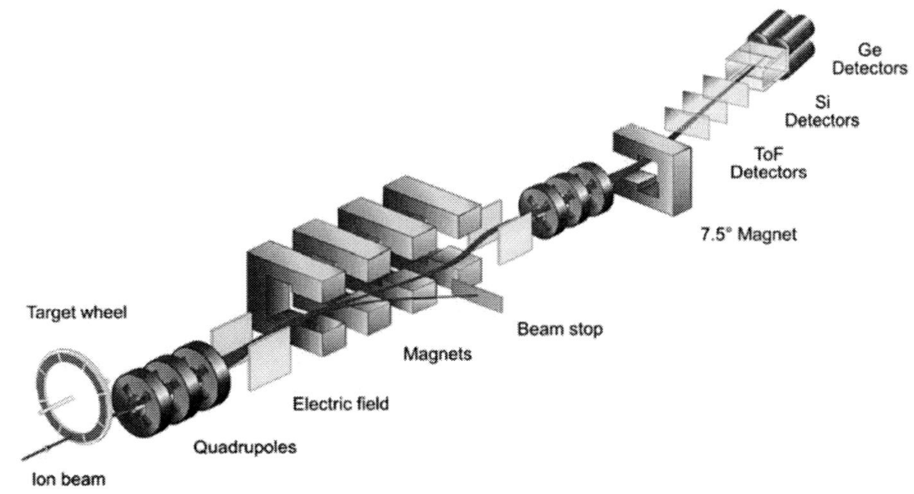

Figure 1. The velocity filter SHIP (Separator for Heavy Ion reaction Products) and its detection system [19, 25, 29]. The figure shows the upgraded version as it has been used since 1994. The drawing is approximately to scale, however, the target wheel and the detectors are enlarged by a factor of two. The length of SHIP from the target to the detector is 11 m. The target wheel has a radius up to the center of the targets of 155 mm. It rotates synchronously with beam macrostructure at 1125 rpm [30]. The target thickness is usually 450 μg/cm^2. The detector system consists of three large area secondary-electron time-of-flight detectors [31], a position-sensitive silicon-detector array, and a germanium clover detector. The flight time of the reaction products through SHIP is 1–2 μs. The filter, originally consisting of two electric and four magnetic dipole fields plus two quadrupole triplets, was later extended by a fifth deflection magnet, allowing for positioning of the detectors away from the straight beam line and leading to further reduction of the background.

traps with half-lives below milliseconds were identified [25, 26]. Of particular interest was the detection of the first proton-emitting isotope beyond the proton drip-line, ^{151}Lu [27].

Properties of fusion reactions were studied by K.H. Schmidt *et al.* [23, 28]. Neutron-deficient nuclei along $N = 126$ were explored. New neutron-deficient isotopes and isomers of Pa were discovered in reactions of a ^{40}Ar beam and targets of ^{181}Ta and ^{184}W [23]. Fusion hindrance was studied using symmetric systems [28].

Appropriate for the tracking of the lifetime of a single atom from the implantation into the detector up to its decay, the new method of plotting the counts on an axis at logarithmically increasing width of bins was introduced [23]. The center of gravity of the so determined universal decay curve marks the lifetime of the isotope. The significance of data in the case of poor statistics was investigated and procedures for determining error bars for lifetime and cross-section at poor statistics were elaborated [32].

Production of nuclei in the region of neutron deficient nuclei along the closed shell $N = 126$ was considered exemplary for the synthesis of spherical SHN. In the focus of the interest was the damping of shell effects at high excitation energies. Systematic data for cross-sections of ERs revealed no local increase from the continuously decreasing values when $N = 126$ was crossed into direction of lighter isotopes. It was concluded that the damping of shell effects at high excitation energy will appreciably reduce the cross-section for production of SHN [32, 33].

Applying the improved techniques in the region of heavy elements, the first result of considerable importance was the identification of ^{257}104 in the reaction ^{50}Ti + ^{208}Pb $\rightarrow$ ^{258}104* [34]. This was the first observation of the emission of only one neutron from the

CN (1*n* channel) and a confirmation of results obtained at Dubna that cold fusion based on Pb or Bi targets provides a possibility for synthesis of heavy nuclei [35]. The also measured half-life of the spontaneously fissioning even-even nucleus ^{256}104 produced in the 2*n* channel confirmed the observation by Yu.Ts. Oganessian *et al.* [35] that the systematics of long half-lives of $N = 152$ isotones is broken at element 104. This phenomenon was successfully explained using Strutinsky's method for calculation of fission barriers. It was shown that at high Z the decreasing liquid drop fission barrier results in a depression of the second hump of the fission barrier below the ground-state, and the shorter SF half-lives result from a narrow single humped barrier [36].

The first new isotopes produced and identified at SHIP in the region of heavy elements were ^{239}Cf as α-decay daughter of ^{243}Fm and ^{247}Md [37]. These nuclei were produced in reactions with a ^{40}Ar beam and targets of ^{206}Pb and ^{209}Bi, respectively.

4. The new elements 107 to 109

Element 107 was the first new element produced and identified at SHIP [38]. The reaction ^{54}Cr + ^{209}Bi $\rightarrow$ ^{263}107* was selected. Two reasons were decisive for that choice. Firstly, reactions with actinide targets were excluded, because beams lighter than ^{40}Ar were forbidden at GSI. There were several accelerators providing beams of light ions in Germany, and it was concluded that GSI should concentrate on reactions with heavy ions. Unfortunately, this decision prevented gaining experience with hot fusion, reactions based on actinide targets and beams of light ions, an experience that would have been very valuable later. Secondly, the preparatory experiments using Pb and Bi targets had delivered promising results in various test reactions before. The circumstance that these targets could be manufactured and handled without radiation protection was another supporting argument.

The irradiations for element 107 were performed in February 1981. The main experiment was preceded by a study of the decay of the expected α-decay daughters, isotopes of elements 105 and Lr, which were not known at that time. For that study a beam of ^{50}Ti ions was used. For the main experiment, beam time with a ^{54}Cr beam of only five days was allocated. Not without surprise, six decay chains were observed, five of them starting with α decays of 10.38 MeV at lifetimes of $\tau = 6.8$ ms and one chain starting with 9.70 MeV at $\tau = 165$ ms. This chain and the longest one of the other five is plotted in Fig. 2. The 9.70-MeV decay was assigned to the decay of the ground state of ^{262}107 and the higher energy decay to an isomeric state. The daughter decays were in agreement with the previously identified isotopes ^{258}105 and ^{254}Lr.

Changing the beam from ^{54}Cr to ^{58}Fe would produce element 109. After emission of one neutron from the CN the produced isotope ^{266}109 was expected to decay into the now known isotope ^{262}107. The irradiation was performed in August 1982. In a three-days preparatory experiment the reaction ^{50}Ti + ^{209}Bi was studied in more detail and a 22% EC branch was determined for ^{258}105. The main experiment lasted two weeks. Only one significant decay chain was measured starting with an α decay of 11.10 MeV [39, 40], see Fig. 2. The daughter decays were in agreement with known decays down to the spontaneously fissioning nucleus ^{258}104.

Attempts to produce element 108 had been postponed. Calculations of SF half-lives of even-N isotopes of element 108 resulted in values close to 1 μs [36]. In that case the produced nuclei would have decayed before they reached the detector. However, it was not only the short half-life which determined the planning of experiments. Also in the case of a partial SF half-life shorter than the α half-life, an identification of the dominantly fissioning nucleus would be hampered. The already well established α-α correlation method could not be applied

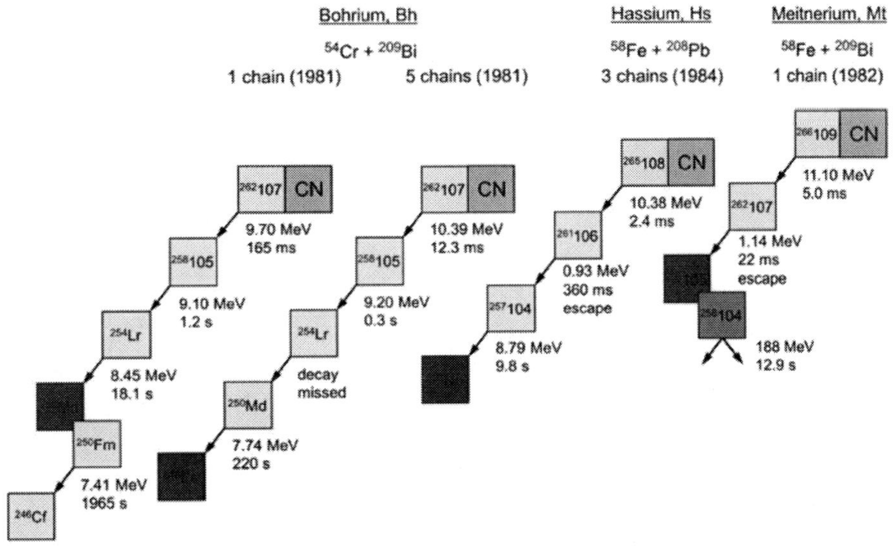

Figure 2. Examples of α-decay chains measured in the search experiments for elements 107, 108, and 109. The first information from the decay chain was obtained from the implantation of the evaporation residue (ER) produced after emission of one neutron from the compound nucleus (CN, blue). For each α-decaying nucleus (yellow) the measured α energy and the lifetime is given. The chains terminate by a known, but unobserved electron capture decay (EC, red), by the unobserved α decay of ^{246}Cf ($T_{1/2} = 36$ h) or, as in the case of ^{266}109, by spontaneous fission (SF, green) after EC. The rows on top show name and symbol of the new element, beam and target of the fusion reaction, number of measured decay chains, and year of the experiment.

in that case. However, after measurement of α decay of ^{266}109 it was expected that also isotopes of element 108 would preferably decay by α emission.

This assumption was corroborated in 1984, when in a preparatory experiment α decay of the isotopes ^{259}106 and ^{261}106 was measured and even for the even-even isotope ^{260}106 the partial SF half-life did not decrease relative to that of the α-decay daughter ^{256}104 [41]. The measured total half-life of ^{260}106 was 3.6 ms and a 1/1 branching ratio b_α/b_{SF} was determined. The reaction for studying element 106 was ^{54}Cr + ^{208}Pb.

It was expected that the reaction ^{58}Fe + ^{208}Pb would produce ^{265}108 in a $1n$-evaporation channel. Three decay chains of this isotope were measured in March 1984 [42]. The first one of these chains is shown in Fig. 2. The mean half-life was 1.8 ms.

Two years later, the even-even nucleus ^{264}108 was identified in a reaction of ^{58}Fe with a ^{207}Pb target [43]. Only one decay chain was observed. Again, α decay was measured but the half-life was now 76 μs.

5. Intermediate conclusions

The realization of the technical concept of UNILAC and SHIP and subsequent development of position sensitive Si detectors has proven to be the right choice for study of rare nuclei and decay modes, in particular those of heavy nuclei. Short separation time, high background suppression and the use of a high-resolution detector system together with fast and low noise signal processing resulted in the identification of three new elements. Admittedly, the fact that both the decay of the nuclei by α emission resulting in decay chains with half-lives between milliseconds and some tens of seconds were within the window of optimal performance of the experimental set-up, was a lucky coincidence.

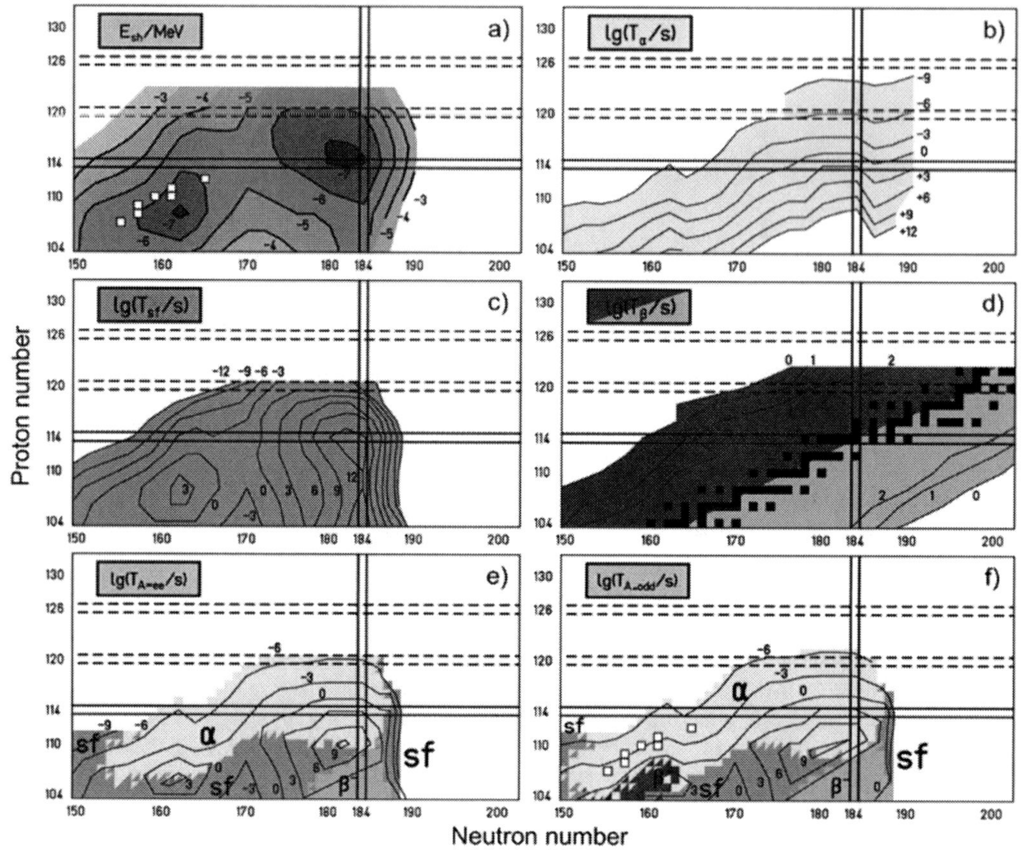

Figure 3. Shell-correction energy (a) and partial half-lives for α decay, SF, and β decay (b)–(d). The calculated values in (a)–(c) were taken from [44, 45] and in (d) from [46]. The filled squares in (d) mark the β stable nuclei. Half-lives and dominating decay modes for α, β^+ decay/EC, β^- decay, and SF are plotted for even-even nuclei in (e) and for odd-A nuclei in (f). Hindrance factors of 10 and 1000 were applied for α decay and SF of odd-A nuclei, respectively. White squares in (a) and (f) mark the produced nuclei $^{262}107$, $^{265}108$, $^{266}109$, $^{269}110$, $^{271}110$, $^{272}111$, and $^{277}112$ of the new elements from 107 to 112.

On the other hand, the aim of the study was to discover the unknown properties of nuclei of new elements not knowing what the result would be. The now available experimental data could be compared with existing theoretical models. It turned out that the measured data could be well explained by the results of calculations by S.G. Nilsson *et al.* [8] already published in 1969 using the MM model. In those calculations, minimal negative SCE were calculated not only for spherical SHN but also for nuclei around $Z = 108$ and $N = 162$. However, there, in contrast to the spherical SHN, the increased stability occurs at relatively large quadrupole deformation due to a relatively high level density for protons and neutrons below level gaps at $Z = 108$ and $N = 162$, respectively.

Later, more refined calculations presented in [46–48] substantiated the early results. SCE values, partial half-lives, and deduced dominant decay modes from these MM model calculations are shown in Fig. 3. In Fig. 3a the two regions of minimal SCE for deformed and spherical nuclei are clearly visible. Both regions are separated by a ridge of weak SCE at about mass number $A = 280$, which separates regions of heavy and super-heavy nuclei. The produced nuclei of the new elements from 107 to 109 and those later identified from elements 110 to 112 marked in Figs. 3a are crossing the region of deformed nuclei. Their dominant decay mode is α emission with half-lives from 0.1 to 100 ms, well in agreement with the results of the theoretical models, see Fig. 3f.

6. The new elements 110 to 112

Measured $1n$ cross-sections for the production of elements from 102 to 109 in reactions with ^{208}Pb and ^{209}Bi revealed an average decrease by a factor of 4.4 per element. An extrapolation to element 110 resulted in a cross-section of 1.5 pb. In contrast to the cross-sections of these so called cold fusion reactions, the extrapolation of known cross-sections of the hot fusion reactions for elements from 102 to 106 based on targets from ^{238}U to ^{249}Cf was worse. A cross-section of 0.1 pb was extrapolated for element 110. For production of one atom of element 109 at a cross-section of 16 pb we needed two weeks of beam time. Under the same conditions, we estimated needing 20 weeks for production of element 110, not taking into account statistical fluctuations.

Such a long beam time could not be allocated at the end of the 1980s. The new synchrotron SIS had been built at GSI and other experiments became the center of interest. In addition, predictions of the reaction theory were not promising. Support for the GSI heavy element program was reduced. Under these circumstances we were forced to improve the experimental set up. The aim was to reach a cross-section of 1 pb in two weeks of beam time.

The improvements were performed during the years 1988 to 1994. The largest gain factor of 2.9 was due to installation of a new high charge-state injector based on an ECR ion source and a 108-MHz RFQ-IH injection accelerator for the Alvarez section of the UNILAC. In this case we profited from the request of other experiments for a stable uranium beam. The injector delivered higher beam intensities at higher stability, an increase of the length of beam pulses to 5.5 ms, and considerably less consumption of isotopically enriched source material. Other improvements were related to an increase of the SHIP acceptance angle by moving the target closer to the first quadrupole, an optimization of the target thickness and new, larger area TOF and position sensitive Si detectors. A fifth deflection magnet reduced the background considerably. The upgraded version of SHIP is shown in Fig. 1.

A serious problem which had to be solved experimentally before starting the main experiment to search for element 110 was caused by reaction theory. It was predicted that with heavier Z projectiles increasing beam energies are needed for fusion of nuclei of high proton number. An extra-push of 30 MeV calculated for the reaction Ni + Pb in [49] belonged to the smaller values predicted. Such an increase of the resulting excitation energy E^* would have reduced the survival probability of the CN considerably.

The quality of excitation functions measured for element 102 to 107 in cold fusion reactions was insufficient for extracting an extra-push energy reliably and thus determining an optimum beam energy solely based on experimental data. In addition, element 108 and 109 were observed only at one relatively high energy which was chosen taking into account a certain amount of extra-push energy.

In order to obtain a reliable value for the optimum beam energy for production of element 110, we decided to measure accurately the excitation functions for the reactions ^{50}Ti + ^{208}Pb and ^{58}Fe + ^{208}Pb. The first measurement should be relatively fast due to the high cross-section. The second was necessary because only one energy of the excitation function was known. In addition, element 108 was the closest approach of an even element to the even element 110. However, for this latter measurement an additional problem arose from the fact, that we needed an amount of approximately 4 g of enriched ^{58}Fe, which cost at that time 500 kDM/g. The necessary amount of money could not be provided by GSI. Therefore, we were extremely happy that our collaborators from the Flerov Laboratory of Nuclear Reactions

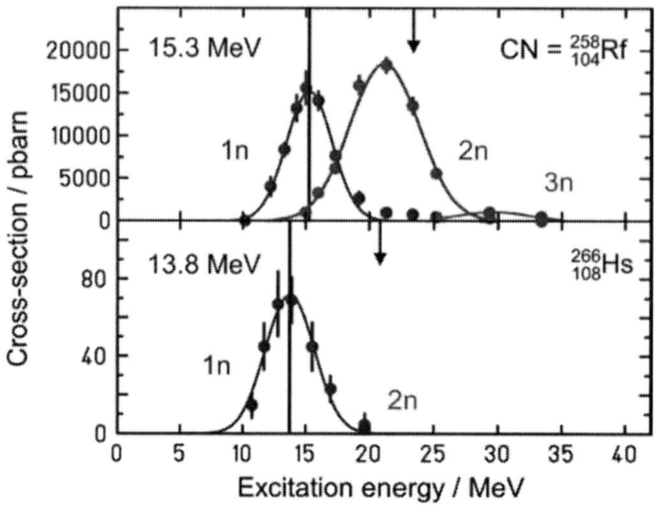

Figure 4. Measured excitation functions for production of isotopes of elements 104 (Rf) and 108 (Hs) by reactions of ^{50}Ti and ^{58}Fe beams with ^{208}Pb targets. The cross-sections are plotted as a function of the excitation energy, E^*, calculated from the beam energy in the middle of the target thickness. The curves represent Gaussian curves fitted to the data points. An extrapolation of the maxima for element 104 and 108 resulted in an optimum value of $E^* = 13.0$ MeV for production of element 110. The arrows mark the interaction barriers of the reactions according to the fusion model of R. Bass [50]. These barriers are defined by a beam energy so that beam and target nuclei are at rest in the center of mass system when their mean radii are at contact.

(FLNR) in Dubna contributed the material. Later, we were able to reimburse part of the efforts by supporting our collaborators at FLNR with detectors and electronics.

The excitation function for element 104 was measured in June, that of 108 in October 1994. The measured data are shown in Fig. 4. Now it was easy to extrapolate the well determined cross-section maxima of the 1*n* curves to element 110, and already on the first day of the irradiation of a ^{208}Pb target with ^{62}Ni ions the first decay chain of ^{269}110 was measured [51]. The experiment lasted twelve days, and a total of four decay chains was accumulated. The longest one consisting of five consecutive α decays is shown in Fig. 5. The isotope ^{62}Ni was chosen with the assumption that the addition of only one α particle to the ^{58}Fe beam used for element 108 would be a minimal change of the reaction and, secondly, the new isotope ^{269}110 could be unambiguously identified by the then well known decay properties of its daughter ^{265}108.

In a second step, the most neutron rich stable Ni isotope, ^{64}Ni, was used as a beam. The yield increased and during another thirteen days a total of nine decay chains was measured of ^{271}110 [52]. One of the two longest chains is shown in Fig. 5. The majority of eight chains started with a half-life of 1.1 ms of ^{271}110, in one case a half-life of 56 ms was measured. Using the argument that high-spin nuclei are populated stronger in fusion reactions, it seems reasonable to assign the eight chains to the decay of a high-spin state which decays preferably into a high-spin state of the daughter nucleus whereas the decay with the long half-life is attributed to a low-spin state, which, however, is hindered by a change of angular momentum. High- and low-spin states near the ground-state of ^{271}110 and its daughter ^{267}108 were obtained in a theoretical work of S. Cwiok *et al.* [53]. The suggested decay scheme is presented in [52].

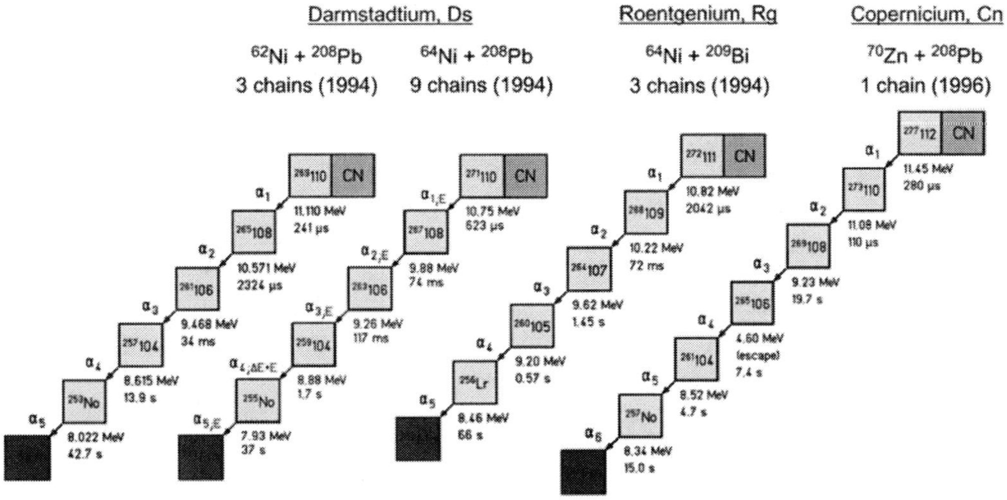

Figure 5. Examples of α-decay chains measured in the search experiments for elements 110, 111, and 112. The first information from the decay chain was obtained from the implantation of the evaporation residue (ER) produced after emission of one neutron from the compound nucleus (CN, blue). For each α-decaying nucleus (yellow) the measured α energy and the lifetime is given. The chains terminate by a known, but unobserved electron capture decay (EC, red). The rows on top show name and symbol of the new element, beam and target of the fusion reaction, number of measured decay chains, and year of the experiment.

Eventually, the target was changed to ^{209}Bi aiming to search for element 111. During an irradiation period of seventeen days we measured three decay chains of ^{272}111 in December 1994 [54]. The longest of the chains is shown in Fig. 5.

Element 112 was searched for using the most neutron-rich stable isotope of Zn, ^{70}Zn, and a ^{208}Pb target in January-February 1996 [55]. Two decay chains were published, the second and longest one is shown in Fig. 5. Already in the publication [55] it was mentioned that the second chain was more complete than the first one. In a confirmation experiment performed in 2000 a third chain was measured, which was in agreement with the second chain down to ^{265}Sg, where the new chain ended by a previously unknown SF branch [56]. A reanalysis of the 1996 data in connection with the confirmation experiment revealed that the first chain published in 1996 did not exist in the original data. This chain had to be retracted [56].

Because of this, reanalysis of all the data on elements 110, 111, and 112 measured at GSI since 1994 logically followed (a total of 34 decay chains was investigated) and revealed that the second chain of ^{269}110 was not recorded in the original data files in 1994. This chain was also retracted [56]. In all other cases the earlier data were exactly reproduced in the reanalysis.

Four further experiments using cold fusion were negative. Nevertheless, they are mentioned here because also the obtained cross-sections limits were of importance.

Firstly, the systematics of $1n$ cross-section maxima of cold fusion reactions revealed a continuous trend to lower excitation energy with increasing proton number of the beam, see Fig. 4 and Figs. 19 and 21 in [52]. Cross-section maxima were measured at beam energies just enough high so that beam and target nuclei come to rest in the center of mass system, when their diffused surfaces are touching. It was noticed that this configuration results in excitation energies below the $1n$ binding energy for reactions with beams of ^{76}Ge and ^{82}Se. Therefore, the possibility of a radiative capture process was tested using the reaction ^{82}Se + ^{208}Pb in

1995. At four beam energies between $E^* = 0$ and 10 MeV cross-section limits of 5 pb were reached [52]. A decay from a possible production of the nucleus $^{290}116$ was not observed.

Secondly, using the reaction ^{68}Zn + ^{208}Pb we tried to produce a lighter isotope of element 112. In that reaction the CN having $N = 164$ is closer to the center of increased stability at $N = 162$ than in the reaction with the ^{70}Zn beam, and a higher cross-section could result, similar to the case of element 110 when the beam was changed from ^{62}Ni to ^{64}Ni, see Fig. 3(a). At two beam energies one-event cross-section limits of 0.7 pb were obtained in 1997 comparable to that of 0.5 pb measured with a ^{70}Zn beam. A strong increase of the cross-section due to shell effects at $N = 162$ can be excluded. More importantly, the higher yield with the ^{70}Zn beam seems to be with the higher number of neutrons.

Thirdly, a search for element 113 started in 1998 and was continued in 2003. The reaction studied was ^{70}Zn + ^{209}Bi. During a total irradiation time of 57 days, a one-event cross-section limit of 160 fb was reached [57]. The experiment had to be stopped due to lack of beam time. However, a successful irradiation was performed at RIKEN. Using the same reaction, three decay chains assigned to $^{278}113$ were measured during a total of 553 days of beam time accomplished in several parts between 2003 and 2012 [58]. The measured cross-section of 28 fb is the smallest one ever achieved in a heavy-ion fusion reaction.

Fourthly, the reaction ^{86}Kr + ^{208}Pb was studied subsequently to reports on positive results of the synthesis of element 118 in 1999 [59]. A one-event cross-section limit of 0.5 pb was reached at SHIP in 24 days of beam time. Although lower than the Berkeley value of 2.2 pb published for three decay chains, our limit did not exclude the correctness of the Berkeley result. Eventually, the first announcement was retracted in 2001 after additional experiments in Berkeley and after a reanalysis of the data of the first experiment [60].

7. Concluding remarks

The half-lives and decay modes of the produced nuclei of the new elements from 107 to 112 are well in agreement with the results of the MM model. Shell effects increase the stability of nuclei around $Z = 108$ and $N = 162$. However, nuclei in that region are deformed in contrast to the spherical SHN. Although minimal SCE values are at –7 MeV in both regions, see Fig. 3a, the calculated partial SF half-lives are considerably shorter in the region of deformed nuclei, 10^3 s instead of 10^{12} s, see Fig. 3c. The more narrow fission barrier and the already elongated shape of the deformed nuclei is responsible for the difference.

Measured α half-lives, although less influenced by shell effects, prove that the region of highest stability is already surpassed with $^{277}112$. High α energies and accordingly short half-lives are measured for the first two decays of the chain, which proceed into direction of increased stability, see Fig. 3. Energies decrease by 2 MeV and half-lives increase by almost five orders of magnitude for the decay of $^{269}108$, see Fig. 5, when the region of highest stability is left into the direction of smaller proton and neutron numbers.

The low cross-section for production of element 113 and the negative results of the search experiments for elements 116 and 118 by cold fusion is understandable by means of Fig. 3a. In addition to increasing re-separation probability of beam and target fragments due to higher proton number of the beam, the locations of the CN are departing the region of strong shell effects and thus high fission barriers of deformed nuclei, but, for the heaviest systems, the CN are not located within the region of strong shell effects of spherical SHN.

At the end of the 1990s, the technical requirements for continuing the experiments at GSI and SHIP using hot fusion reactions were excellent. An intensive and stable beam of ^{48}Ca was achievable from the ECR ion source. The experiments on identification of new elements using cold fusion have convincingly demonstrated the high efficiency and background suppression of the separator SHIP and the reliable identification of the produced nuclei by the position-time correlation method.

However, rather pessimistic perspectives were published for a production of SHN by hot fusion reactions [61, 62]. Nevertheless, when we developed a medium-range plan for experiments at SHIP at the end of 1998, which was requested by the GSI management, we also suggested to study hot fusion reactions based on ^{48}Ca beams for synthesis of SHN, knowing that a larger program for such an investigation was already started at FLNR in Dubna. In our proposal we also attempted to stimulate a discussion on the option of a new accelerator delivering high intensity DC beams dedicated to low-energy experiments, in particular to research on SHN. This happened at the time, when an extension of the GSI accelerator facilities were discussed, aiming at higher currents at relativistic energies. With our proposal we hoped to prepare a bright future for SHN research at GSI. We attempted to make available our suggestions to the public in form of a report [63]. However, already at the beginning of 1999, we were forced to retract that report. Distribution and thus a public discussion was prohibited. The study of hot fusion reactions at SHIP had to be postponed.

The decisive experiments on exploration of the island of spherical SHN were performed at FLNR in Dubna. During the years 1999 to 2010 nuclei of the new elements 113 to 118 were produced and identified [64]. Cross-sections increased with approach of the CN to the center of the island of stability, see Fig. 3a. Values of 10 pb were measured for production of isotopes of elements 114 and 115 in reactions of a ^{48}Ca beam with targets of ^{244}Pu and ^{243}Am, respectively. Decay modes, α-decay half-lives and in particular the systematics of partial SF half-lives confirmed the existence of a region of increased stability due to a double shell closure at $Z = 114$ and $N = 184$, as it was predicted already in the 1960s by theoretical models.

The excellent performance of SHIP also for the study of nuclei produced in hot fusion reactions was demonstrated later when targets of ^{238}U and ^{248}Cm were irradiated with ^{48}Ca ions [65, 66]. Results, previously obtained at FLNR, were confirmed, and energies of the α decays were improved. In a continuation of the experiments based on a ^{248}Cm target we attempted to produce element 120 using a beam of ^{54}Cr. The results of this search are presented in a recent review article [67], and a related study on fission barriers deduced from α-decay chains is published in [68]. Experiments to search for element 120 will be continued at FLNR [69] and at RIKEN [70]. At GSI, the UNILAC which was so important for a successful low-energy program and in particular the study of SHN, is presently optimized for injection of short beam pulses into the accelerators of the new Facility for Antiproton and Ion Research (FAIR). It will not be available for a heavy element program in the future.

References

[1] M. Goeppert Mayer, *Nobel Lecture: The Shell Model*, `http://www.nobelprize.org/nobel_prizes/physics/laureates/1963/mayer-lecture.html`.
[2] V.M. Strutinsky, Nucl. Phys. **A95**, 420 (1967).
[3] W.D. Myers and W.J. Świątecki, Nucl. Phys. **81**, 1 (1966).
[4] A. Sobiczewski, F.A. Gareev, B.N. Kalinkin, Phys. Lett. **22**, 500 (1966).
[5] H. Meldner, Ark. Fys. **36**, 593 (1967).
[6] S.G. Nilsson *et al.*, Nucl. Phys. **A115**, 545 (1968).
[7] S.G. Nilsson, S.G. Thompson, C.F. Tsang, Phys. Lett. **B28**, 458 (1969).
[8] S.G. Nilsson *et al.*, Nucl. Phys. **A131**, 1 (1969).
[9] U. Mosel, W. Greiner, Z. Phys. **222**, 261 (1969).
[10] J. Grumann *et al.*, Z. Phys. **228**, 371 (1969).

[11] E.O. Fiset and J.R. Nix, Nucl. Phys. **A193**, 647 (1972).
[12] A. Sobiczewski, Physica Scripta A **10**, 47 (1974).
[13] J. Randrup *et al.*, Phys. Rev. C **13**, 229 (1976).
[14] S.M. Polikanov *et al.*, Sov. Phys. JETP **15**, 1016 (1962).
[15] T. Sikkeland, Ark. Fys. **36**, 539 (1967).
[16] G.N. Flerov *et al.*, JINR P7-3808 (1968).
[17] A. Ghiorso *et al.*, Phys. Rev. Lett. **24**, 1498 (1970).
[18] F. Beck *et al.*, *Memorandum zur Errichtung eines gemeinsamen Ausbildungszentrum für Kernphysik Hessischer Hochschulen Darmstadt, Frankfurt und Marburg* (1966).
[19] G. Münzenberg *et al.*, Nucl. Instr. Meth. **161**, 65 (1979).
[20] P. Armbruster, Proc. 3rd Int. Conf. on Nuclei far from Stability, Report **CERN 76-13**, 3 (1976).
[21] P. Armbruster *et al.*, GSI Jahresbericht 1976, GSI Report **GSI-J-1-77**, 74 (1977).
[22] G. Münzenberg *et al.*, GSI Jahresbericht 1977, GSI Report **GSI-J-1-78**, 75 (1978).
[23] K.H. Schmidt *et al.*, Nucl. Phys. **A318**, 253 (1979).
[24] F.P. Heßberger *et al.*, GSI Scientific Report 1981, GSI Report **GSI-82-1**, 223 (1982).
[25] S. Hofmann *et al.*, Z. Phys. **A291**, 53 (1979).
[26] S. Hofmann *et al.*, Nucl. Instr. Meth. **223**, 312 (1984).
[27] S. Hofmann *et al.*, Z. Phys. **A305**, 111 (1982).
[28] C.C. Sahm *et al.*, Z. Phys. **A319**, 113 (1984).
[29] S. Hofmann, G. Münzenberg, Rev. Mod. Phys. **72**, 733 (2000).
[30] H. Folger *et al.*, Nucl. Instr. Meth. **A362**, 64 (1995).
[31] S. Saro *et al.*, Nucl. Instr. Meth. **A381**, 520 (1996).
[32] K.H. Schmidt *et al.*, Z. Phys. **A316**, 19 (1984).
[33] K.H. Schmidt, J.G. Keller, D. Vermeulen, Z. Phys. **A315**, 159 (1984).
[34] G. Münzenberg *et al.*, GSI Scientific Report 1979, GSI Report **GSI-80-3**, 59 (1980).
[35] Yu.Ts. Oganessian *et al.*, Nucl. Phys. **A239**, 157 (1975).
[36] A. Baran *et al.*, Proc. 3rd Int. Conf. on Nuclei far from Stability, Report **CERN 76-13**, 537 (1976).
[37] G. Münzenberg *et al.*, Z. Phys. **A302**, 7 (1981).
[38] G. Münzenberg *et al.*, Z. Phys. **A300**, 107 (1981).
[39] G. Münzenberg *et al.*, Z. Phys. **A309**, 89 (1982).
[40] G. Münzenberg *et al.*, Z. Phys. **A315**, 145 (1984).
[41] G. Münzenberg *et al.*, Z. Phys. **A322**, 227 (1985).
[42] G. Münzenberg *et al.*, Z. Phys. **A317**, 235 (1984).
[43] G. Münzenberg *et al.*, Z. Phys. **A324**, 489 (1986).
[44] R. Smolanczuk, J. Skalski, A. Sobiczewski, Phys. Rev. C **52**, 1871 (1995).
[45] R. Smolanczuk and A. Sobiczewski, Proc. XV. Nuclear Physics Divisional Conference on Low Energy Nuclear Dynamics, St.Petersburg, Russia, 1995, World Scientific, Singapore, p. 313 (1995).
[46] P. Möller, J.R. Nix, K.L. Kratz, At. Data Nucl. Data Tables **66**, 131 (1997).
[47] A. Sobiczewski and K. Pomorski, Prog. Part. Nucl. Phys. **58**, 292 (2007).
[48] P. Möller *et al.*, At. Data Nucl. Data Tables **59**, 185 (1995).
[49] P. Fröbrich, Phys. Lett. **B215**, 36 (1988).
[50] R. Bass, Nucl. Phys. **A231**, 45 (1974).
[51] S. Hofmann *et al.*, Z. Phys. **A350**, 277 (1995).
[52] S. Hofmann, Rep. Prog. Phys. **61**, 639 (1998).
[53] S. Cwiok, S. Hofmann, W. Nazarewicz, Nucl. Phys. **A573**, 356 (1994).
[54] S. Hofmann *et al.*, Z. Phys. **A350**, 281 (1995).

[55] S. Hofmann *et al.*, Z. Phys. **A354**, 229 (1996).
[56] S. Hofmann *et al.*, Eur. Phys. J. **A14**, 147 (2002).
[57] S. Hofmann *et al.*, GSI Scientific Report 2003, GSI Report **GSI-2004-1**, 1 (1980).
[58] K. Morita *et al.*, J. Phys. Soc. Jpn. **81**, 103201 (2012).
[59] V. Ninov *et al.*, Phys. Rev. Lett. **83**, 1104 (1999).
[60] V. Ninov *et al.*, Phys. Rev. Lett. **89**, 039901 (2002).
[61] K.H. Schmidt and W. Morawek, Rep. Prog. Phys. **54**, 949 (1991).
[62] P. Armbruster, Rep. Prog. Phys. **62**, 465 (1999).
[63] S. Hofmann, *SHIP-2000 – A proposal for the study of superheavy elements*, GSI-Report **GSI-99-02**, 1 (1999).
[64] Yu.Ts. Oganessian and V.K. Utyonkov, Nucl. Phys. **A944**, 62 (2015).
[65] S. Hofmann *et al.*, Eur. Phys. J. **A32**, 251 (2007).
[66] S. Hofmann *et al.*, Eur. Phys. J. **A48**, 62 (2012).
[67] S. Hofmann *et al.*, Eur. Phys. J. **A52**, 180 (2016).
[68] S. Hofmann *et al.*, Eur. Phys. J. **A52**, 116 (2016).
[69] V.K. Utyonkov, in *Int. Symposium 'Super Heavy Nuclei', College Station, Texas, USA, 2015*, `http://cyclotron.tamu.edu/she2015/assets/pdfs/presentations/Utyonkov_SHE_2015_TAMU.pdf`.
[70] K. Morita, *Int. Symposium 'Super Heavy Nuclei', College Station, Texas, USA, 2015*, `http://cyclotron.tamu.edu/she2015/assets/pdfs/presentations/Morita_SHE_2015_TAMU.pdf`.

How good are superheavy element *Z* and *A* assignments?

Kenneth Gregorich[a]

Lawrence Berkeley National Laboratory, Berkeley, CA, USA

Abstract. Proton number, *Z*, and mass number, *A*, assignments for newly discovered heavy element nuclides have historically been made by observing α decay to a daughter with well-established *Z* and *A*, and then observing the well-know decay of that daughter. For all of the new superheavy element isotopes observed in ^{48}Ca irradiations of actinide targets, this correlation technique has not been possible, because the α-decay chains end in spontaneous fission of previously unknown isotopes. Consequently, *Z* and *A* assignments have been made by less-direct means. The superheavy element *Z* and *A* assignment methods are summarized, and possibilities for how they may be incorrect are explored. While it is highly likely that most of the superheavy element *Z* and *A* assignments are correct, there is a real need for a direct proof.

1. History about heavy element *Z* and *A* assignments

For the discovery of elements 104–113 [1–12] proton number (*Z*) and mass number (*A*) assignments have been made by observing the α decay of the parent (new element) nucleus followed by the α decay of daughter nuclei with well-established *Z* and *A*. These daughter α decays have characteristic decay energies and half-lives. With the assumption of no intervening, unseen decays (such as electron-capture or β decay), the parent *Z* and *A* assignment was made by adding two to each of the *Z*and *A* of the first daughter. The assumption of no intervening decays could be supported by observation of a smooth trend in the α-decay energies, matching that expected by well-established α-decay systematics [13].

For all of the superheavy element (SHE) isotopes produced by ^{48}Ca irradiations of actinide targets [14, 15] the series of α decays (α-decay chains) terminate with spontaneous fission (SF) of the last chain member. The SF decay mode provides little information, other than half-life, to identify the *Z* and *A* of the decaying nucleus. SF lacks the characteristic decay energy seen in α decay. Thus, the *Z* and *A* of the α-decay chain members cannot be "anchored" by identification of the later chain member.

2. How SHE *Z* and *A* assignments have been made

For the SHE isotopes produced in ^{48}Ca irradiation of actinide targets, and all of the daughter isotopes resulting from successive α-decays, experimenters have resorted to less-direct means of *Z* and *A* assignment. These less-direct means are excitation functions, cross bombardments, and decay systematics.

[a] e-mail: kegregorich@lbl.gov

2.1 Excitation functions

An excitation function for formation of superheavy element isotopes by compound nucleus formation followed by de-excitation by evaporation of neutrons is a record of the production cross section for evaporation residue (EVR) formation as a function of projectile energy. Projectile energies are chosen from near the Coulomb barrier between projectile and target nuclei to a few tens of MeV above the barrier. The production cross sections are typically modelled as a three-step process,

$$\sigma_{EVR} = \sigma_{cap} \cdot P_{CN} \cdot P_x \cdot \prod_{i=1}^{x} \left(\Gamma_n / \Gamma_{tot}\right)_i, \tag{1}$$

where σ_{EVR} is the EVR production cross section. The first step of the process, represented by σ_{cap}, is the cross section for capture of the projectile and target nuclei into a minimum of their mutual Coulomb + nuclear potential to form a di-nuclear complex. The second step is the evolution this di-nuclear complex to an equilibrated compound nucleus, represented in Eq. (1) as P_{CN} (the probability for compound nucleus formation). The third and final step is the de-excitation of this excited compound nucleus by neutron evaporation in competition with fission. This third step is shown in Eq. (1) as the probability of evaporation of exactly x neutrons, P_x, times the product of the probabilities for emitting a neutron, rather than fissioning, in each of x neutron evaporation steps.

At the lowest projectile energies, σ_{EVR} is limited by the Coulomb barrier. Compound nucleus formation ($\sigma_{cap} \cdot P_{CN}$) increases monotonically from zero at far sub-barrier energies (although it may decrease at the highest projectile energies due to critical angular momentum effects). The CN excitation energy results from the center-of-mass kinetic energy and the mass difference of the target + projectile and the CN. The CN is "cooled" during each neutron evaporation step by an amount defined by the neutron separation energy and the kinetic energy of the neutron.

The end result for the formation of SHE in ^{48}Ca irradiations of actinide targets is near Gaussian excitation functions for $2 \leq x \leq 5$ with FWHM $\approx$ 10 MeV. The experimental excitation functions for SHE formation have been summarized in a recent review [15], and are reproduced here in Fig. 1. In Fig. 1, horizontal error bars indicate the range of projectile energies subtended in the target layer, and vertical error bars indicate statistical errors. Systematic errors in energies and cross sections are not represented.

2.2 Cross bombardments

Cross bombardments, where the same nuclides are produced using different target/projectile combinations, also provide information used in Z and A assignments. Evidence that the same nuclide was produced in each reaction includes observation of decay mode, decay energy, and half-lives for the nuclide in question, and for all α-decay daughters. Cross bombardment information for the SHE has been concisely summarized in the recent review of SHE formation and decay by Oganessian and Utyonkov [15]. Their cross bombardment summary figures are reproduced here in Fig. 2 (even-Z) and Fig. 3 (odd-Z).

Cross bombardments can produce the same nuclide directly. For example, ^{287}Fl (element 114) was produced directly in both the ^{244}Pu(^{48}Ca,5n)^{287}Fl and ^{242}Pu(^{48}Ca,3n)^{287}Fl reactions. When considering these two excitation functions (Fig. 1), it can be seen that ^{287}Fl was produced at the highest CN excitation energies in the ^{244}Pu(^{48}Ca,5n)^{287}Fl reaction and at the lowest CN excitation energies in the ^{242}Pu(^{48}Ca,3n)^{287}Fl reaction. This provides strong evidence that the exit channels (determination of x) have been assigned consistently.

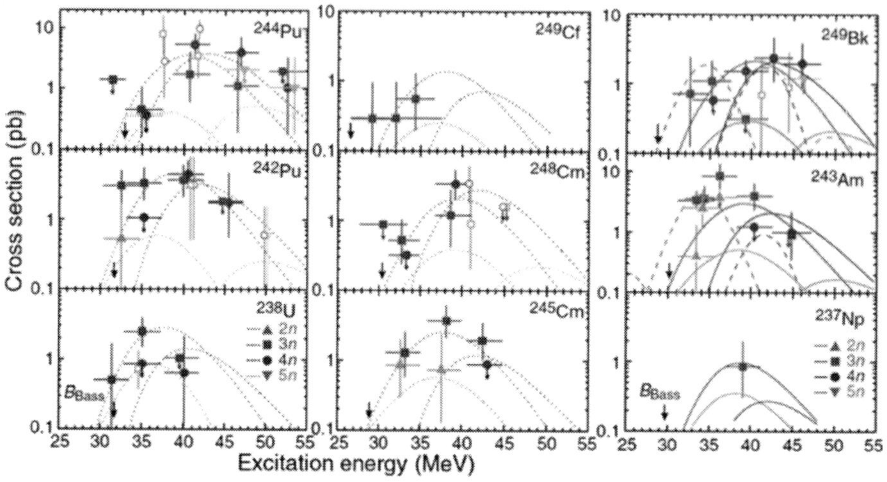

Figure 1. Excitation functions for even-Z SHE (left, center) and odd-Z SHE (right) (reproduced from Ref. [15]).

Another class of cross bombardments has been used, where the same nuclide is produced directly, and as an α-decay daughter. Continuing the example with ^{287}Fl, this element 114 nuclide has also been produced as the α-decay daughter of ^{291}Lv (element 116). This element 116 parent nuclide was produced in both the ^{248}Cm(^{48}Ca,5n)^{291}Lv and ^{245}Cm(^{48}Ca,2n)^{291}Lv reactions. Once again, observing that $^{291}\text{Lv} \rightarrow ^{287}\text{Fl} \rightarrow ^{283}\text{Cn} \rightarrow ^{279}\text{Ds}$ was produced at the highest CN excitation energies in the ^{248}Cm(^{48}Ca,5n)^{291}Lv reaction, and at the lowest excitation energies in the ^{245}Cm(^{48}Ca,2n)^{291}Lv reaction provides strong evidence that these exit channels (determination of x) have been assigned consistently.

Cross-bombardment data indicates that the Z and A assignments have been made consistently: either consistently correct, or consistently incorrect (always off by the same ΔZ and/or ΔA). As may be expected when making Z and A assignments based on a limited number of decay chains, there may be some assignments for individual decay chains, or for certain nuclides, which are in error. As an example, recent data and more comprehensive statistical assessments question the originally suggested α-decay link between the isotopes ^{293}117 and ^{289}115 (cf. Fig. 3) [16–18]. It is expected that these discrepancies will be cleared up as more SHE data become available.

2.3 Decay systematics

Alpha-decay energies tend to increase with increasing Z of the emitting nucleus because the increased Z causes a Coulomb separation energy increase. Additionally, α-decay energies tend to decrease with increasing neutron number, N, because the increased nuclear radius of the α-decay daughter causes a Coulomb separation energy decrease. These $E_\alpha(Z, N)$ systematics have been quantified empirically [13]. The same $E_\alpha(Z, N)$ trends result from calculation of α-decay Q-values, Q_α, using mass models [19]. Q_α trends for even-even nuclides (with even Z and N) are clear, because the decay is dominated by decay from the 0^+ ground-state in the parent to the 0^+ ground state in the daughter. The situation is somewhat less clear in the α-decay of odd and odd-odd isotopes, because the dominant α-decay usually proceeds to an excited analogue-state in the daughter nucleus (with the odd particle(s) in the same Nilsson state as in the parent).

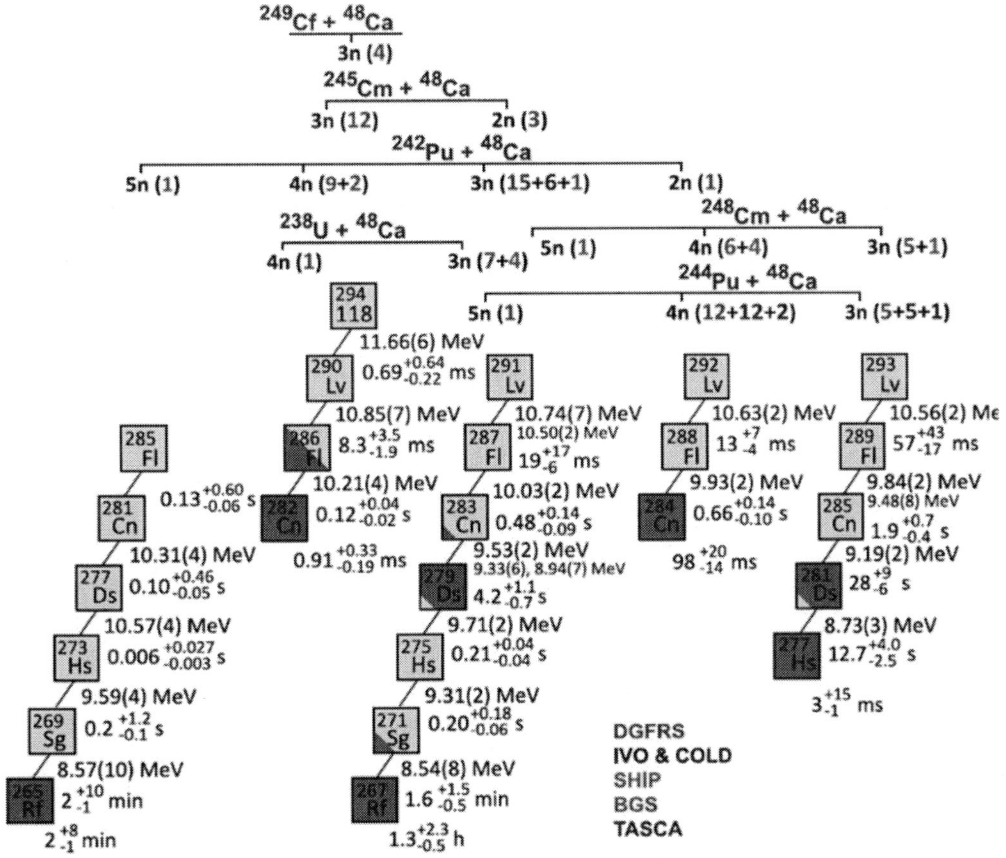

Figure 2. Cross bombardment summary for even-Z SHE (reproduced from Ref. [15]).

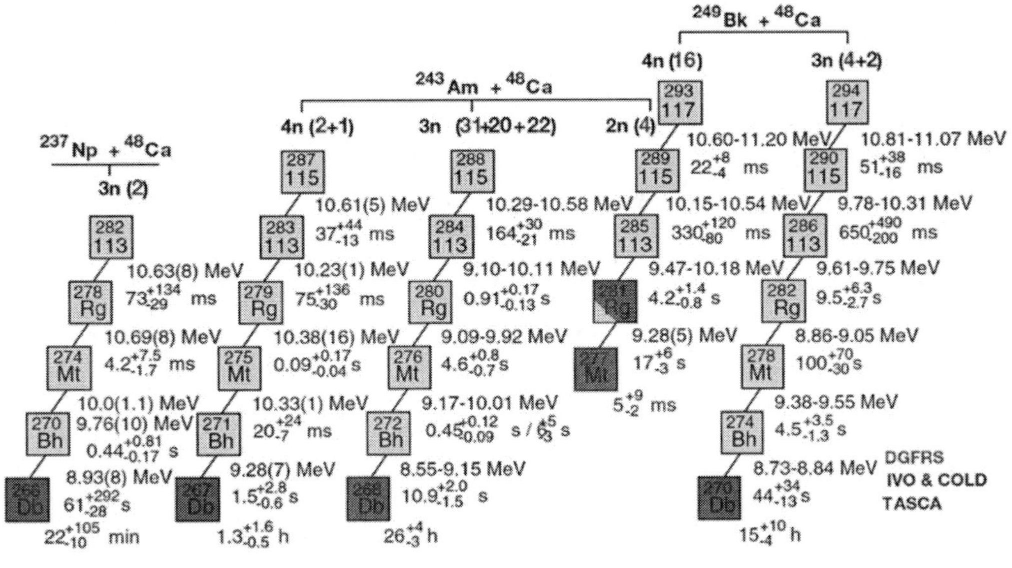

Figure 3. Cross bombardment summary for odd-Z SHE (reproduced from Ref. [15]).

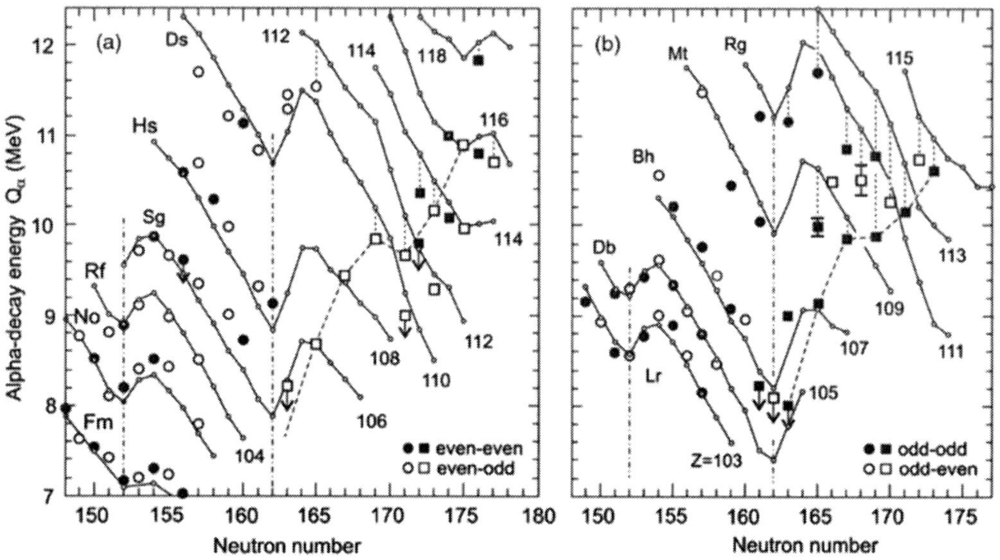

Figure 4. Comparison of α-decay energies with predictions [19] (taken from Ref. [14]).

The maximum α-decay energies for even-Z SHE isotopes are plotted as a function of N in the left panel of Fig. 4. These decay energies have been corrected for the recoil energy imparted to the α-decay daughter nucleus, resulting in α-decay Q-values (assuming a ground-state to ground-state transition). They show a smoothly decreasing α-decay energy with increasing N, giving confidence that the *relative* neutron numbers have been assigned correctly. In addition, the systematic trend of increasing Q_α with increasing Z is readily apparent, giving confidence that the *relative* proton numbers have been correctly assigned.

The maximum α-decay energies (corrected for daughter recoil) for odd-Z SHE isotopes are plotted as a function of N in the right panel of Fig. 4. Once again, the expected trends with increasing Z and N are apparent. The *relative* Z and N for these odd-Z SHE isotopes appear to have been assigned correctly. Also, it should be noted that the odd-Z Q_α values at a given N fall between those for the adjacent even Z. This shows that the Zs of these isotopes do, in fact,fall between those assigned for the even-Z SHE isotopes.

Spontaneous fission (SF) half-life systematics can also be used to test the accuracy of SHE Z assignments. SF is strongly hindered by the presence of odd protons and/or odd neutrons. Even-even isotopes decay by SF with half-lives according to regularly trending spontaneous fission systematics. The SF decays of isotopes with odd numbers of protons or neutrons are hindered by several orders of magnitude compared to even-even neighbouring isotopes. A review of the spontaneous fission half-lives reported for SHE isotopes shows that the even-Z vs. odd-Z and even-N vs. odd-N assignments have been made correctly.

2.4 Summary of accuracy of existing assignments

Evidence provided by measurement of excitation functions, implementation of cross bombardments, and analysis of decay systematics, provides a consistent picture of formation and decay of SHE. The Z and A assignments result in a contiguous set of isotopes and isotones. It has been demonstrated conclusively that, with very few exceptions, the *relative* positions in Z and A are assigned correctly.

Given the strong evidence supporting the contiguous nature of the present Z and A assignments, we must now consider if the whole set of assignments could be wrong. Odd-even effects in the SF half-lives give strong evidence that the set of assignments is not off by one unit in Z or and/or one unit in A. If the set of assignments is incorrect, the Z and/or A assignments must be off by an even number of protons and/or neutrons.

3. Uncertainties in *A* assignments

Mass number, A, assignments have been made based on the energetics of the compound nucleus evaporation residue reaction. The number of neutrons evaporated from the compound nucleus is calculated by summing the center-of-mass frame reaction energy and the Q-value for target plus projectile going to an EVR and x free neutrons. This is further corrected by an estimate for the kinetic energy of the evaporated neutrons. Since the mass defect of the EVR is not experimentally known, predictions from mass models are used.

If these reactions actually proceed via compound nucleus formation followed *exclusively* by evaporation of neutrons, and the mass number is incorrect by at least two neutrons (as shown in the previous section), the mass models must be incorrect by nearly 20 MeV to arrive at incorrect A assignments. Could the mass models, which have been shown to be accurate in the No ($Z = 102$) region, be inaccurate by ~20 MeV in the $Z = 114$ SHE region? If SHE nuclei behave similarly to the lighter heavy elements, cumulative errors in α-decay Q-values, stepping up from the No region (where masses have been measured) to the SHE region could result in 5–10 MeV errors in SHE masses. To reach a 20-MeV mass error, some important new physics, not included in the mass models would be required. The study of SHE is a study at the extreme limits of Z and A. One may expect that the nucleus "finds" innovative ways to stabilize itself. Examples of such new physics (not included in many mass models) could be unusual nucleon density distributions within the nucleus, such as a "bubble nucleus" with a reduced proton or nucleon density at its centre. The result of a calculation including such nucleon density effects [20] is shown in Fig. 5. Another possibility is that the SHE nuclei take on different shapes to increase stability. Many of the mass models consider only axially-symmetric shapes. Non-axially symmetric shapes such as octahedral (stabilized by six Z~20, A~48 clusters) or tetrahedral (stabilized by four $Z = 28$ clusters) could result in enhanced SHE stability, and correspondingly reduced mass. Some plots of these octahedral and tetrahedral deformations [21] are presented in Fig. 6.

As an example, the ^{243}Am(^{48}Ca,xn)$^{291-x}$115 reaction is considered. If element 115 masses are 20 MeV smaller than predicted by mass models, the ^{291}115 compound nucleus will be formed with an excitation energy 20 MeV larger than expected. With this increased excitation energy, approximately two extra neutrons will be evaporated from the compound nucleus to form the fully de-excited evaporation residue. What has been assigned as the ^{243}Am(^{48}Ca,2n)^{289}115 reaction would actually be ^{243}Am(^{48}Ca,4n)^{287}115. Similarly, what has been assigned as ^{243}Am(^{48}Ca,3n)^{288}115 would actually be ^{243}Am(^{48}Ca,5n)^{286}115, and ^{243}Am(^{48}Ca,4n)^{287}115 would be ^{243}Am(^{48}Ca,2n)^{285}115. Conversely, if the element 115 masses are 20 MeV larger than predicted by the mass models, the ^{291}115 compound nucleus will be formed at an excitation energy 20 MeV lower than expected. With this decreased excitation energy, approximately two fewer neutrons will be evaporated from the compound nucleus to form the fully de-excited evaporation residue. What has been assigned as the ^{243}Am(^{48}Ca,2n)^{289}115 reaction would actually be ^{243}Am(^{48}Ca,0n)^{291}115. Similarly, what has been assigned as ^{243}Am(^{48}Ca,3n)^{288}115 would actually be ^{243}Am(^{48}Ca,1n)^{290}115, and ^{243}Am(^{48}Ca,4n)^{287}115 would be ^{243}Am(^{48}Ca,2n)^{289}115. There are two arguments supporting the present assignments, rather than those shifted by plus-minus two neutrons. First, the 0n

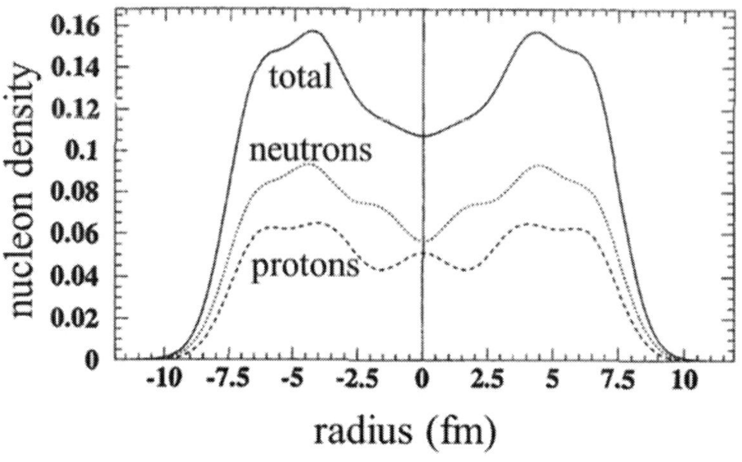

Figure 5. Calculation of radial nucleon density for $^{292}120$ (from Ref. [20]).

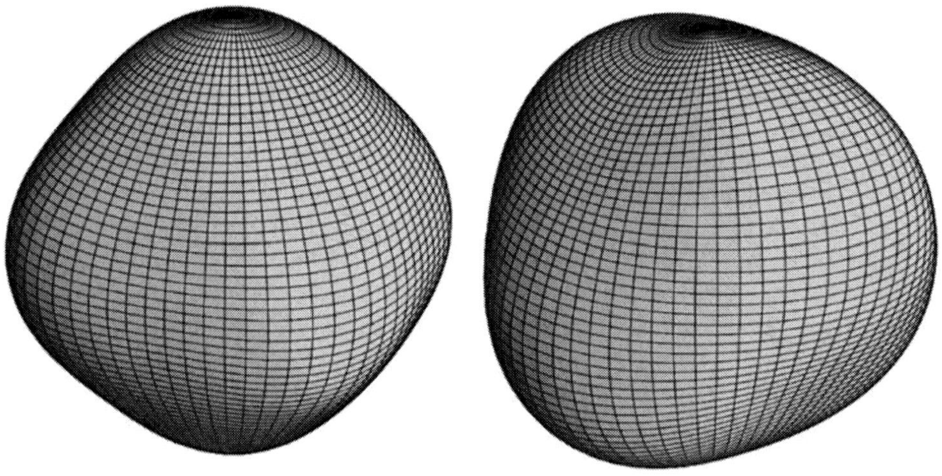

Figure 6. Non-axial nuclear deformations (left: octahedral, right: tetrahedral) (taken from Ref. [21]).

exit channel in the $^{243}Am(^{48}Ca,0n)^{291}115$ reaction is expected to have a vanishingly-small cross section. However, the sub-picobarn cross sections reported for what has been assigned as SHE produced by two-neutron exit channels may qualify as "vanishingly-small" with this heretofore unobserved exit channel becoming observable. Second, the α-decay systematics presented in Fig. 4 show the effect of the $N = 162$ deformed shell, supporting the present set of A assignments. However, the position of this deformed shell could easily be shifted by plus-minus two neutrons with one neutron level in the Nilsson diagram having a shape dependence different than calculated.

4. Uncertainties in Z assignments

Z-assignments for the contiguous set of SHE isotopes have been made largely on the assumption that SHE formation proceeds by compound nucleus formation, followed by de-excitation *exclusively* by neutron emission. There are other possibilities. In the formation of neutron-deficient isotopes of lighter elements (Z = 50–80) by compound nucleus – evaporation reactions, de-excitation by emission of α-particles (and neutrons) is quite common, accounting for a large fraction of the EVR cross section. Such αxn exit channels have not been found to be significant in the production of elements with $102 \leq Z \leq 113$ in "cold fusion" reactions using projectiles from ^{48}Ca to ^{70}Zn with Pb and Bi targets. However, αxn exit channels could become more important at even higher Z for formation of SHE. Also, processes, such as αxn exit channels could have been proceeding at picobarn levels in the $102 \leq Z \leq 110$ region, which are insignificant compared to the microbarn – to – nanobarn cross sections found for the formation of elements 102–110. These charged-particle evaporation processes could dominate the picobarn cross sections found for formation of SHE at even higher Z. For example, the energetics of (^{48}Ca, $3n$) reactions are similar to those for (^{48}Ca, αn) reactions, and what has been interpreted as ^{248}Cm(^{48}Ca, $3n$)^{293}116 could, in fact, be ^{248}Cm(^{48}Ca, αn)^{291}114.

There is an argument against dominance of αxn exit channels in the production of SHE. SHE produced in αxn exit channels will have broader recoil velocity and angular distributions than SHE produced in xn exit channels. Similar SHE production cross sections have been measured with three gas-filled recoil separators and a velocity filter. These separators, are, respectively, The Dubna Gas Filled Recoil Separator at Dubna, the Berkeley Gas-Filled Separator in Berkeley, and the TransActinide Separator and Chemistry Apparatus at GSI, and the Separator for Heavy Ion Products (SHIP) at GSI. These four separators have different angular acceptance characteristics, and therefore have different relative efficiencies for SHE produced in xn vs. αxn exit channels. A consistent set of SHE production cross sections have seemingly been measured at all four separators, suggesting that αxn exit channels do not dominate SHE production.

One must also consider evaporation of protons during de-excitation of SHE compound nuclei. The same arguments apply here. Processes such as pxn exit channels may become more important at the high-Z of SHE. While pxn exit channels have been insignificant compared to the microbarn – to – nanobarn cross sections found for the formation of elements 102–110, they could have been proceeding at picobarn levels. These picobarn levels could dominate the picobarn level cross sections for formation of SHE. For example, the energetics of (^{48}Ca,$3n$) reactions are similar to those for (^{48}Ca,$2p$) reactions, and what has been interpreted as ^{248}Cm(^{48}Ca,$3n$)^{293}116 could, in fact, be ^{248}Cm(^{48}Ca,$2p$)^{294}114. Once again, the similarity in SHE cross sections measured with four different recoil separators provides some evidence against the predominance of $2pxn$ reactions in SHE production, although this evidence is weaker than in the case of αxn reactions, because the kinematics of $2pxn$ recoils are more similar to those of xn recoils.

Finally, compound nucleus formation followed by neutron evaporation has been the assumed reaction mechanism for SHE formation using ^{48}Ca projectiles with actinide targets. The possibility of pre-compound nucleus particle emission has been ignored. Pre-compound emission of high-kinetic-energy neutrons could be occurring at picobarn cross section levels. Because of the high neutron kinetic energy, the compound nucleus will be cooled much more than expected. Similar arguments can be made for emission of high-kinetic energy pre-compound proton or pre-compound α emission.

5. Conclusion

Excitation functions, cross bombardments, and decay systematics show that the Z- and A-assignments for SHE produced in ^{48}Ca irradiations of actinide targets, and all of their decay daughters form a contiguous block of isotopes. If the formation mechanism, nuclear masses, and radioactive decay of SHE isotopes behave as in the lighter heavy elements, the position of this contiguous block of isotopes in Z and A is – by and large – correctly assigned.

If mass models diverge from true masses when extrapolating from regions of Z and A with well-known masses to the SHE isotopes, excitation functions could be improperly interpreted, leading to incorrect A-assignments. This divergence between model masses and true masses would have to be unexpectedly large: 20 MeV. Such a large divergence would indicate interesting new physics in nuclear shapes and/or nucleon density distributions.

The production mechanism for SHE isotopes is formation of an excited compound nucleus, followed by de-excitation by neutron evaporation in competition with fission. The Z and A assignments have assumed there is no charged particle emission during the de-excitation process. CN de-excitation involving charged particle emission has been observed in other parts of the chart of nuclides, but has had an insignificant impact on formation of heavy element isotopes close to $Z = 102$. If insignificant charged particle evaporation continues to the SHE, the SHE Z-assignments are correct. However, the SHE lie at the high-Z limit of nuclear stability, and a resurgence of charged particle emission during CN de-excitation is a possibility.

A strong case has been made for the present set of SHE Z and A assignments. Excitation functions, cross bombardments, and decay systematics show that, with very few exceptions, the present Z and A assignments are correct relative to each other. This results in a contiguous set of isotopes. There is a small probability that the Z and/or A assignments for the whole group of SHE isotopes is incorrect by an even number (–2) of protons and/or neutrons.

Before using the measured Q_α values from SHE decay to adjust mass models, the Z and A assignments must be proven to be correct. Otherwise, the mass models could be adjusted to produce mass defects that support the incorrect assignments. If the group of Z and A assignments for SHE is incorrect, there is important and exciting new physics to be learned. If the A assignments are incorrect, this new physics could be new variations in nuclear shapes or nucleon density distributions. If the Z assignments are incorrect, new heavy element formation reaction mechanisms which become important in the heaviest elements on the picobarn cross section scale could be explored.

6. Future work

Experiments are being carried out to measure the Z of SHE by looking for X-ray emission with characteristic energies coincident with SHE α decay [16, 22]. Alpha decay from an odd or an odd-odd isotope typically proceeds to an excited analogue state with odd particle(s) in the same state(s) as in the parent. If the excitation energy of this analogue state is above the K-shell edge, de-excitation by internal conversion of a K electron usually results in a K Xray. These K X rays have well-known energies and relative intensities which are unique for each Z. The first direct proof of SHE Z will probably come from a future α-K-X-ray coincidence experiment in a SHE isotope where the α-decay analogue state in the daughter lies above the K-shell edge and decays by internal conversion with a $M1$ multipolarity.

There are also some possibilities for a direct proof of SHE Z and A by observing α-decay chains that end in isotopes with well-characterized Z and A. Target-projectile combinations that are more neutron-deficient than those that have been successful so far are required. Unfortunately, these reaction combinations seem to have sub-picobarn cross sections, and the α-decay chains tend to terminate by SF before reaching daughters with well-characterized Z and A. Another difficulty results because electron capture decay may become dominant in these neutron-deficient decay chains, shifting the decay chain away from the known daughters.

Chemical separations will play only a minor confirmatory role in the determination of SHE proton numbers. Chemical separations can be performed with SHE to show similar chemical behaviour to periodic table homologs, but the chemical information gained will not be sufficient for assignment of Z. An example is the gas-phase chemistry with what is presumed to be element 112, where atoms of 112 was shown to be volatile and to adsorb strongly onto a gold surface [23, 24]. This behaviour was claimed to be similar to that for the periodic table homolog element, Hg, and different from noble gas behaviour of Rn. However, this chemistry experiment cannot be used to assign Z, as it used atoms with uncertain Z in a chemical separation system that has not been proven to be selective for element 112 over other nearby SHE elements.

An apparatus for direct measurement of SHE mass numbers is being constructed at the Lawrence Berkeley National Laboratory's 88-Inch Cyclotron. A RF Gas Catcher is being installed at the Berkeley Gas-filled Separator focal plane to stop the SHE recoils in ultra-pure He, where they will retain a positive charge state. RF fields and a DC gradient will move the SHE ions toward an exit orifice. SHE ions will be swept through the orifice and into a radiofrequency quadrupole trap where they will be trapped and cooled. The SHE ions will then be ejected from the trap, accelerated to $\sim$10 kV, and passed through a mass separator. The device will have a mass resolution of $A/\Delta A = 1000$, sufficient to determine SHE A from single events.

Financial support was provided Office of High Energy and Nuclear Physics, Nuclear Physics Division, Office of Basic Energy Sciences, of the U.S. Department of Energy under contract DE-AC02-05CH11231.

References

[1] A. Ghiorso *et al.*, Phys. Rev. Lett. **22**, 1317 (1969).
[2] G.N. Flerov *et al.*, Phys. Lett. **13**, 73 (1964).
[3] A. Ghiorso *et al.*, Phys. Rev. Lett. **24**, 1498 (1970).
[4] G.N. Flerov *et al.*, Atomnaya Énergiya **29**, 243 (1970).
[5] A. Ghiorso *et al.*, Phys. Rev. Lett. **33**, 1490 (1974).
[6] G. Münzenberg *et al.*, Z. Phys. **A300**, 107 (1981).
[7] G. Münzenberg *et al.*, Z. Phys. **A317**, 235 (1984).
[8] G. Münzenberg *et al.*, Z. Phys. **A309**, 89 (1982).
[9] S. Hofmann *et al.*, Z. Phys. **A350**, 277 (1995).
[10] S. Hofmann *et al.*, Z. Phys. **A350**, 281 (1995).
[11] S. Hofmann *et al.*, Z. Phys. **A354**, 229 (1996).
[12] K. Morita *et al.*, J. Phys. Soc. Jpn. **73**, 2593 (2004).
[13] Y. Hatsukawa, H. Nakahara, D.C. Hoffman, Phys. Rev. C **42**, 674 (1990).
[14] Y. Oganessian, J. Phys. G **34**, R165 (2007).
[15] Yu.Ts. Oganessian and V.K. Utyonkov, Rep. Prog. Phys. **78**, 036301 (2015).
[16] J.M. Gates *et al.*, Phys. Rev. C **92**, 021301(R) (2015).
[17] U. Forsberg *et al.*, Nucl. Phys. **A953**, 117 (2016).
[18] U. Forsberg *et al.*, Phys. Lett. **B760**, 293 (2016).
[19] I. Muntian, Z. Patyk, A. Sobicziewski, Phys. At. Nucl. **66**, 1015 (2003).
[20] J.-F. Berger *et al.*, Nucl. Phys. **A685**, 1c (2001).
[21] J. Dudek, A. Goizdz, N. Schunck, Acta Phys. Pol. **B34**, 2491 (2003).
[22] D. Rudolph *et al.*, Phys. Rev. Lett. **111**, 112502 (2013).
[23] R. Eichler *et al.*, Nature **447**, 72 (2007).
[24] R. Eichler *et al.*, Angew. Chem. Int. Ed. **47**, 3262 (2008).

The discovery of elements 113 to 118

Vladimir Utyonkov[1,a], *Yuri* Oganessian[1], *Sergey* Dmitriev[1], *Mikhail* Itkis[1], *Kenton* Moody[2], *Mark* Stoyer[2], *Dawn* Shaughnessy[2], *James* Roberto[3], *Krzysztof* Rykaczewski[4], and *Joseph* Hamilton[5] for the collaboration[1,2,3,4,5,6,7]

[1] Flerov Laboratory of Nuclear Reactions, Joint Institute for Nuclear Research, 141980 Dubna, Russian Federation
[2] Nuclear and Chemical Sciences Division, Lawrence Livermore National Laboratory, Livermore, California 94551, USA
[3] Science and Technology Partnerships Directorate, Oak Ridge National Laboratory, Oak Ridge, Tennessee 37831, USA
[4] Physics Division, Oak Ridge National Laboratory, Oak Ridge, Tennessee 37831, USA
[5] Department of Physics and Astronomy, Vanderbilt University, Nashville, Tennessee 37235, USA
[6] Department of Physics and Astronomy, University of Tennessee, Knoxville, Tennessee 37996, USA
[7] Research Institute of Atomic Reactors, 433510 Dimitrovgrad, Russian Federation

Abstract. Discovery and investigation of the "Island of stability" of superheavy nuclei at the separator DGFRS in the ^{238}U-^{249}Cf+^{48}Ca reactions is reviewed. The results are compared with the data obtained in chemistry experiments and at the separators SHIP, BGS, TASCA, and GARIS. The synthesis of the heaviest nuclei, their decay properties, and methods of identification are discussed and compared with the criteria that must be satisfied for claiming the discovery of a new chemical element. The role of shell effects in the stability of superheavy nuclei is demonstrated by comparison of the experimental results with empirical systematics and theoretical data.

1. Introduction

For half a century, various theoretical models have predicted the existence of an area of enhanced stability of heaviest (superheavy) nuclei (SHN). A group of nuclei in the vicinity of the predicted doubly magic spherical nucleus $^{298}114_{184}$ were named superheavy. Just after the appearance of the first theoretical predictions of this area of SHN, numerous attempts were undertaken to synthesize them artificially; however, all the efforts were in vain. Later on, it became clear that in order to synthesize short-lived ($\leq$ 1 min) SHN the sensitivity of experiments should be increased by two to three orders of magnitude.

In 1998, experiments aimed at the synthesis of SHN were initiated in FLNR, JINR. In these experiments, we employed the Dubna gas-filled recoil separator (DGRFS), which allows the separation of the products of complete fusion reactions from the beam of bombarding ions, elastically-scattered nuclei, and products of incomplete fusion; with the subsequent collection of the complete fusion products in a detection system. The detection system includes proportional chambers used to measure the time of flight (TOF) of particles and several semiconductor detectors; the latter are separated in position-sensitive strips.

[a] Corresponding author: utyonkov@jinr.ru

Table 1. Production and radioactive properties [4] used for identification of nuclei.

Production properties[a]	**Radioactive properties**[a]
Energy of bombarding particles, C	Kind of decay, C
Cross section, C	Branching ratio, C
Yield curve, C,A(A,Z)	Half-life, C
Cross bombardments, C,A(A,Z)	Energy of α-particles, C
Angular distribution[b], A(Z)	Maximum energy of β-particles[c], C
Angular selection, A(Z)	Energy of γ-radiations[c], C
Mass separation[d], A(A)	X-ray spectrum (K or L)[c], C,A(Z)
Velocity filter (separator), A(Z)	Fission characteristics, C
Time of flight selection, A(Z)	Genetic relation between ancestor and n[th] generation descendant[b], C,A(Z)
Chemistry, A(Z)	

[a] C – characterization property, A – assignment property: A(Z) for Z, A(A) for A, and A(A,Z) for both.
[b] If combined with chemistry experiments.
[c] Was not applied.
[d] With limited mass resolution.

During 15 years of experiments with the DGFRS, six new superheavy elements located on the shore of the "Island of stability" and having atomic numbers 113 through 118 were synthesized for the first time. Radioactive properties of more than fifty new heaviest nuclides were explored in the experiments carried out in collaboration with the laboratories of the USA in Livermore (LLNL), Oak Ridge (ORNL), Knoxville (UT), Nashville (VU) and with Russian centers RIAR (Dimitrovgrad) and RIEPh (Sarov) [1–3].

Radioactive properties of the synthesized nuclides demonstrate a substantial increase of nuclear stability with increase of neutron number and closer approach to the predicted spherical shells $Z = 114$ (or at even higher Z, such as 120, 124 or 126) and $N = 184$. New observations establish a consistent pattern of nuclear properties in the area of the heaviest nuclides. They demonstrate the decisive role of nuclear shells and provide experimental proof of the existence of the predicted "Island of stability" of superheavy elements.

2. Experimental results

Decay chains of all of the nuclei synthesized hitherto in the ^{238}U–^{249}Cf+^{48}Ca reactions are terminated by spontaneous fission (SF). Thus, the region of SHN is not yet linked to known isotopes; here, the method of the detection of consecutive α decays that lead to the known nuclei can be applied for the identification of the new ones only after independent identification of at least one isotope of a given element in the decay chain, e.g., by chemical identification of its atomic number. However, in addition to this method, approximately 20 other criteria were developed by IUPAC/IUPAP in 1991 "that must be satisfied for the discovery of a new chemical element to be recognized" [4]. The criteria were subdivided into *Production properties* and *Radioactive properties*: "The first establishes physical and/or chemical properties of samples suspected of containing the new element and that are sufficient to categorize it. The second extends to properties that are used to demonstrate that the "characterization properties" are indeed those of an unknown element. Some properties can be used for both purposes." Most of these criteria demonstrate, beyond a reasonable doubt, the synthesis of the nuclides with atomic numbers $Z = 113$–118 (see Table 1).

Characteristics of the DGFRS that is used in the experiments meet the following requirements. The principle of operation of the separator is selection of products of the complete-fusion reaction by their charge state (q) in a rare gas and kinematic characteristics

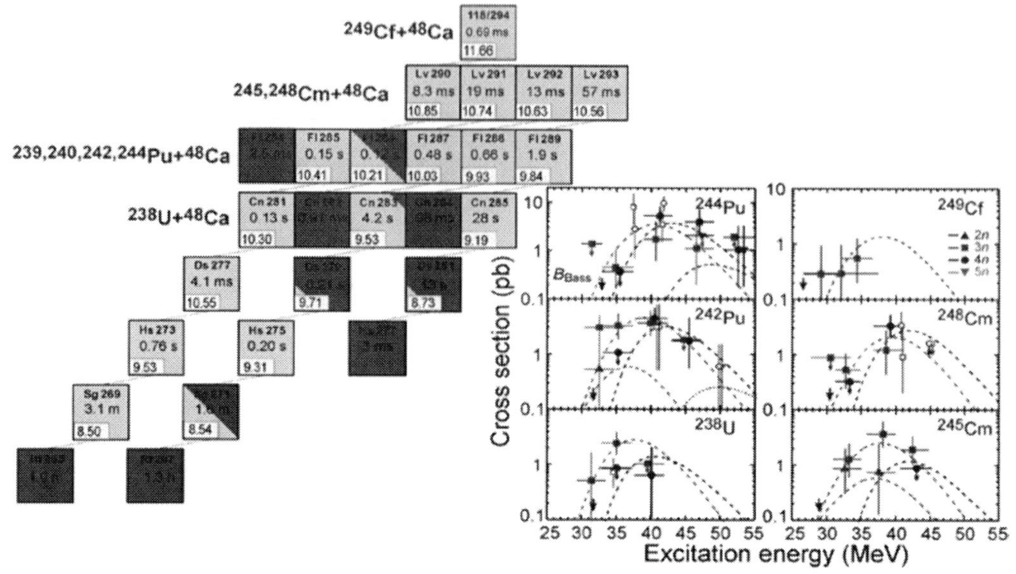

Figure 1. Left-hand side: summary of decay properties of isotopes of even-Z elements synthesized in the given reactions. The average energies of α particles and half-lives are given for α emitters (yellow squares). For spontaneously fissioning nuclei marked by green squares the half-lives are listed. Right-hand side: Excitation functions for the $2n$- to $5n$-evaporation channels (see explanations in text) from the complete-fusion reactions ^{238}U–^{249}Cf+^{48}Ca measured at DGFRS (solid symbols) and SHIP, BGS, and TASCA (open symbols). Dotted lines show results of calculations [7]. Vertical error bars correspond to statistical uncertainties of the DGFRS experiments and available data from other experiments. Upper cross-section limits are shown by arrows. Horizontal error bars represent the range of excitation energies corresponding to a given beam energy. For reference purposes we show the energy at the Bass barrier ($B_{\rm Bass}$) [8] (black arrow).

(mass of recoil nucleus (m) and its velocity (v)) in accordance with the separator magnetic rigidity $B\rho = mv/q$. Note that q depends linearly on v. These values are calculated for the xn-reaction channel when setting the separator's parameters. The setup (separator) strongly separates forward-peaked evaporation residues (ER), products of complete-fusion reactions, within a narrow angle $\pm 2 - 3°$ ("angular selection") with a huge suppression of the products of the transfer reactions and even incomplete fusion, e.g., αxn reactions ("mass separation" with limited resolution). The TOF selection in the existing separators may be substituted by the equivalent combined measurement of recoil energy and TOF. Note, the production properties discussed here were called "assignment properties" in [4].

The first superheavy nucleus ^{289}Fl was discovered in 1999 in the ^{244}Pu(^{48}Ca,3n) reaction studied at DGFRS (here and after we refer to reviews [1–3] containing references to most of earlier experimental data). The decay properties of ^{289}Fl and descendant nuclei ^{285}Cn, ^{281}Ds, and ^{277}Hs, later observed at the TASCA separator [5], are shown in Fig. 1.

With an increase of the excitation energy E^* of the compound nucleus (CN), the yields of different nuclides vary ("at increasingly higher energy values for increasing values of x" [4]). At the lowest E^* values, the nuclides with relatively lower α-particle energies and longer half-lives $T_{1/2}$ are produced (viz., ^{289}Fl). At larger projectile energies, the yield of these nuclei decreases, but the yield of other nuclides (^{288}Fl and ^{287}Fl), with higher E_α and shorter $T_{1/2}$ values (by about 0.1 MeV and by factor of about two, respectively, for parent nuclei), increases and then decreases again with the further increase of the excitation energy, see Fig. 1. The decay properties of the nuclides, consecutively appearing with the rise of E^*, are different (compare the E_α and $T_{1/2}$ values and decay modes of all of the nuclei in the chains in Fig. 1). In particular, the decay chains of even-Z, odd-N nuclei are longer indicating the hindrance of fission process as compared to the even-even nuclei decays. However, the difference in E_α and $T_{1/2}$ values of parent nuclei $^{287-289}$Fl is rather small; this observation excludes the assignment of the observed nuclides to differing types of reactions (e.g., considering that one isotope is produced in the *xn* channel and others – in the *pxn* or *αxn* channels).

The production cross-section values as well as energies of bombarding particles (see items in Table 1) are comparable for all of the studied reactions, which might indicate the identity of the mechanism of all of the reactions. In Fig. 1, the excitation energy values were calculated from mass tables [9]. One could use other mass tables which would result in a shift of all of the data points by a couple of MeV. However, the excitation energy relative to the Bass barriers (*a value independent of mass predictions*) is firmly fixed.

For synthesis of the lighter isotopes of Fl, the reactions with lighter target nuclei were studied at DGFRS [1–3]. In the ^{242}Pu+^{48}Ca reaction, the isotope ^{287}Fl was predominantly observed at low excitation energies while the lighter isotope (^{286}Fl) was obtained at higher projectile energies. Further increase of ^{48}Ca energy resulted in observation of the next isotope ^{285}Fl, first at BGS [10]. The same isotope was also produced at DGFRS in the reaction with a lighter Pu isotope as the product of the ^{240}Pu(^{48}Ca,3n)^{285}Fl reaction [11]. In the same experiment carried out at higher ^{48}Ca energy as well as in the ^{239}Pu+^{48}Ca reaction, a spontaneously fissioning nuclide was observed and assigned to the new isotope ^{284}Fl.

The excitation functions (item "yield curve" in Table 1) have been measured for the reactions with ^{244}Pu and ^{242}Pu in the E^* interval of about 32–53 MeV and at $E^* = 39$ and 43 MeV in the reaction with ^{240}Pu (Fig. 1), which, together with decay properties of the observed nuclides (items "kind of decay", "branching ratio", "half-life", and "α-particle energy" in Table 1), demonstrate *the production of the neighbouring isotopes of the same element* in each reaction.

Two Fl isotopes as well as their descendant nuclei were observed in cross bombardments ^{244}Pu(^{48}Ca,5n)^{287}Fl, ^{242}Pu(^{48}Ca,3n)^{287}Fl, ^{242}Pu(^{48}Ca,5n)^{285}Fl, and ^{240}Pu(^{48}Ca,3n)^{285}Fl. Moreover, Cn isotopes ^{282}Cn and ^{283}Cn, daughter nuclei of ^{286}Fl and ^{287}Fl, were produced in another cross-bombardment reaction ^{238}U+^{48}Ca. This proves that these nuclei were produced in *the same type of reaction – xn, pxn, αxn*, and so on. Indeed, with increase of the atomic (by two protons, e.g., from ^{238}U to ^{242}Pu or ^{244}Pu) and mass numbers of the target nuclei, heavier parent nuclides are produced; their α decay leads to nuclei that were synthesized in reactions with lower-Z and lighter target nuclei. Another example of the cross bombardments is the observation of the same nuclides in reactions with target isotopes of the same Z but with different mass numbers. If the target isotopes differ by several neutrons, e.g., ^{242}Pu and ^{244}Pu (or ^{240}Pu and ^{242}Pu), then the observed parent nuclide, e.g., ^{287}Fl (or ^{285}Fl), can be produced in the 3n-evaporation channel of the reaction, with ^{242}Pu (or ^{240}Pu) at a lower excitation energy, and in the 5n channel of the reaction, with ^{244}Pu (or ^{242}Pu) at higher excitation energy. Here, the identity of the decay properties observed in cross bombardments definitely proves that the same nucleus was produced in two reaction channels that differ merely by the number of evaporated neutrons.

The first nucleus of element 116, ^{293}Lv, was synthesized at DGFRS in 2000 in the ^{248}Cm+^{48}Ca reaction [1–3]. This isotope was observed on the low-energy side of the excitation function. Increasing the excitation energy of the CN resulted in the observation of two isotopes: one already observed at lower energy ^{293}Lv and a new lighter isotope ^{292}Lv. Two even lighter isotopes, ^{291}Lv and ^{290}Lv, were produced in the reaction of ^{48}Ca with the lighter target isotope ^{245}Cm followed by evaporation of two and three neutrons: ^{245}Cm(^{48}Ca,2n)^{291}Lv and ^{245}Cm(^{48}Ca,3n)^{290}Lv. The decay properties of descendants of the

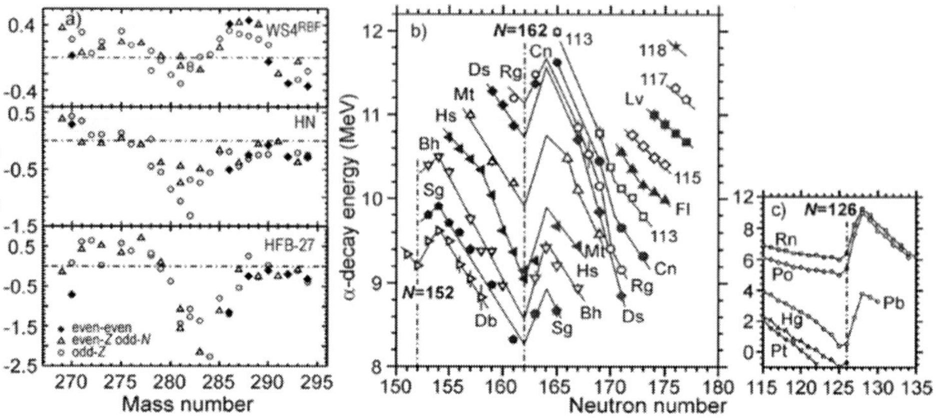

Figure 2. (a) Deviation between experimental and calculated (see Fig.12 in [3]) α-decay energies for even-even, even-Z and odd-N, and odd-Z nuclei. (b) Measured α-decay energy vs. neutron number for the isotopes of elements 106–118 (filled and open symbols refer to even-Z and odd-Z nuclei, respectively; Q_α values for nuclei produced in the Ra-Cf+^{48}Ca reactions are shown in red; other data (blue symbols) are taken from [12]. The lines are drawn to guide the eye. (c) The Q_α values for isotopes of even-Z elements Pt-Rn [12] are shown for comparison.

isotopes ^{290}Lv-^{293}Lv were in full agreement with those found in the ^{244}Pu+^{48}Ca, ^{242}Pu+^{48}Ca, and ^{238}U+^{48}Ca cross-bombardment reactions.

The first decay chain of element 118 was found in 2002 in the ^{249}Cf(^{48}Ca,3n)^{294}118 reaction studied at DGFRS [1–3]. Here one should note that the descendant SF isotope ^{282}Cn was produced in the following cross reactions: ^{238}U(^{48}Ca,4n)^{282}Cn, ^{242}Pu(^{48}Ca,4n)^{286}Fl-(α)-^{282}Cn, ^{245}Cm(^{48}Ca,3n)^{290}Lv-(α,α)-^{282}Cn, and ^{249}Cf(^{48}Ca,3n)^{294}118-(α,α,α)-^{282}Cn.

Assignment of all of the synthesized nuclides to the products of the xn-reaction channel is justified by characteristics of the DGFRS, excitation-function measurements, production of most of the nuclides in cross bombardments, as well as by decay properties of nuclei, especially their measured α-particle energies [see Fig. 2(b)].

The population of the relatively high-energy levels in the parent nucleus followed by α decay to the low-energy levels or the ground state of the daughter nucleus leads to a higher α-particle energy E_α than that corresponding to the ground-state-to-ground-state α-decay energy Q_α. However, the following α decay starts from a low-energy level. Thus, the observation of decays from high- to low-energy levels for all of the nuclei in the long decay chains (e.g., ^{285}Fl, ^{291}Lv, or odd-Z nuclei, see below) is absolutely excluded. Vice versa, α decay can go to excited states of the daughter nucleus, followed by the emission of γ rays, which results in a lower E_α than that corresponding to the Q_α value. The even-Z isotopes investigated so far do not indicate α decays from the excited (isomeric) states (see Fig. 1 and [2, 3]); their α-particle energies are identical and form single peaks in both cases – after production of nuclei in primary reactions and after α decays of parent nuclei (with several lower-energy peaks observed for even-odd ^{291}Lv, ^{289}Fl, and ^{283}Cn). Calculation of the Q_α values from α peaks appears to be reasonable for even-even isotopes as well as for even-Z odd-N nuclei because the energies of odd isotopes have intermediate values between those of even-even nuclides, which is in agreement with the observations for the lighter nuclei (see Figs. 21–26 in [12]).

An important feature observed in these experiments (see Fig. 2) is the fact that $Q_\alpha(N)$ systematics for isotopes of neighbouring elements never intersect (see also Figs. 25–26 in [12]). Moreover, the Q_α values for isotopes of Sg and Hs perfectly follow the trend

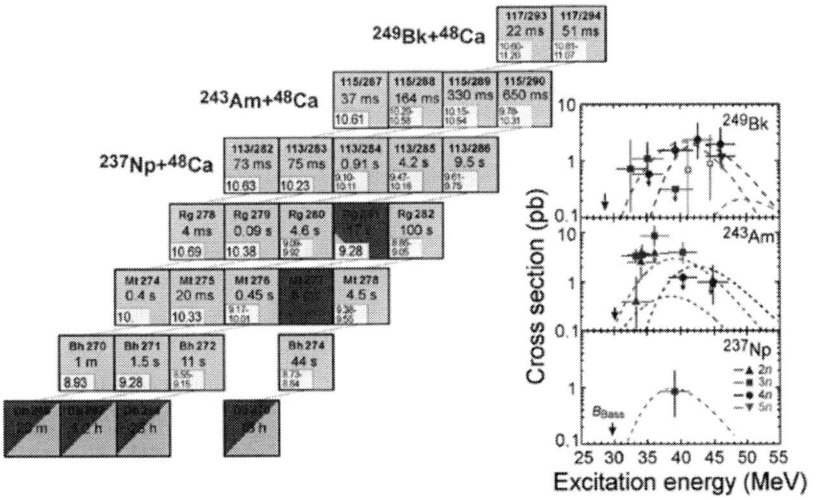

Figure 3. Same as Fig. 1, but for the isotopes of odd-Z elements. For Db isotopes only SF was observed however, electron capture (EC) cannot be excluded.

of variation of $Q_\alpha(N)$ at the crossing of the magic neutron number $N = 162$ (identical to what is observed for the isotopes of Ds and Rg as well as for numerous nuclides with $N \approx 152$ and 126, see Fig. 2c. Therefore, the systematics of α-decay energies of even-Z elements and their descendants provides proof of identification of the Z and N of these elements. In addition, the mass numbers of nuclei can be easily established by considering their decay properties (e.g., SF is much more probable for even-N and/or Z isotopes, see Figs. 1 and 3 below) and excitation functions (e.g., products of the ^{244}Pu+^{48}Ca might be assigned to the 1n- to 3n-evaporation channel instead of 3n- to 5n, but in this assumption, the product of the ^{245}Cm(^{48}Ca,2n)^{291}Lv reaction should originate from the radiative-capture channel, which was not registered even in the cold-fusion reactions of heavy ions with ^{208}Pb and ^{209}Bi).

Furthermore, the chemical characterization of Cn, based on the direct comparison of the adsorption characteristics of atoms of ^{283}Cn to that of Hg and Rn, was obtained. The results of these experiments "establish element 112 as a typical element of group 12" [15]. Thus, the atomic numbers of all of the isotopes in the decay chain, starting with ^{291}Lv and ending with ^{267}Rf, were determined by the genetic relation (Table 1).

In all of the reactions with even-Z target nuclei, two to three different parent nuclides were produced in the 2n- to 5n-evaporation channels, except for the ^{249}Cf+^{48}Ca reaction which was studied at $E*$ values lower than those corresponding to the predicted cross-section maximum. Thus, it would be reasonable to expect similar observations for the reactions with odd-Z target nuclei. With the aim of synthesis of isotopes of elements 113, 115 and 117 three reactions using odd-Z actinide target materials were investigated: ^{237}Np, ^{243}Am, ^{249}Bk+^{48}Ca [1–3] (see Fig. 3).

The heaviest odd-Z nucleus was synthesized in the ^{249}Bk(^{48}Ca,3n)^{294}117 reaction at low projectile energies in 2009 [2, 3]. At higher excitation energy we registered shorter decay chains of the even-N isotope ^{293}117; each consisted of three consecutive α decays terminated by SF of ^{281}Rg or 5-ms ^{277}Mt instead of α decay observed for their odd-N neighbours ^{280,282}Rg and ^{276,278}Mt. The nuclei ^{277}Mt ($N = 168$) and ^{281}Rg ($N = 170$) belong to the region with the lowest stability with respect to SF observed for even-Z nuclei (Fig. 1). Accordingly, even the high hindrance gouverned by the characteristic of odd particles does not "save" these odd nuclei (^{281}Rg, ^{277}Mt) from SF, which is caused by the weakening of the stabilizing effect of the neutron shells at $N = 162$ and $N = 184$ (see Refs. in [1–3]).

The observations of elements 113 and 115 were first reported in 2003–2004 [1–3]. In the reaction ^{243}Am+^{48}Ca, these two new elements were simultaneously synthesized for the first time. In 2010–2012, new series of experiments with ^{243}Am were performed at DGFRS aimed at the measurement of the excitation function in a wider energy range. At the lowest ^{48}Ca energies, we detected decay chains consisting of two consecutive α decays ending in SF. Similar α-α-SF decay chains were observed for descendant nuclei in the chain starting with the isotope ^{293}117. Therefore, assignment of these chains to the product of the $2n$-reaction channel, ^{289}115, seems to be the most reasonable. Note, the product of the $2n$-reaction channel was evidently observed with the same yield in the ^{245}Cm(^{48}Ca,$2n$)^{291}Lv reaction [1–3]. At higher projectile energies, decay chains of ^{288}115, the product of evaporation of three neutrons, were registered that undergo five α decays followed by SF. At the highest bombarding energy, we detected decay chains of ^{287}115.

The same decay chains were later produced in experiments with ^{243}Am which were performed at TASCA [16] and BGS [17]. In [18], the summary of decay times of nuclei in the short decay chains were re-analyzed which led to the interpretation that some of these chains might start from the isotope ^{288}115 and proceed through either SF or electron capture decay branches of ^{284}113 and ^{280}Rg. However, authors concluded that "clearly, more ... data ... are needed to verify *any* proposed decay scenario of ^{288,289}115".

For investigation of the region of neutron-deficient odd-Z SHN, we studied the ^{237}Np(^{48}Ca,$3n$)^{282}113 reaction. The decay properties of this isotope and its descendant nuclei ^{278}Rg, ^{274}Mt, ^{270}Bh and ^{266}Db are in agreement with those expected from the extrapolation of the radioactive decay properties of the heavier neighbouring isotopes ^{283,284}113 and, respectively, their descendants.

Similar to even-Z elements, the assignment of all of the odd-Z nuclides to the products of the xn-reaction channel is based on characteristics of the DGFRS, excitation-function measurements for the reactions with ^{243}Am and ^{249}Bk ($2n$- to $4n$-evaporation channels were observed), and production of ^{289}115 in cross bombardments, as well as decay properties of nuclei. First of all, the decay properties of all of the odd-Z nuclei evidently differ from those observed for $^{290-293}$Lv, $^{285-289}$Fl, etc. (potential products of the pxn-reaction channels or EC of parent nuclei).

For odd-Z nuclei, the α spectra are more complex, which makes estimations of Q_α more difficult. However, the calculation of their α-decay energies from the highest measured α-particle energies [1–3] appears to be justified. For example, the Q_α values for odd-odd nuclei ^{288}115-^{272}Bh, estimated or measured from the observed α-γ coincidences [16, 17] are in good agreement with such calculations. For all of the nuclides, these values are lower than those from [16, 17] by 0.03–0.24 MeV [3], which might indicate that transitions occur to the low-lying exited states of the descendants. Thus, the calculated Q_α values for odd-Z nuclei (Fig. 2) can represent their lower limits, which can differ from the exact Q_α values by a few hundred keV at most. The systematics shown in Fig. 2b demonstrates that the α-decay energies of all of the odd-Z nuclei are located in between and are in good agreement with the interpolations/extrapolations from neighbouring even-Z nuclides with ($Z \pm 1$) and the same neutron number. The Q_α values for Bh isotopes are located between $Q_\alpha(N)$ systematics for Sg and Hs and perfectly correspond to the pattern of variation of $Q_\alpha(N)$ for Bh isotopes at the crossing of $N = 162$.

Moreover, the terminal SF nuclide in the decay chain of ^{288}115 produced in the ^{243}Am+^{48}Ca reaction was chemically characterized as a transactinide element [1–3]. In these experiments, the products from the irradiation of ^{243}Am by ^{48}Ca projectiles, having the same

^{48}Ca energy as in the DGFRS experiments, were collected within a larger acceptance angle of $\pm 12.5°$ that would result in increase of collection efficiency for transfer-reaction products by a factor of more than twenty. Nevertheless, after chemical extraction of transactinide elements, the SF nuclide was observed with the same decay properties – decay mode, half-life, and total kinetic energy (TKE) – and with the same cross section as at the DGFRS. All of these factors allowed for the conclusion that the same isotope was observed in both the physical and chemical experiments [1–3]. Simultaneously, all of the precursors (Z = 107, 109, 111, 113, and 115) discovered in [1–3] were identified by establishing the genetic link between the ancestor and the descendant.

In addition one could mention that the properties of α-decaying even-even SHN are in agreement with the empirical Geiger-Nuttall relationship (see Fig. 26 in [1]). For SF nuclei it is observed that, with the transition $Z \geq 110$, the TKE increases with increasing Z in agreement with the previously established dependence of TKE vs. parameter $Z^2/A^{1/3}$, typical for the asymmetric fission of lighter nuclei (see Fig. 29 in [1]).

Therefore, conformity with all of the above-mentioned criteria demonstrates that the observed SHN originate from the complete-fusion reactions followed by the evaporation of the neutrons. Furthermore, the decay properties of ^{294}117, $^{291-293}$Lv, $^{287-289}$115, $^{285-289}$Fl, and ^{283}Cn and their descendants were determined in different laboratories with use of different techniques (separators DGFRS, SHIP, TASCA, BGS, GARIS, and chemistry setups) that is in accordance with demand of reproducibility for recognition of the discovery of new elements (see [1, 3, 5, 10, 11, 15–18], references therein, and proceedings of this Symposium).

3. Discovery of superheavy nuclei

As a result of the experiments performed with use of a beam of ^{48}Ca ions, the heaviest elements with atomic numbers 113–118 were synthesized. These elements fill the seventh period of the table by D.I. Mendeleev. In these investigations, more than fifty new nuclides, isotopes of elements 104–118 having the largest number of neutrons, were produced for the first time. Their decay properties were determined. The chart of the nuclei was essentially extended up to the nuclides with $Z = 118$ (^{294}118) and $N = 177$ (^{294}117 and ^{293}Lv).

These nuclei demonstrate amazing vitality. The increase of the neutron number in nuclei with $N \geq 165$ results in a decrease of the Q_α energy (Fig. 2) and a considerable increase of their half-lives (Fig. 4). An especially strong growth of $T_\alpha(N)$ with increasing N is observed for the isotopes of elements 109–113.

The maxima of the total cross sections, σ_{xn}, for the fusion-evaporation ^{48}Ca-induced reactions reach about 10 pb for $Z = 114$ and 115 nuclei, exceeding the values from extrapolation of σ_{xn} systematics for hot-fusion reactions by about three orders of magnitude. Such relatively high cross sections are caused by the large survivability of the CN, which is directly related to the high fission barriers in SHN that appear due to nuclear shell effects.

Even-even isotopes of Cn and Fl with $N = 170$ and 172 are located in a region where a steep rise of $T_{SF}(N)$ is predicted. The difference of two neutrons in these isotopes increases the SF partial half-life by two orders of magnitude. Similarly, the addition of two neutrons to the nucleus ^{286}Fl leads to an increase of the stability of ^{288}Fl with respect to SF by at least a factor of fifteen. For heavier even-even nuclei, ^{288}Fl, ^{290}Lv, ^{292}Lv, and ^{294}118, SF was not detected due to the more considerable rise of stability with regard to SF compared to α decay, with the neutron number approaching the magic number $N = 184$. On the other hand, the considerable drop of the evaporation cross sections observed for the reactions with neutron-deficient isotopes ^{239}Pu and ^{240}Pu (by a factors of about fifty and four, respectively,

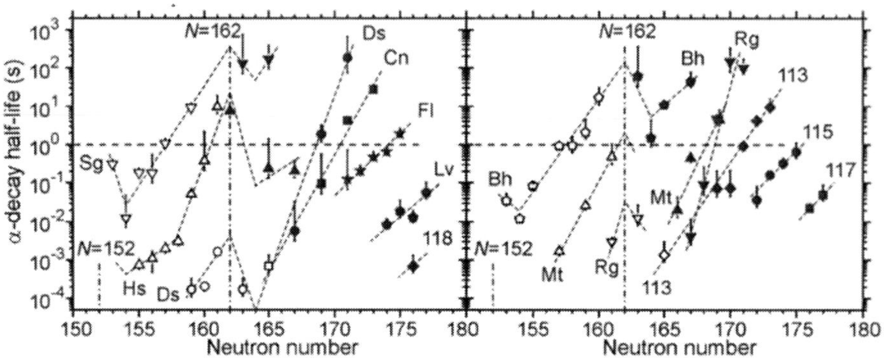

Figure 4. Half-lives vs. neutron number for the isotopes of even-Z (left panel) and odd-Z (right panel) elements with $Z = 106$–118 [3] (results from Ra-Cf+^{48}Ca reactions are shown by full blue symbols; other data are taken from [12]). Lines are drawn to guide the eye.

in comparison with that for the reaction with the heavier isotope ^{244}Pu [11]) indicates one is approaching the neutron-deficient border of stability of SHN.

In spite of the disagreement between different theoretical models in predicting the proton magic number which varies from 114 to 126 for SHN, the resulting calculated α-decay energies, Q_α, which are determined by nuclear masses, do not strongly deviate from experimental values. In Fig. 2a, the deviation between experiment and theory, $\Delta Q_\alpha = Q_\alpha^{exp} - Q_\alpha^{\text{th}}$, for even-even nuclei in all of the three cases is within approximately ±0.5 MeV; however, the HFB-27 data systematically exceed the experimental Q_α values by approximately 0.5 MeV. The production cross-section values, calculated within different models (see Figs. 1, 2 and references given in [1–3]), as well as SF lifetimes of SHN (see, e.g., Fig. 5 in [11]) which are largely dependent on fission barriers of nuclei, are in reasonable agreement with experimental data. Therefore, the fundamental outcomes of the microscopic theory concerning the predictions of the "Island of stability" of SHN were validated by experimental evidence.

We acknowledge the support by the Russian Foundation for Basic Research, including recent Grants Nos. 13-02-12052, 13-03-12205, and 16-52-55002. Research at ORNL was supported by the U.S. DOE Office of Nuclear Physics under DOE Contract No. DE-AC05-00OR22725 with UT-Battelle, LLC. Research at LLNL was supported by LDRD Program Project No. 08-ERD-030, under DOE Contract No. DE-AC52-07NA27344 with Lawrence Livermore National Security, LLC. This work was also supported by the U.S. Department of Energy through Grant No. DE-FG-05-88ER40407 (Vanderbilt University). Most of the actinide target materials used in the DGFRS experiments were made at ORNL and provided through the DOE Isotope Program and at RIAR.

References

[1] Yu.Ts. Oganessian, J. Phys. G: Nucl. Part. Phys. **34**, R165 (2007).
[2] Yu.Ts. Oganessian and V.K. Utyonkov, Rep. Prog. Phys. **78**, 036301 (2015).
[3] Yu.Ts. Oganessian and V.K. Utyonkov, Nucl. Phys. **A944**, 62 (2015).
[4] A.H. Wapstra, Pure Appl. Chem. **63**, 879 (1991).
[5] Ch.E. Düllmann *et al.*, Phys. Rev. Lett. **104**, 252701 (2010)
[6] J.M. Gates *et al.*, Phys. Rev. C **83**, 054618 (2011).
[7] V.I. Zagrebaev and W. Greiner, Nucl. Phys. **A944**, 257 (2015).

[8] R. Bass, in: W. von Oertzen (ed.), Proc. Symp. *Deep Inelastic and Fusion Reactions with Heavy Ions*, in: Lecture Notes in Physics, vol. 117 (Springer-Verlag, Berlin, 1980) p. 281.
[9] W.D. Myers and W.J. Swiatecki, Nucl. Phys. **A601**, 141 (1996).
[10] P.A. Ellison *et al.*, Phys. Rev. Lett. **105**, 182701 (2010).
[11] V.K. Utyonkov *et al.*, Phys. Rev. C **92**, 034609 (2015).
[12] M. Wang *et al.*, Chin. Phys. **36**, 1603 (2012).
[13] G. Audi *et al.*, Chin. Phys. **36** 1287 (2012).
[14] G. Audi *et al.*, Chin. Phys. **36** 1157 (2012).
[15] R. Eichler *et al.*, Nature **447**, 72 (2007).
[16] D. Rudolph *et al.*, Phys. Rev. Lett. **111**, 112502 (2013).
[17] J.M. Gates *et al.*, Phys. Rev. C **92**, 021301(R) (2015).
[18] U. Forsberg *et al.*, Nucl. Phys. **A953**, 117 (2016).

EPJ Web of Conferences **131**, 06004 (2016) DOI: 10.1051/epjconf/201613106004
Nobel Symposium NS160 – Chemistry and Physics of Heavy and Superheavy Elements

Validation of new superheavy elements and IUPAC-IUPAP joint working group

Cecilia Jarlskog[a]

Div. Math. Phys., Physics Department, LTH, Lund University, Sweden

Abstract. The great chemist Glenn Seaborg has written a delightful little book "Man-made Transuranium Elements", published in 1963, in which he points out that: "The former basic criterion for the discovery of a new element – namely, chemical identification and separation from all previously-known elements – had to be changed in the case of lawrencium (element 103). This also may be true for elements beyond lawrencium."

Indeed this is what has happened. The elements with $Z \geq 103$ are produced in nuclear reactions and are detected by counters. The detectors have undergone substantial refinement. For example one uses multiwire proportional chambers [for which George Charpak received the 1992 Nobel Prize in Physics] as well as solid state micro-strip detectors. In spite of this remarkable shift from chemistry to physics, the managerial staff of the International Union of Pure and Applied Chemistry (IUPAC) does not seem to be aware of what has been going on. The validation of superheavy elements should be done by physicists as the chemists lack the relevant competence as I will discuss here below.

This article is about a collaboration between International Union of Pure and Applied Chemistry (IUPAC) and its sister organization International Union of Pure and Applied Physics (IUPAP), to deal with discovery of superheavy elements beyond $Z = 112$. I spent a great deal of time on this issue. In my opinion, the collaboration turned out to be a failure. For the sake of science, which should be our most important concern (and not politics), the rules for the future collaborations, if any, should be accurately defined and respected. The validation of new elements should be done by people who have the relevant competence – the physicists.

1. Introduction in a nutshell: Physicists discover – IUPAC gets the credit

The organizers of this Nobel Symposium have asked us, the speakers, to be

"brave, simple & bold"

and that is what I intend to be, hoping for a better future.

My talk concerns the latest IUPAC-IUPAP Joint Working Group (JWG) which was created in 2011 to deal with issues concerning discoveries of superheavy elements beyond element 112. As the president-designate of IUPAP, and soon-to-become its president, I was asked to negotiate with IUPAC to establish the Group. Being naive and believing that scientists are impartial and logical I thought the task was going to be easy. I could not imagine that there would be "political" aspects in it. It turned out that the managerial staff of IUPAC

[a] e-mail: cecilia.jarlskog@matfys.lth.se

cared primarily about getting credit rather than being fair and giving credit to those who deserved it. This behaviour is, in my opinion,

old-fashioned, wrong, bad, and immoral.

I wish to emphasize that I am not talking about the scientists, the chemists in IUPAC, but about their managerial staff.

After starting my work I realized that the responsibility put on my shoulders was heavier than I could have imagined. While discussing this matter in IUPAP, Alan Astbury (who served IUPAP for nine years, 2002–2011, as president-designate, president, and former president), in response to an email by me, sent the following lines in a message addressed to several of us (in September 2011, i.e., about two months before his appointment in IUPAP ended):

> "... For historical reasons the Chemists cling to this process dearly. Judy and I spent hours on this, and look what we achieved!!"

Here Judy refers to Judy Franz who was the Executive Officer of the American Physical Society (APS). She was also the Secretary-General of IUPAP. Alan had a very high opinion of Judy and their collaboration was very successful. When I was asked to take charge of negotiations with IUPAC the office had already moved from APS to Institute of Physics (IOP) in London. People I consulted would tell me that chemists are "impossible to deal with" (especially the managerial staff of IUPAC, who were in addition arrogant and didn't answer emails). I simply could not believe this as I had met many chemists, for example at the Royal Swedish Academy of Sciences where I had been a member since 1984, i.e., by then for more than 27 years, and had found chemists to be just fine. So what was going on? It was clear that the roots of the problem were deep.

2. Discoveries of superheavy elements fall into domain of nuclear physics

Research papers on discovery of superheavy elements are almost all published in physics journals such as Physical Review Letters, Physical Review, Europhysics Journal, and so on. One may sometimes find a few articles in the Journal of Pure and Applied Chemistry, but these are in general not written by discoverers but by managerial staff of IUPAC or a working group reporting on what others have done.

The leading role of physicists in the discovery of the superheavy elements is not a new phenomenon. In 1951 the Nobel Prize in Chemistry was awarded jointly to Edwin Mattison McMillan and Glenn Theodore Seaborg. McMillan, who in 1940 had discovered the first transuranium element, declared in his Nobel Lecture:

> "in spite of what the Nobel Committee may think, I am not a chemist."

He had used the cyclotron built by the 1939 Nobel Laureate in Physics, Ernest Orlando Lawrence, who was awarded the Prize

> "for the invention and development of the cyclotron and for results obtained with it, especially with regard to artificial radioactive elements".

The Prize was also justified by the fact that

> "... only with the cyclotron was it possible to produce active isotopes in large quantities. This was, among other things, an essential condition for the use of active elements for biological and medical purposes."

Glenn Seaborg who succeeded McMillan, and was involved in discovery of about ten transuranium elements, described in his Nobel lecture the research environment at Berkeley as follows:

> "The work on the transuranium elements which I am about to describe was made possible by this circumstance and because of the unusual and excellent spirit of cooperation which exists at the Radiation Laboratory, under the direction of Professor E.O. Lawrence, where the cyclotron bombardments so crucial to the final results were made."

In an interview Seaborg has said that he loved physics but became a chemist because there were more jobs for chemists than for physicists. The important point is that there was an admirable spirit of collaboration among these great scientists, irrespectively of which department they belonged to.

As the years passed, elements up to and including $Z = 101$ were produced using neutron, deuteron, or He ions from the cyclotron, with the exception of $Z = 99$ and $Z = 100$, which were discovered in thermonuclear (bomb) explosions. Elements beyond $Z = 101$ have been produced via collision of heavier projectiles such as carbon and calcium or zinc ions with transuranium elements or lead or bismuth, respectively. Seaborg expected transuranium elements to play an important role in energy production. However, this has not happened yet. Elements beyond $Z = 100$ are produced in very small amounts in particle accelerators and decay rapidly. However, getting to know their properties is important for basic science.

3. Why does it matter that the physicists get the credit they deserve?

The area of superheavies is without any doubt very interesting. Physicists, whenever possible, try to study physics under extreme conditions hoping to learn something new and interesting about nature. That is why some profitless-looking enterprises are extremely interesting, such as studying black holes, even though we "pray" and hope that we will never meet any of them. Black holes constitute a laboratory for testing some fundamental ideas, such as the information paradox.

Superheavy elements fall into a more tangible extreme domain of physics. An obvious question is whether we will reach the so called, up to now hypothetical, island of stability, inhabited by long-lived superheavies, and if so what will we learn? I remember that many years ago there was a proposal called Lee-Wick vacuum, named after two outstanding theorists, T.D. Lee (Nobel Laureate in Physics, 1957) and his colleague Gian Carlo Wick. These authors were concerned with the structure of the "vacuum", the "nothing" which all along has enjoyed enormous popularity among theorists as this "nothing" may actually be something subtle. The familiar hypotheses of ether and of the Dirac sea, are two early versions of non-trivial vacua. In the Lee-Wick model, the "hyper-superheavies" were predicted to form not just an island but a continent. The nuclear matter was thus predicted to be highly "abnormal".

The superheavies, beyond element $Z = 100$, that have been discovered up to now, are short-lived. However, they all have isotopes, many of them have been discovered but many more have not been accessible. Generally, one expects some neutron-rich isotopes to have much longer lifetimes and perhaps take us to the so called "island of stability". This makes the physics of the superheavies particularly interesting. Moreover, the decays of these objects involve a host of nuclear phenomena (α, β, γ decays as well as electron capture and spontaneous fission) that provide information about nuclei. The study of electronic structure of the superheavies is of interest to atomic physicists. However, these facts are not generally appreciated.

Nowadays, in several countries, the inhabitants are afraid of the word "nuclear". They do not realize that almost 100% of their own mass comes from the nuclei of hydrogen and heavier elements in their bodies. This fear has sometimes had ridiculous consequences. For example, in Sweden at hospitals the letter N in NMR (nuclear magnetic resonance) has been removed. One uses instead the acronym MR which people erroneously call "magnetic Röntgen", which means "magnetic X-ray". Another acronym used is MI (magnetic imaging). In fact research in nuclear physics is not rated highly. When the researchers apply for funding to produce and study superheavy elements, there is an eminent danger that the evaluators at funding agencies do not understand or appreciate their work and therefore discriminate against them. After all this field is rather "narrow" and cannot count on having supporters on the relevant panels. Superheavy hunters tend not to get the beam time that they deserve. In some cases, their project is canceled due to "funding priorities". Being **bold** I would add:

It certainly does not help that in addition IUPAC steals credit from them!

4. IUPAC-IUPAP joint working group and what went wrong

Some facts about early history of the IUPAC-IUPAP collaboration on transuranium elements can be found in a report by Barber *et al.* [Prog. Part. Nucl Phys., Vol. 29 (1992) pp. 453]. According to this report, the first attempt to form such a collaboration was made in 1974. A committee of nine people was appointed but according to the report "This committee never completed its work, nor issued a report nor, indeed, met as a group, though there were some correspondence between members". The next round started in 1985. I will not go into what happened but have to say, with great regret, that some of the leading scientists in this committee have already passed away, among them its chair Denys Wilkinson. From the statement made by Alan Astbury, that I quoted earlier ["... For historical reasons the Chemists cling to this process dearly. Judy and I spent hours on this, and look what we achieved!!"] I conclude that there were problems with the collaboration as recently as in 2011. Sadly, also Alan has passed away. In IUPAP, we have been particularly unfortunate when it comes to our history. We have had no permanent office anywhere. I have contacted a few people but nobody seems to know what was going on, further back in time than 1970s.

In 1987 a new Joint Committee was created which had seven members nominated by IUPAP [D.H. Wilkinson (UK) Chairman, R.C. Barber (Canada), A. Hrynkiewicz (Poland), M. Lefort (France), M. Sakai (Japan), I. Ulehla (Czechoslovakia), A.H. Wapstra (Netherlands)] and two nominated by IUPAC [N.N. Greenwood (UK), Y.P. Jeannin (France)]. The Committee took its task seriously and produced the Barber *et al.* report quoted above. Note that at that time none of these members was from a claimant country. This was supposed to be a rule to be followed. Unfortunately, this rule was violated later on.

Taking a leap in time, the 2011 IUPAP General Assembly took place in London, 2–4 November. Two documents on IUPAC-IUPAP collaboration were presented to the delegates at this meeting. One was the request from the Inter-Union Group that the General Assembly approve the names and symbols for elements 110, 111 and 112. The second document was called *Joint IUPAC/IUPAP Working Groups to Validate Claims for the Discovery of New Elements*. It was information on *Guidelines for the Appointment of Members* as well as a list of names of the nominees from IUPAC and IUPAP. In retrospect, this document was already obsolete when it was presented.

The IUPAP members of the latest Joint Working Group were suggested by the IUPAP Commission C12 – Nuclear Physics, at their annual meeting at MIT on 25 July 2011! IUPAP followed their recommendation as the Nuclear Physics Commission is the only one in IUPAP which has the relevant competence. It would have been "silly" if such a recommendation had come from the managerial staff! The guidelines included the statements that the members should not come from a claimant country and that each Union be represented by at least two members. And then just about everything went wrong.

Committing errors is said to be human, but repeating them is stupid. Instead one should learn from them, in order to avoid them. When charged with the above duty I took for granted that the two Unions will collaborate on **all issues** having to do with the discovery of superheavy nuclei, i.e., also on validations, and making announcements. Everyone agrees that the discovery of the new superheavy elements, a process which takes ages, lies in the "camp of physicists". Once discovered, the atomic physicists and chemists can deal with the identification of the electronic structure of these elements. As noted earlier, our physicists deserve and need to be publicly acknowledged for their achievements. However, it has turned out that the IUPAC managerial staff has gone behind our back and violated the "ethical rules", in spite of not having the competence to validate the discoveries.

In connection with the latest announcements by IUPAC, it is said that IUPAC collected referee reports but lacking the sufficient competence how can you choose appropriate referees and how can you evaluate their reports? We in IUPAP have been kept out of this process; we have not seen these reports. However, I know from private sources that they were not all positive. A major idea was to avoid having members from claimant countries (Germany, Japan, Russia, and USA). This was not followed. In fact IUPAC took the liberty of proposing a chemist from the USA as the chair! This worried me. I contacted Alan Astbury who wrote back to me "I presume he [meaning the Chair] feels that being God-like he can rise above all conflicts." This turned out to be very unfortunate. I should have protested. I was too naive to take for granted that the two sister Unions would operate together, in the spirit of heroes of the past such as Lawrence and Seaborg. The Chair was reluctant to communicate. As an example, at the end of his report to the 2014 IUPAP General Assembly in Singapore (which was supposed to be on behalf of the Joint Working Group) he had added:

> "Disclosure: This informal progress statement has been prepared entirely by the Chair of the JWP and has not been reviewed by the membership."

I do not know what was going on. The great surprise was that IUPAC, unexpectedly and without consultation with us, released the results of "our" Joint Working Group on 30 December 2015 (note the date!). On that occasion one of the members of the Working Group sent me the following message:

> "... I know you are now Past President of IUPAP, but I am writing because we corresponded about the IUPAP/IUPAC Joint Working Group in the past. As you know our reports on elements 113, 115, 117, and 118 have been accepted for publication in Pure and Applied Chemistry as is the case for past reports. Yesterday IUPAC released a press release, which I have attached. I am just letting you know in case there is anything IUPAP should do. My apologies in case I should have done anything sooner to let you know this was happening. I was not sure of the timing of the publication and thought it was going to be next year. **I also had assumed the IUPAC and IUPAP were working jointly on this** and there might be a joint announcement. ..." [my bold-facing].

This is not a good sign. It only shows that there has been something fundamentally wrong in this Inter-Union Collaboration. Not even (all) the members of the committee were informed about what was going on!

5. Looking ahead

Times have changed and the old rules and guidelines are no longer appropriate. As Seaborg predicted, more than 50 years ago, the former basic criterion for the discovery of superheavy elements moved a long time ago from chemistry to physics. Superheavies are created by physicists at accelerators and are identified by using physical measuring devices. Only physicists have the competence to check the validity of the results. It is natural that IUPAP should take the leading role in this enterprise.

Another point of concern is that the managerial staff of IUPAC as well as IUPAP keep on changing. Even if say the two presidents would happen to know each other well and trust each other their terms are limited to two years in IUPAC and three years in IUPAP. The superheavies take much longer time to discover. Moreover, the presidents, and other managerial staff, should humbly be aware of the fact that they do not have the competence to pass judgment on superheavies.

We all hope that new superheavy elements will be found. If so, IUPAC and IUPAP may decide to form a new Joint Working Group, even though the contribution of IUPAC cannot possibly be significant. IUPAP, through its Commissions C2 (Commission on Symbols, Units, Nomenclature, Atomic Masses and Fundamental Constants) and specially C12 (Nuclear Physics) possesses sufficient competence to deal with the most important aspects of these discoveries. Since these elements are expected to have very short lifetimes, the determination of their electronic structure could be a challenge.

In my opinion, if there is going to be a new Joint Working Group, it is essential that IUPAP should insist on a well-defined mandate for the group. IUPAP should insist on openness. The referee reports should be ordered by IUPAP and be available for scrutiny by eligible nuclear physicists.

It is irritating when IUPAC "proudly announces" the discovery of new elements. What have they done to be proud of? They should congratulate the discoverers. You should not steal someone's discovery and be proud of it. The articles in the non-scientific journals do not help either. They give the credit to IUPAC, by making statements such as "The International Union of Pure and Applied Chemistry, which verified their existence, announced the entry of the four elements ...". Well, I am sorry to have to say that those who made the announcement did not have the competence to verify "their existence".

First of all, I wish to thank the organizers of this Nobel Symposium for giving me the opportunity to be "brave, simple & bold". I wish to thank Prof. Dan-Olof Riska who was the Chair of the Nuclear Physics Commission of IUPAP, 2008-2011, and his successor Prof. Hideyuki Sakai. Dan-Olof, with whom I had an extensive email exchange, answered all my questions and ensured that his commission, C12, efficiently proposed members to be included in the Joint Working Group. Hide-san remained in constant contact with me. On this occasion, my thoughts go to the late Alan Astbury, a brave & bold man who truly cared for IUPAP. I wished I could have expressed my deep sense of gratitude to him. Finally, and above all, I wish a successful future for the superheavy community.

EPJ Web of Conferences **131**, 07001 (2016) DOI: 10.1051/epjconf/201613107001
Nobel Symposium NS160 – Chemistry and Physics of Heavy and Superheavy Elements

Advances in chemical investigations of the heaviest elements

Andreas Türler[1,2,3]

[1] Paul Scherrer Institut, Department Nuclear Energy and Safety, 5232 Villigen PSI, Switzerland
[2] University of Bern, Department of Chemistry and Biochemistry, 3012 Bern, Switzerland
[3] University of Bern, Albert Einstein Center for Fundamental Physics, 3012 Bern, Switzerland

Abstract. Although somewhat in the shadow of the discoveries of new elements, experimental chemical investigations of the heaviest elements have made tremendous progress in the last decades. Indeed, it was possible to experimentally determine thermochemical properties of heavy transactinide elements such as copernicium or flerovium. But will it be possible to chemically study all currently known elements of the periodic table up to element 118? While it is experimentally feasible to work with single atoms, the short half-lives of even the longest currently known isotopes of elements 115 through 118 call for new experimental approaches.

1. Introduction

Recently, the discovery of elements with atomic numbers 113, 115, 117, and 118 has been validated by a joint IUPAC and IUPAP working group and attribution of priority of discovery to a laboratory or to a collaborating group of laboratories has been made [1, 2]. It should be noted that these findings were not in all parts unanimously welcomed by the scientific community. The reservations concern the hitherto proposed cross-reaction link between alpha-decay chains associated with the isotopes $^{293}117$ and $^{289}115$, which was shown to not withstand a rigorous statistical treatment [3, 4]. Nevertheless, the respective discoverers were invited to propose a name and a symbol to the IUPAC Inorganic Chemistry in accordance with IUPAC recommendations on how to name a new element, which was recently revised [5]. The most important adjustment was that the names of new elements should have an ending that "reflects and maintains historical and chemical consistency. This would be in general "-ium" for elements belonging to groups 1–16, i.e. including the f-block elements, "-ine" for elements of group 17 and "-on" for elements of group 18." The positioning of a newly discovered element in a group of the periodic table is not considered or discussed in the published discovery profiles [1, 2] and IUPAC proceeded to place elements with atomic numbers 113, 115, 117 and 118 into groups 13, 15, 17, and 18 respectively, without offering any further considerations. But interestingly, IUPAC has no recommendation for a specific form of the periodic table. Indeed, an IUPAC project has been recently initiated [6] to resolve the question whether group 3 consists of the elements Sc, Y, Lu, and Lr or, of the elements Sc, Y, La and Ac? This dispute has no influence on the name of the involved elements; however, the misplacement of an element in groups 17 and 18 would definitely affect their suffixes. Currently, the discovery profiles of the four new elements concentrate mainly on the nuclear aspects in order to prove that indeed a new element with a previously unreported atomic number was synthesized. The place an element occupies in the Periodic Table is not only defined by its atomic number, i.e. the number of protons in the nucleus, but also by its electronic configuration, which defines its chemical properties. Strictly speaking, a new element is assigned its proper place only after its chemical properties have been sufficiently investigated. No chemical information was obtained whatsoever. Is it then conceivable, that the shape of the periodic table may be altered in the future and newly discovered elements are moved away from their current position to different groups? Not very likely for elements with atomic numbers 113, 115, 117, and 118, as recent theoretical calculations have shown (see Ref. [7] for a review). Nevertheless, strong relativistic effects significantly influence the chemical behaviour of the heaviest elements and, therefore, experimental chemical investigations of these elements are warranted and of decisive importance.

The study of the chemical properties of the heaviest known elements in the Periodic Table is an extremely challenging task and requires the development of unique methods, but also the persistence to continuously improve all the processes and components involved in order to achieve the ultimate goal of chemically isolating one single atom that lives for only a few seconds (or even less). Due to the low production rates of the heaviest elements, chemical information obtained from experiments is limited to the knowledge of very few basic properties. It mainly answers the question whether a new element behaves similarly to its lighter congeners in a chemical group, or whether some deviations from the trends occur due to very strong relativistic effects on its valence electron shells. Indeed, in some cases it has been possible to experimentally investigate chemical properties of transactinide elements and even synthesize simple compounds. The half-life limitation of current chemical systems allow only investigations of transactinide nuclides with $T_{1/2} > 1$ s. Thus the heaviest element whose chemical properties have been investigated experimentally is flerovium ($Z = 114$, Fl) [8, 9]. While the body of experiments and data on the chemistry of transactinide elements is already impressive, pushing the limits to investigations of even heavier and shorter-lived elements will require the improvement of existing technologies and the development of new techniques. For the not yet studied elements, like Mt, Ds, and Rg, isotopes with $T_{1/2}$ suitable for chemical studies have already been identified. Currently, also element 115 may come within reach of present day technologies.

2. Basic considerations and requirements

A chemistry experiment with a transactinide element can be divided into three basic steps: first, synthesis of the transactinide nuclide, second, rapid transport of the synthesized nuclide to the chemical apparatus, third, fast chemical isolation of the desired nuclide, preparation of a sample suitable for nuclear spectroscopy, and detection of the nuclide via its characteristic nuclear decay properties.

The synthesis of suitable, sufficiently long-lived transactinide nuclei is accomplished in the same manner as in the nuclear physics discovery experiments. High intensity heavy-ion beams such as ^{18}O, ^{22}Ne, ^{26}Mg, ^{36}S, and mainly ^{48}Ca are directed at exotic, highly radioactive target nuclides such as ^{244}Pu, ^{243}Am, ^{248}Cm, ^{249}Bk, or $^{249,251}Cf$. On the one hand as intense beams as possible are to be used; on the other hand the destruction of the very valuable and highly radioactive targets has to be avoided. Production rates are proportional to both the target thickness and the beam intensity. For nuclear reactions involving the lighter projectiles, target thickness is limiting the compound nucleus recoil ranges (<1 mg/cm^2), because the separation techniques used require the compound nucleus products to recoil out of the target. With the higher-Z projectiles, the recoil ranges are larger, but because of the higher rate of projectile energy loss in the target material and the narrow projectile energy range effective for heavy-element production, the effective target thicknesses are once again limited to approximately 1 mg/cm^2.

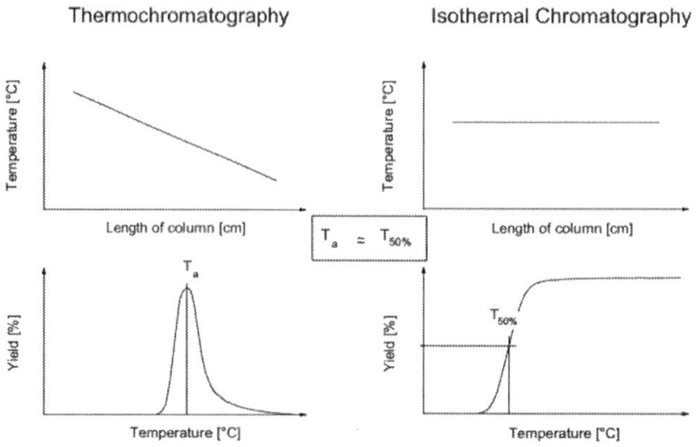

Figure 1. Upper panel: temperature profiles employed in TC and IC; lower panel: deposition peak and integral chromatogram resulting from TC and IC, respectively.

Since transactinide nuclei are inevitably produced among many orders of magnitude larger amounts of "background" activities, which interfere with the detection and identification of the decay of transactinide atoms, there is nowadays a recognized need for physical pre-separation of the transactinide atoms before chemical investigation. Except for the most asymmetric synthesis reactions, the recoil energy of the fusion reaction products imparted by the momentum of the beam is sufficient to allow the separated nuclei to pass through a thin Mylar® window, which separates the low pressure in the separator (i.e. 1 mbar or vacuum) from the stopping gas cell (i.e. 1 bar). The thin windows are supported by honeycomb grids. Thus, a recoil chamber can be coupled to a kinematic pre-separator instead of a focal plane detector array, making transactinide atoms available for chemical experiments. Since the separated transactinides are usually thermalized in a gas, rapid gas-phase chemical reactions and separations are the method of choice. Gas-phase separation procedures are fast, efficient and can be performed continuously, which is highly desirable in order to achieve high overall yields. Two different types of gas chromatographic separations have been developed for the experimental investigation of volatile transactinide elements or compounds, isothermal chromatography (IC) and thermochromatography (TC). A schematic of the operating principle of IC and TC is provided in Fig. 1.

In IC, a carrier gas is flowing through a chromatography column of constant, isothermal temperature. Depending on the temperature and on the enthalpy of adsorption, ΔH_{ads}, of the species on the column surface, the species travel slower through the length of the column than the carrier gas. This retention time can be determined by continuously introducing a short-lived nuclide into the column and detecting the fraction of nuclides that have decayed at the exit of the column. A characteristic quantity is the temperature at which half of the introduced nuclides are detected at the exit ($T_{50\%}$). In this case, the retention time in the column is equal to $t_{1/2}$ of the introduced nuclide. Since volatile species rapidly emerge at the exit of the column, they can be condensed and assayed with nuclear spectroscopic methods. This is done by turbulent mixing of the gas stream exiting the chromatography column (containing the volatile species under investigation) with a second gas stream containing aerosol particles. Volatile species will condense on the surface of these aerosol particles and can be transported to a counting device, i.e. a rotating wheel system or a tape system, where they are collected by impaction and subsequently assayed by α-particle and SF-spectroscopy.

In TC, a carrier gas is flowing through a chromatography column, to which a negative longitudinal temperature gradient has been applied. Species that are volatile at the starting point are transported downstream of the column by the carrier gas flow. Due to the decreasing temperature in the column, the time the species spend in the adsorbed state increases exponentially. Different species form distinct deposition peaks, depending on their ΔH_{ads} on the column surface and are thus separated from each other. A characteristic quantity is the deposition temperature (T_a), which depends on various experimental parameters. The mixture of species to be separated can be injected continuously. The deposition position (and thus T_a) of each single radioactive decay in the column contributes chemical information about ΔH_{ads} of the investigated species. In recent years, TC detectors proved enormously successful. The carrier gas containing volatile atoms or molecules is flowing through a narrow channel formed by a series of planar silicon diodes. Due to the close proximity of the silicon diodes facing each other, the probability to register a complete decay chain consisting of a series of α-particle decays is very high. Serious limitations constitute the required operating conditions of silicon diodes, which sustain only temperatures from +35 °C to about −180 °C.

A new development concerns vacuum chromatography, where, under molecular flow conditions, atoms or molecules can, in principle, be isolated within about 100 milliseconds. In conjunction with diamond detectors that can be operated up to 180 °C, so far inaccessible elements beyond Fl can be accessed in the future. These developments will be presented later in the form of a proof-of-principle experiment with relatively short-lived Tl nuclides as homologs of element 113.

3. Recent accomplishments and selected results

With the introduction of silicon based cryo-thermochromatography detectors a number of breakthroughs were made possible. This was mainly due to the much improved overall efficiency of the chemical systems provided by these detectors. While the overall efficiency to detect a three member decay chain in IC-experiments with Bh ($Z = 107$) [10] was estimated at a meager 4%, this number increased by more than one order of magnitude to about 50% in experiments with Hs ($Z = 108$) [11]. This accomplishment also allowed for the discovery of two new Hs isotopes, ^{270}Hs and ^{271}Hs, and their hitherto unknown daughter nuclei, ^{266}Sg and ^{267}Sg and ^{263}Rf, respectively. The success of these experiments paved the way to first ever chemical investigations of copernicium ($Z = 112$, Cn) [12] and Fl ($Z = 114$) [8, 9] in the elemental state, but also to the first characterization of a volatile $Sg(CO)_6$ compound [13]. The results of these experiments have been reviewed already extensively [14] and will also be the topic of several contributions to the current conference proceedings.

An interesting development concerns investigations of the interaction of superheavy elements Cn and Fl with Se surfaces. The stability of Fl–selenides is expected to be the highest in group 14, whereas it is the lowest for Cn-selenides in group 12. Both selenides are expected to be the most volatile within their respective groups. First experiments revealed surprising results for the behavior of Cn versus its lighter homolog Hg. Further experiments are in progress to elucidate the behavior of Cn- and Fl-selenides.

The recent introduction of carbonyl complex chemistry to the heavy element region opens up new possibilities to synthesize and study other transactinide containing carbonyl compounds with the elements Bh, Hs, and Mt. Furthermore, the thermal stability towards decomposition of $Sg(CO)_6$ was studied in a recent experiment by an international collaboration at RIKEN, Japan. The data of these experiments is still under analysis.

4. Vacuum chromatography with heavy transactinide elements

Based on trends of the standard enthalpies of monoatomic gaseous elements in the periodic table, all transactinide elements in groups 12 through 17 and groups 1 and 2 are expected

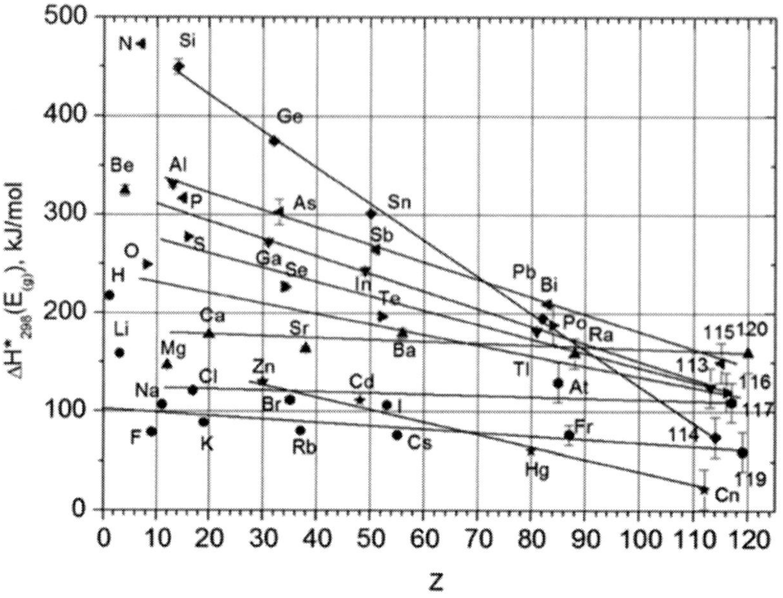

Figure 2. Extrapolation of the standard enthalpies of monoatomic gaseous elements $\Delta H^*_{298}(E_{(g)})$ along the groups of the Periodic Table based on the atomic number Z (adapted from [15]).

to be the most volatile ones in the group. A simple and very early extrapolation of the standard enthalpies of monoatomic gaseous elements, $\Delta H^*_{298}(E_{(g)})$ over the atomic number Z, predicts the heavy transactinides Cn through element 120 [15] to be rather volatile (Fig. 2), since the standard enthalpies of monoatomic gases are mostly equal to the standard sublimation enthalpy ΔH_S of these elements. Those linear extrapolations should be used with care, as there is no solid theoretical basis for them. However, the high volatility of Cn and Fl has been demonstrated experimentally and was predicted by much more sophisticated relativistic theoretical calculations. One might therefore argue that elements such as Tl or At serve as suitable homologs for the development of new chemistry setups for investigations of elements 113, 115, Lv, and 117. Depending on the ΔH_S of these elements, column temperatures of up to about 700 °C will be required to rapidly transmit them through an open chromatography column at vacuum conditions.

A suitable isothermal vacuum chromatography (IVAC) setup was constructed at Paul Scherrer Institute, Switzerland and used to investigate the volatility of Tl on SiO_2 at the single-atom level in an experiment at JAEA, Japan [16]. A mass separated low-energy ion beam of $^{184}Tl^+$ ions was provided by the JAEA-ISOL system at the tandem accelerator. A 3D construction drawing of the IVAC system is provided in Fig. 3. The low energy $^{184}Tl^+$ beam enters the Hf hot catcher (HC) through a small hole. Neutralized ^{184}Tl atoms exit the hot catcher and enter the kinked quartz chromatography column in the isothermal oven. The kink prevents Tl atoms from traversing the chromatography column in a straight flight path without surface interactions. At the exit of the column a four-fold diamond detector is registering α-particle decays of condensed ^{184}Tl. As the transport of the released atoms is governed exclusively by random walk, the number of surface interactions is directly proportional to the geometric area. This means that the probability for an atom to leave the HC containment through the 6-mm exit hole into the chromatography column is four times greater (80%) than for it to exit back through the 3-mm entrance orifice (20%). However, as the conductance of the chromatography column is rather low, the bulk of the species under observation re-enters

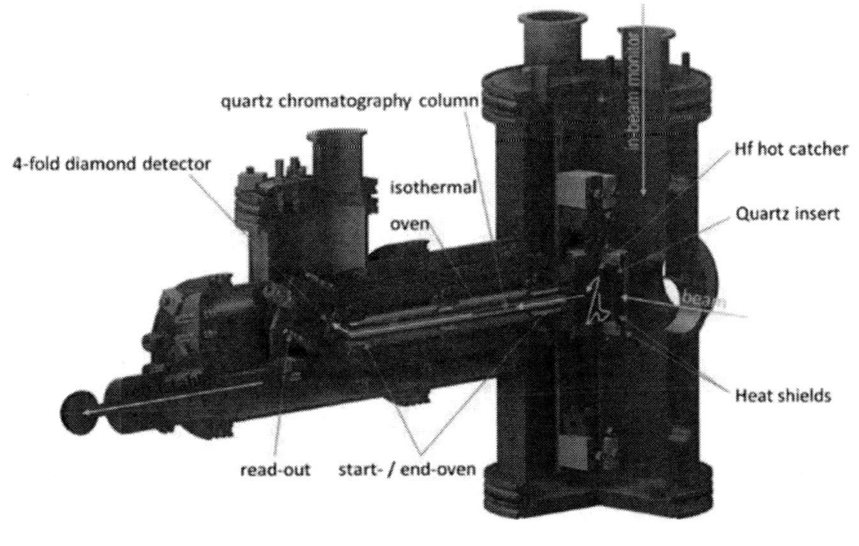

Figure 3. Construction drawing of the IVAC setup.

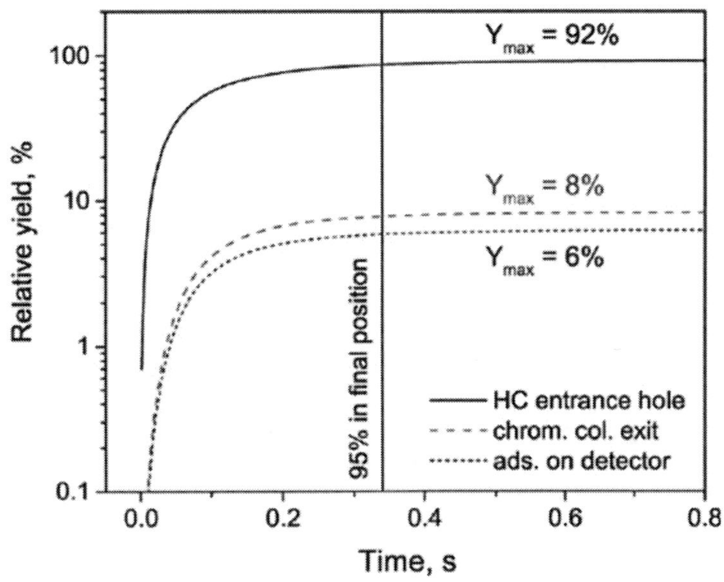

Figure 4. Simulated transmission efficiency through the IVAC setup assuming a temperature independent adsorption behavior as a function of time. Figure from [16].

the HC containment multiple times and eventually leaves it through the small entrance orifice, leading to its ultimate loss.

In Fig. 4 the transmission efficiency through the setup simulated with the COMSOL multiphysics mathematical particle tracing module is shown (assuming no temperature dependent retention). A major portion (92%) of all inserted Tl atoms escape through the HC entrance orifice (black solid line), whereas 8% (red dashed line) reach the end of the chromatography column, and 6% (blue dotted line) adsorb on the active area of the detector.

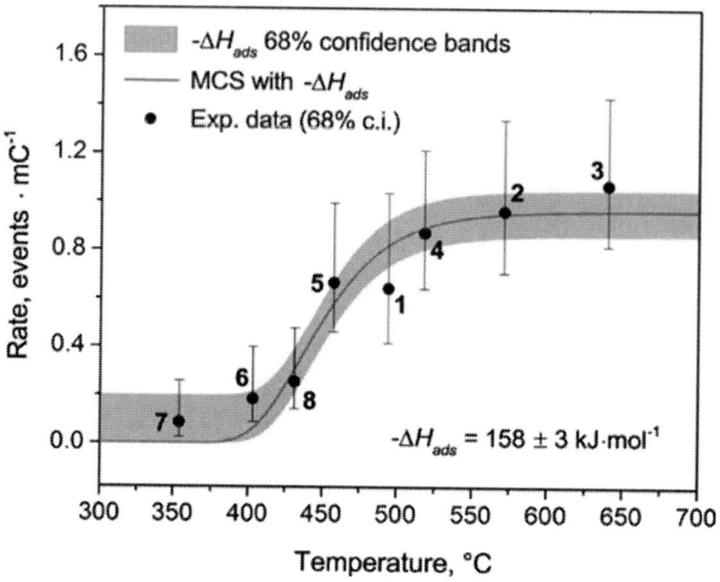

Figure 5. Measured event rate of ^{184}Tl decays in the diamond detector at the exit of the column as a function of isothermal temperature. For a detailed description see text. Figure from [16].

After 100 ms 50% of all atoms that will reach the column exit have already reached it. The vertical line at 340 ms indicates the time at which 95% of all atoms are in their final position. The measured transmission efficiency between the in-beam monitor and the exit detector amounted to only 2.2% (compared to a theoretical yield of 6%). This low yield would clearly be insufficient for an experiment with transactinide elements. A mechanism has to be developed to transmit the product beam through a thin hot window into the HC. This way the huge losses through the entrance orifice (92%) could be avoided and the yield would be increased significantly. This should be feasible as the product beam from a gas-filled separator or a velocity filter has a much higher energy (20–50 MeV) compared to the one from the JAEA-ISOL system.

In Fig. 5, the experimentally measured external chromatogram (black dots, error bars: 68% c.i.) for Tl atoms passing through a quartz column of the IVAC setup is shown. The data was modelled using a Monte Carlo simulation procedure assuming an adsorption enthalpy of $-\Delta H_{\mathrm{ads}}^{\mathrm{SiO_2}}(\mathrm{Tl}) = (158 \pm 3)\,\mathrm{kJ \cdot mol^{-1}}$ (red line) and the corresponding confidence bands (red area); the bold black numbers indicate the sequence of the conducted measurements.

The determined adsorption enthalpy differed noticeably from previously reported values of $-\Delta H_{ads} = 137\,\mathrm{kJ \cdot mol^{-1}}$ that could eventually be attributed to the formation of TlOH. In this experiment, due to vacuum conditions, there were negligible particle-particle interactions in the molecular flow regime. Furthermore, high temperatures of $\approx$1000 °C were applied all along the quartz chromatography column prior to the experiment, so that reactions with O_2 and H_2O were virtually excluded. A further indication that indeed the adsorption of elemental Tl on SiO_2 was investigated was that the newly determined adsorption enthalpy value now perfectly fell in line with the empirical correlation function between the sublimation enthalpies of various elements and their respective adsorption enthalpies on a quartz surface [16], which can be described as:

$$-\Delta H_{\mathrm{ads}}^{\mathrm{SiO_2}} = (0.84 \pm 0.07) \cdot \Delta H_{\mathrm{subl}}^{\circ} - (2.8 \pm 8.9)\,\mathrm{kJ \cdot mol^{-1}}. \quad (1)$$

5. Conclusions

Applying vacuum chromatography in conjunction with diamond detectors may open up new avenues of transactinide research and will allow chemical studies of nuclides with half-lives as short as 100 ms or even less. However, the modest yields of first "proof-of-principle" experiments clearly demonstrate the need for further significant improvements, before experiments with trans-flerovium elements can be attempted with a realistic chance of success. While looking at possible chemical experiments with the heaviest known elements, one should not forget about the fact, that so far no chemical studies of elements Mt, Ds, and Rg have been conducted, even though isotopes with half-lives >1 s are already known. Also, results on the chemical properties of element 113 are not convincing so far [14, 17].

References

[1] P.J. Karol *et al.*, Pure Appl. Chem. **88**, 139 (2016).
[2] P.J. Karol *et al.*, Pure Appl. Chem. **88**, 155 (2016).
[3] U. Forsberg *et al.*, Nucl. Phys. **A953**, 117 (2016).
[4] U. Forsberg *et al.*, Phys. Lett. **B760**, 293 (2016).
[5] W.H. Koppenol *et al.*, Pure Appl. Chem. **88**, 401 (2016).
[6] E. Scerri *et al.*, Chemistry International **38**, 22 (2016); see also: `www.iupac.org/project/2015-039-2-200`.
[7] A. Türler and V. Pershina, Chem. Rev. **113**, 1237 (2013).
[8] R. Eichler *et al.*, Radiochim. Acta **98**, 133 (2010).
[9] A. Yakushev *et al.*, Inorg. Chem. **53**, 1624 (2014).
[10] R. Eichler *et al.*, Nature **407**, 63 (2000).
[11] Ch.E. Düllmann *et al.*, Nature **418**, 859 (2002).
[12] R. Eichler *et al.*, Nature **447**, 72 (2007).
[13] J. Even *et al.*, Science **345**, 1491 (2014).
[14] A. Türler, R. Eichler, A. Yakushev, Nucl. Phys. **A944**, 640 (2015).
[15] B. Eichler, Kernenergie **19**, 307 (1976).
[16] P. Steinegger *et al.*, J. Phys. Chem. C **120**, 7122 (2016).
[17] S.N. Dmitriev *et al.*, Mendeleev Commun. **24**, 253 (2014).

Theoretical chemistry of superheavy elements: Support for experiment

Valeria Pershina[a]

GSI Helmholtzzentrum für Schwerionenforschung GmbH, 64291 Darmstadt, Germany

Abstract. Spectacular developments in the relativistic quantum theory and computational algorithms in the last few decades allowed for accurate calculations of properties of the superheavy elements (SHE) and their compounds. Often conducted in a close link to the experimental research, these investigations helped predict and interpret an outcome of sophisticated and expensive experiments with single atoms and to reveal the magnitude and importance of relativistic effects.

1. Importance of chemical theory for SHE

Conceptually, it is the atomic number Z and ground state electronic configuration of an element that define its position in the Periodic Table. However, information on chemical behaviour of a new element with respect to its homologs often serves as an additional proof of the proper assignment [1, 2]. This is particularly important for superheavy elements (SHE): It is meanwhile well established that relativistic effects on their electron shells are so strong that deviations from trends in chemical properties observed within lighter homologs in the groups can occur [3]. Important properties to assess similarity of SHE with the lighter congeners are atomic/ionic radii, redox potentials and complex formation.

The atomic number of many SHE can be determined by physical methods, while the electron configuration can currently be provided only by theory. Experimental studies of chemical properties of short-lived SHE isotopes are also limited to measurements of very few properties, mainly, volatility of atoms or molecules using gas-phase chromatography [4], or complex formation using liquid-phase separation techniques [5]. (See also the contribution of A. Türler in this volume). Knowledge of many other important properties, such as chemical composition, stability, geometrical configuration, ionization potentials (IP), or electron transitions, can only be provided by theory. Thus, in the area of SHE, theoretical studies become extremely important and are often the only source of useful chemical information [6–9]. They are also invaluable for predicting and/or interpreting the outcome of sophisticated and demanding experiments with single atoms. Moreover, since nature is relativistic, it is only theory that can demonstrate the importance and magnitude of relativistic effects by comparing the observed behaviour with that predicted on the basis of relativistic vs. non-relativistic calculations.

[a] e-mail: v.pershina@gsi.de

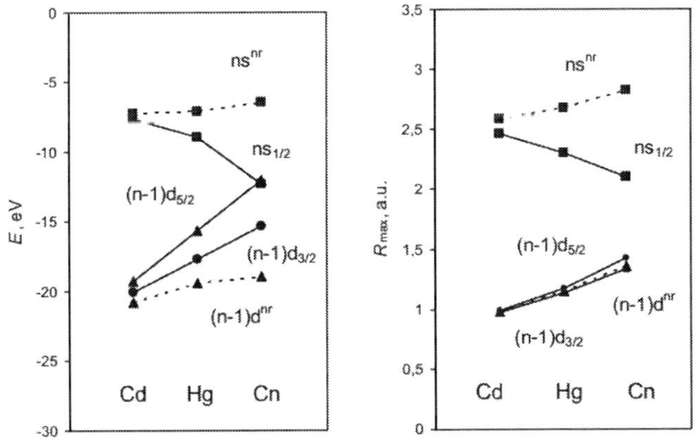

Figure 1. Relativistic (solid line) and nonrelativistic (dashed line) energy, E, and the maximum of the radial charge density, R_{max}, of the group-12 valence ns and (n-1)d AOs [6].

2. Relativistic and QED effects for SHE

With increasing Z, the velocity of an electron increases. The Bohr model for a hydrogen-like species gives it as $v = (2\pi e^2/nh)Z$. As a result, the relativistic mass of an electron also increases

$$m = m_0\left[(1-(v/c)^2\right]^{-1/2}, \tag{1}$$

where m_0 is the mass at zero velocity (rest mass) and c is the speed of light. From the sixth row onwards, the m/m_0 ratio exceeds by 10%, so that relativistic effects cannot be neglected anymore. For example, for Fl, $m/m_0 = 1.79$, and it is 1.95 for element 118. As a result of the relativistic mass increase, the electron (negative) energy increases and the orbital radius shrinks according to

$$E = -(2\pi^2 e^4/n^2h^2)mZ^2, \tag{2}$$

$$r = Ze^2/mv^2, \tag{3}$$

where n is the principal quantum number, e is the charge of the electron, and h is Planck's constant. The contraction (Eq. (3)) and stabilization (Eq. (2)) of the hydrogen-like s and $p_{1/2}$ electrons is a direct relativistic effect and it was shown to originate from the inner K- and L-shell regions. The second (indirect) relativistic effect causes an expansion and destabilization of the $p_{3/2}$, d, f, and g atomic orbitals (AOs) due to increased screening of the nucleus by the relativistically contracted s and $p_{1/2}$ AOs. The third relativistic contribution comes from the spin-orbit (SO) splitting of the AOs with $l > 0$. All the three effects change approximately as Z^2 for the valence shells down a column of the Periodic Table.

Figure 1 shows, e.g., relativistic effects on group-12 elements: the ns AO stabilization reaches 5.8 eV for Cn, which is the maximum in the 7th row of the Periodic Table, and its contraction is 25%; the SO splitting of the 5d AOs is also large, e.g., 3.3 eV for Cn. The SO splitting of the 7p AOs reaches 11.8 eV at element 118. For the heavier elements, relativistic effects on their valence orbitals are even more pronounced and could lead to properties very different from those of the lighter homologs.

Breit effects (accounting for magnetic and retardation interactions) on valence orbital energies and IPs of the heaviest elements are small, for example, only 0.02 eV for element 121. Quantum electrodynamic (QED) effects such as vacuum polarization and electron self-energy can reach a few % in IP, electron affinities (EA), or transition energies for SHE (see, for instance, Refs. [10, 11]).

3. Methods to calculate electronic structures of SHE

There are several issues devoted to the relativistic methods (see, for example, [12]). It is meanwhile established that for highly relativistic SHE systems, the most appropriate approaches are those that treat both relativity and electron correlation at the highest possible level of theory: both effects were shown to contribute to more than 50% to binding energies and other properties of the 7th-row elements. Presently the highest theoretical level for the many-body methods is the Dirac-Coulomb-Breit (DCB) Hamiltonian

$$h_{DCB} = \sum_i h_D(i) + \sum_{i<j}(1/r_{ij} + B_{ij}) \quad (4)$$

where B_{ij} is the Breit term and the Dirac-Coulomb one-electron Hamiltonian is

$$h_D(i) = c\vec{\alpha}_i\vec{p}_i + c^2(\beta_i - 1) + V^n(i). \quad (5)$$

It contains the one-electron Dirac Hamiltonian plus the nuclear potential, V^n, and the operator $1/r_{ij}$ for the Coulomb interaction between electrons. The wave function is a four-component (4c) vector (spinor). The V^n includes the effect of the finite nuclear size, while some finer effects, like QED, can be added to h_{DCB} self-consistently for atoms, or perturbatively for molecules. Correlation effects are taken into account by configuration interaction (CI), many-body perturbation theory (MBPT) or, presently at the highest level of theory, coupled cluster theory with single-double (and perturbative triple) excitations (CCSD(T)) or Fock-Space (FS) CC theory.

Wavefunction based (ab-initio) methods solving the Dirac many-electron Eq. (4) are the most accurate. Nevertheless, the problems of electron correlation and proper basis sets make the usage of the 4c-DF methods very limited in molecular calculations. These methods are still too computer time intensive. Therefore, almost equally accurate 2c-methods have recently been developed.

Effective core potentials (ECP) allow for more economic calculations within the DHF schemes by replacing the inner core orbitals, which do not take part in the bond formation, by a special (effective core) potential. In this way, the number of basis functions and, therefore, two-electron integrals, is drastically diminished. There are two main types of the ECPs, pseudo potentials (PP) and model potentials (MP).

Relativistic density functional theory (DFT). Due to the high accuracy and efficiency, computational schemes based on the 4c- and 2c-DFT methods are among the most popular in theoretical chemistry, especially for extended systems, large molecules, liquids, or solids. The accuracy depends on the adequate choice of the exchange-correlation potential, E^{ex}, whose exact form is unknown. There is quite a number of the GGA E^{ex} and their choice is dependent on the system. For solid-state and adsorption calculations of SHE, the ADF BAND program is very promising [13].

4. Atomic properties of SHE

4.1 Electronic configurations

Earlier predictions of chemical properties of elements $Z = 104$ through $Z = 172$ were made on the basis of Dirac-Fock (DF) and Dirac-Slater (DS) atomic calculations (see [3] for a review). Later, atomic properties of SHE were calculated using the MCDF and DCB

Table 1. Ground states predicted using various relativistic atomic methods[a].

Element	Madelung principle	DS	DF	MCDF	DCB CC
	Seaborg (1968)	*Waber et al.* (1968)	*Fricke et al.* (1971)	*Nefedov et al.* (2006)	*Eliav et al.* (1998)
121	$7d$	$7d$	$8p_{1/2}$	$8p_{1/2}$	$8p_{1/2}$
122	$5g^2$	$7d^2$	$7d8p_{1/2}$	$7d8p_{1/2}$	$7d8p_{1/2}$
123	$5g^3$	$6f^17d^2$	$6f7d8p_{1/2}$	$6f^28p_{1/2}$	...

[a] See [3, 6, 7, 9] for exact references.

FSCC/CCSD(T) methods, with the heaviest element treated in this way being $Z = 122$ [9]. Table 1 shows the progress in the predictive power of the atomic methods upon their development for ground-state configurations of elements 121, 122, and 123.

One can see that the calculated states deviate from those predicted via the simple Madelung's principle. In the area of $Z > 122$, the elements are characterized by mixing of configurations coming from the partially filled $8p_{1/2,3/2}$, $7d_{3/2,5/2}$, $6f_{5/2,7/2}$ and $5g_{7/2,9/2}$, etc. shells. The proximity of the valence SO bands makes the search for the correct ground state very difficult. The usual classification on the basis of a simple electronic configuration and the placement of these elements in this part of the Periodic Table becomes, therefore, problematic, so that the shape of the table at these high Z is still under discussion and debate [3, 14, 15].

4.2 Spectroscopic properties and predictions of SHE transport

Ionization potentials in the DF and DS approximations are available for SHE up to $Z = 166$ [3], and in the MCDF and DCB CC approximation till $Z = 122$ [9]. Multiple IPs were also calculated at the MCDF level of theory for elements Rf through Hs. They were used for estimates of stabilities of various oxidation states of these elements (see [6]).

Electron affinities were also calculated for quite a few of the heaviest elements at the DCB CC level of theory. Fl, has, e.g., no EA. On the contrary, element 118 has a positive EA of 0.058 eV, shown by the DCB FSCC + QED calculations. Both unusual results were shown to be due to the influence of relativistic effects.

Atomic/ionic/covalent radii and polarizabilities of SHE are now also available at various levels of theory [2, 6, 7, 14]. An important finding is smaller covalent radii of the 6*d* elements than those of the 5*d* ones in group 9 through 12, called a "transactinide break" [14, 15].

The atomic data were useful in predicting chemical behaviour of SHE [3, 6, 7], such as transport through Teflon capillaries from an accelerator to the chemistry set up, and/or their interaction with inert materials of gas-phase chromatography column. For that purpose, the following adatom-slab model for the adsorption energy, E_{ads}, was used:

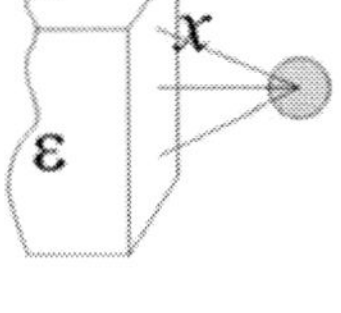

$$E(x) = -\frac{3}{16}\left(\frac{\varepsilon - 1}{\varepsilon + 2}\right)\frac{\alpha_{mol}}{\left(\frac{1}{IP_{slab}} + \frac{1}{IP_{mol}}\right)x^3}. \qquad (6)$$

Here, ε is the dielectric constant of the adsorbent material and x is the adatom-surface distance related to the van der Waals radius of the species. Table 2 shows two examples of such predictions for very different cases: for Fl with a small α and a high IP, and for element 120 with a large α and a small IP. The obtained low value of E_{ads} of Fl is indicative of its easy transport through Teflon capillaries to the chemistry set up, which was not obvious for

Table 2. DC CCSD(T) calculated atomic properties of Fl ($7s^27p_{1/2}{}^2$) and of element 120 ($8s^2$), and their adsorption enthalpies, $-\Delta H_{ads}$, on Teflon [7].

Element	α, a.u.	IP, eV	R_{vdW}, Å	$-\Delta H_{ads}$, kJ/mol
Fl	29.5	8.539	3.94	10.4
120	29.5	8.539	3.94	10.4

element 120. Using the calculated properties and equation 6, a yield N/N_0 (N_0 is the number of atoms entering the column or capillary, N is the number at the exit) of a hypothetical $^{300}120$ of 90% at the end of a Teflon capillary of 1 m was then determined. Thus, this element should also be transported, and the limiting factor is its half-life [7].

5. Predictions of experimental behaviour of SHE

5.1 Gas-phase experiments

Group 4-8 molecules. Gas-phase chromatography experiments provide measurements of volatility of SHE as adsorption temperature, T_{ads}, on a surface of detectors of the chromatography column usually covered with SiO_2 or gold layers [2, 4]. Obtained on their basis the adsorption enthalpy, ΔH_{ads}, is then used to deduce the sublimation enthalpy, ΔH_{sub}, being a measure of volatility in macrochemistry, using an empirical correlation between these quantities. The task of the theory was to predict both ΔH_{ads} and ΔH_{sub} and to see whether such a correlation is valid in the area of the heaviest elements.

For weak interactions of closed-shell SHE atoms or molecules with inert surfaces, no modern DFT methods can provide reliable values of E_{ads}: there are no dispersion-corrected E^{ex} for SHE, as they should be based on empirical data. Thus, for those systems, adatom/molecule-slab models proved to be useful. For molecules without dipole moments, Eq. 6 was used, while for those with dipole moments, μ, the following equation:

$$E(x) = -\frac{2Qe\mu_{mol}{}^2}{x^2} - \frac{Q^2e^2\alpha_{mol}}{2x^4} - \frac{3}{2}\frac{\alpha_{mol}\alpha_{Cl}}{\left(\frac{1}{IP_{mol}} + \frac{1}{IP_{Cl}}\right)x^6}. \tag{7}$$

All the properties of group-4 through 8 molecules used in Eqs. (6) and (7) have accurately been calculated via 4c-DFT and PP CC methods (see Tables 8 and 9 in [6]). Thus, e.g., for the prediction of adsorption of group-7 oxychlorides MO_3Cl (M = Tc, Re, and Bh) on a chlorinated quartz surface, equation 7 was used [6]. For an effective charge $Q = -0.4$ of the surface chlorine ions, $-\Delta H_{ads} = (48 \pm 2)$ kJ/mol and $-\Delta H_{ads} = (78 \pm 5)$ kJ/mol for TcO_3Cl and BhO_3Cl, respectively, were obtained with respect to the measured $-\Delta H_{ads}$ (ReO_3Cl) = (61 ± 2) kJ/mol. These $-\Delta H_{ads}$ of the Tc and Bh oxychlorides were confirmed by experiment giving (51 ± 2) kJ/mol and (75^{+6}_{-9}) kJ/mol, respectively (see [2]). The sequence in volatility, $TcO_3Cl > ReO_3Cl > BhO_3Cl$, was revealed and explained by increasing μ in this row.

Another example deals with predictions of volatility of group-8 tetroxides, MO_4 (M = Ru, Os, and Hs). Their adsorption on a SiO_2, or a silicon nitride surface was predicted with the use of Eq. (6) and 4c-DFT calculated molecular properties (Table 3).

Thermochromatography gas-phase experiments have, indeed, shown that $-\Delta H_{ads}(HsO_4)$ is only 6 kJ/mol larger than $-\Delta H_{ads}(OsO_4)$ (see [2] for the reference), in excellent agreement with the calculations (Table 3). Such a reversal of the trend in volatility in the group could not be predicted from a simple extrapolation, but came out as a result of the accurate calculations and precise measurements.

Table 3. Ionization potentials, IP (in eV), polarizabilities, α (in a.u.), bond lengths, R_e (in Å), and adsorption enthalpies, $-\Delta H_{ads}$, (in kJ/mol), of MO_4 (M = Ru, Os, and Hs) on quartz[a].

	Meth.	RuO_4	OsO_4	HsO_4
IP	calc.	12.21	12.35	12.29
	exp.	12.19	12.35	–
α	calc.	58.07	55.28	65.99
	exp.	58.64	55.13	–
R_e	calc.	1.712	1.719	1.779
	exp.	1.706	1.711	–
$-\Delta H_{ads}$	calc.	41.0 ± 1	39.0 ± 1	45.4 ± 1
	exp.	–	39 ± 1	46 ± 2

[a] See [2, 6, 7] for references.

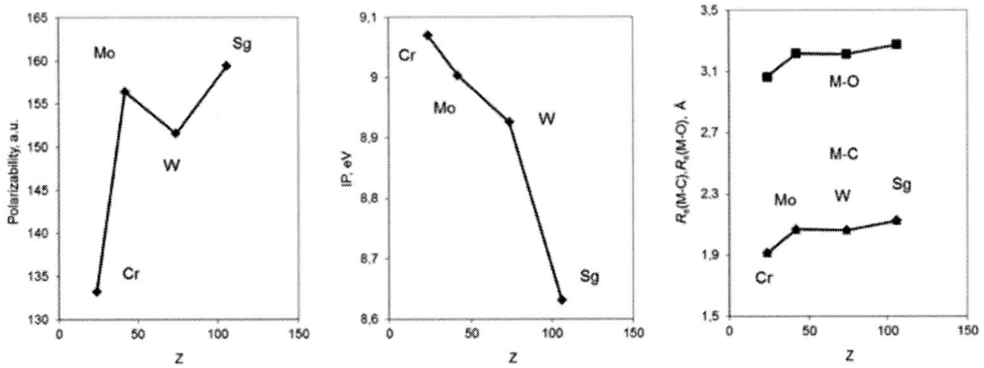

Figure 2. Calculated ionization potentials (IP), polarizabilities (α) and equilibrium bond lengths (R_e) of $M(CO)_6$ (M = Cr, Mo, W, and Sg) [17].

Another example deals with recent successful predictions of volatility of group-6 carbonyls. First of all, stability of $Sg(CO)_6$ had to be known. A work [16] based on the CCSD RECPs calculations have shown that the first bond dissociation energy (FBDE), $M(CO)_6 \rightarrow M(CO)_5 + CO$, should increase in group 6, reaching 2.20 eV for $Sg(CO)_6$. The $Sg(CO)_6$ was, indeed, synthesised and experiments are on the way to measure its FBDE.

To predict volatility of $Sg(CO)_6$, the 4c-DFT calculations were performed for the group-6 carbonyls and E_{ads} were estimated using the adatom-slab model (Eq. (6)) [17]. The trends in the calculated molecular properties are shown in Fig. 2 reflecting, e.g., peculiarities in R_e and α caused by the lanthanide contraction from Mo to W. The $-\Delta H_{ads}$ of (46.2 ± 2.5) kJ/mol of $Sg(CO)_6$ was predicted to be almost equal to the measured $-\Delta H_{ads}$ of $W(CO)_6$ of (46.5 ± 2.5) kJ/mol. The experimental ΔH_{ads} have, indeed, indicated an almost equal volatility of the W and Sg carbonyls giving $-\Delta H_{ads}$ of $Sg(CO)_6$ of (50 ± 4) kJ/mol [18], in very good agreement with the calculations [17].

Elements 112 and 114. Another exciting case was a study of volatility of Cn and Fl, having closed-shell ($6d^{10}7s^2$) and quasi-closed shell ($7s^2 7p_{1/2}{}^2$) ground states, respectively. Due to the strong relativistic effects on their AOs (Fig. 1), these elements were expected to be extremely inert and volatile. Also their extrapolated ΔH_{sub} are the smallest in the corresponding groups.

Experimentally, volatility of Cn in comparison with that of Hg and Rn was to be studied as adsorption process on a gold surface of detectors of a gas-phase thermochromatography column [19–21]. The questions to the electronic structure theory, therefore, were: are Cn and Fl metallic in the solid state, or are they more like a solid noble gas? How volatile and reactive towards gold are Cn and Fl atoms in comparison with the lighter homologs and with Rn?

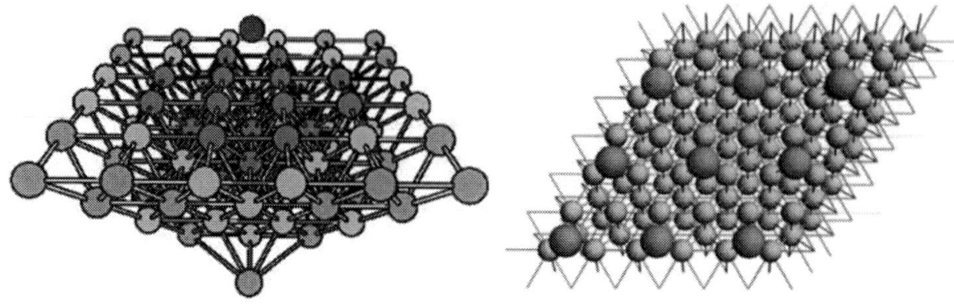

Figure 3. Models simulating adsorption of M on a Au(111) surface. The left panel is a cluster model treated via a molecular code. The right panel is a super-cell one treated via a periodic code.

Table 4. Adsorption enthalpies, $-\Delta H_{ads}$ (in kJ/mol), of Hg, Cn, Pb and Fl on a hydroxylated quartz and a gold (111) surface at very low coverage calculated using the ADF BAND periodic and 4c-DFT cluster approaches for single species in comparison with experiment.

Surface	Hg	Cn	Pb	Fl	Method	References
SiO_2	54	26	152	21	ADF BAND	[25]
	42 ± 2	–	165 ± 4	–	Experiment	see [2]
Au	78[a]	45	240	68	BAND/4c-DFT	[23, 24]
	98 ± 3	52^{+4}_{-3}	234	34^{+54}_{-11}; ≥ 48	Experiment	[19, 21]

[a] Preliminary ADF BAND value for the non-ideal surface.

The 4c-DFT calculations of the binding energies of group-12 homonuclear dimers [6, 7] and the scalar relativistic (SR) DFT periodic ones of cohesive energies [22] have shown Cn-Cn and the bulk of Cn to be more bound than the corresponding Hg systems. Thus, a decrease in ΔH_{sub} in group 12 should not continue with Cn. In contrast to the M-M-bonding, the M-Au one should be weaker for Cn than for Hg due to the stronger relativistic stabilization of the $7s$(Cn) AO [23]. This means that $-\Delta H_{ads}(\text{Hg}) > -\Delta \text{H}_{ads}(\text{Cn})$, which is opposite to $\Delta H_{sub}(\text{Hg}) < \Delta H_{sub}(\text{Cn})$. Thus, there is no correlation between $-\Delta H_{ads}$ and ΔH_{sub} in group 12. The Fl-Au bond was also shown to be weaker than the Pb-Au one, but stronger than the Cn-Au one. Numerous calculations for the Cn/Fl-Au$_n$ systems, where Au$_n$ are clusters simulating a gold surface (Fig. 3, left), have come to the same result, $E_b(\text{Fl-Au}) > E_b$ (Cn-Au) (Table 4).

Experiments [19], being in good agreement with the theoretical predictions for adsorption of Cn on gold [23], turned out to be in some disagreement with each other and partially with the theory for adsorption of Fl [20, 21] (Table 4). Therefore, further experiments are required to shed more light on this interesting case.

For element 113, to be experimentally studied next, the calculations have shown that it should be much more reactive than Cn and Fl due to one unpaired $7p_{1/2}$ electron, so that its $-\Delta H_{ads}$ on gold of 150 kJ/mol should be larger than $-\Delta H_{ads}$ of Cn and Fl (Table 4) [7]. In group 13 the trend to an increase in volatility should continue with element 113: its weaker adsorption on gold is a result of the $7p_{1/2}$ AO SO stabilization and, therefore, weaker bonding.

Calculations of E_{ads}(M-gold) using the ADF BAND code are in progress [24] (Fig. 3, right). Modelling non-ideal surfaces (with vacancies, steps) allowed for reaching even better agreement of the calculated E_{ads} of Hg and Pb with their experimental $-\Delta H^{Au}_{ads}$ (see Table 4). Results for SHE will follow.

Assistance was also rendered to experiments on volatility of Hg/Cn and Pb/Fl with the use of SiO_2 covered detectors located ahead of the gold-plated ones in a chromatography column (allowing for differentiating between ΔH_{ads} on surfaces of these different materials).

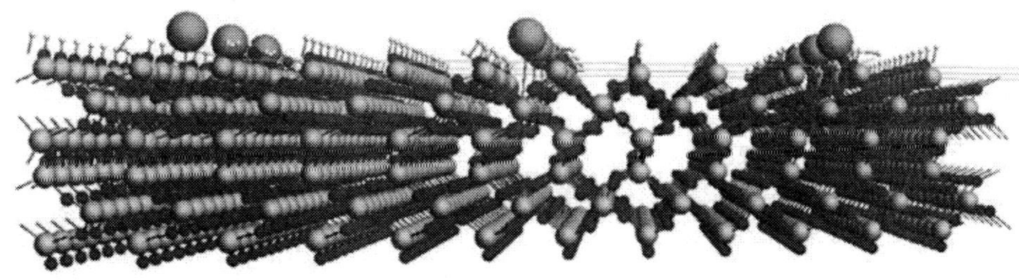

Figure 4. A supercell modelling adsorption of M on vicinal silanols for low adsorbate coverage. Reproduced from [25] with permission from the PCCP Owner Societies.

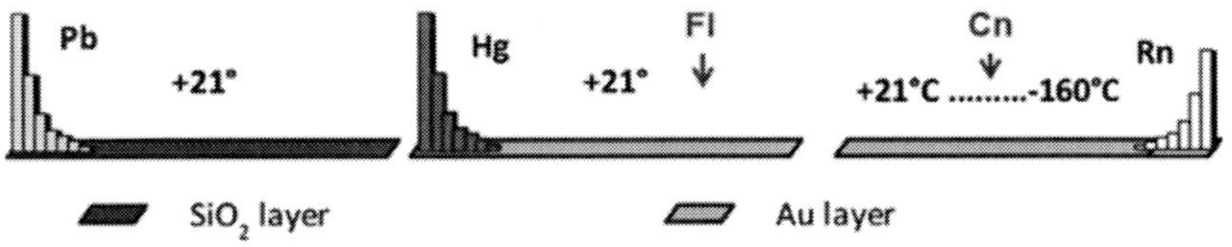

Figure 5. Predicted adsorption of Cn and Fl on quartz and gold surfaces of the chromatography column. Reproduced from [25] with permission from the PCCP Owner Societies.

Accordingly, periodic SR and SO calculations were performed on E_{ads} of these elements on a quartz surface using the ADF BAND code [25]. Two modifications of the hydroxylated SiO_2 surfaces were considered: vicinal (80%) and germinal (20%) silanols. A (4 × 4) super-cell modelling adsorption on a vicinal silanol for very low coverage is shown in Fig. 4.

Thus, according to the calculations [25], Pb should adsorb on the SiO_2 surface at room temperature, while Hg, Cn, and Fl should not. The latter three elements should adsorb, however, on gold, with Hg revealing much stronger interaction (Table 4 and Fig. 5). The reason for the non-interaction of Cn and Fl with quartz and weaker interaction with gold than that of Hg and Pb, respectively, is the strongest relativistic effects on their valence electrons making these elements rather inert. Experiments have, indeed, revealed that Cn and Fl do not adsorb on SiO_2 (ΔH_{ads} to be published).

The theory has also considered interaction of even heavier elements such as 115 through 120 with gold and other metals of interest on the basis of Au-containing dimers [6, 7]. These results might be of interest for future vacuum chromatography experiments, provided sufficiently long-living isotopes of these elements are found.

Some molecular species of even heavier elements are suggested in [14], whose stability may be proven by detailed calculations.

5.2 Aqueous chemistry experiments

Complex formation is known to increase in the groups at the beginning of the 3*d*-5*d*- series. It was, therefore, important to establish whether this trend is continued with the 6*d*-elements. Experimental investigations were accordingly devoted to the study of the complex formation of Rf, Db, and Sg (elements with sufficiently long half-lives) in aqueous solutions of HF, HCl, HBr, and H_2SO_4 acids. Liquid-liquid extraction, or sorption chromatography separations were applied to the studied species [2, 5]. To render assistance to those experiments, predictions of complex-formation constants and distribution coefficients, K_d, between an aqueous and organic phases (or resins) were made with the use of a model that treats a free energy change of a complex formation reaction as a sum of the ionic and covalent constituents. The latter are obtained via 4c-DFT electronic structure calculations of the systems of interest [6, 7].

The results have demonstrated that even though the heaviest elements are homologs of their lighter congeners in the chemical groups, trends in the complex formation are not necessarily continued with them. (See, e.g., predictions of an unexpected trend in the K_d values of the group-5 complexes by extraction from the HCl solutions into amines, confirmed by experiments [5].) The calculations have also shown that simple models based on the ratio of the cation charge to its size, or extrapolations are unreliable in the area of SHE. Only by performing relativistic calculations of the real chemical equilibria, can the complex formation, or hydrolysis be correctly predicted.

A summary of the predicted trends in hydrolysis, complex formation and extraction of the group-4 through 6 elements including the heaviest is given in [6, 7]. As one can see there, most of the predictions have been confirmed by the experiments, while some of them, like predictions for Sg in HF solutions, are still awaiting confirmation.

6. Summary

A large number of theoretical chemical studies were performed on the heaviest elements and their compounds such as molecules, aqueous complexes, and solid-state systems. They were based on the accurate calculations with the use of the most advanced relativistic methods, such as 4c/2c-DFT, PP and DF CC ones. With many of them carried out in a close link to the experimental research, these studies have contributed to better understanding of the chemistry of these exotic elements, and the role and magnitude of relativistic effects. A synergy between theoretical and experimental research has been achieved.

The results have demonstrated that even though the heaviest elements are homologs of their lighter congeners in the chemical groups, trends in properties are not necessarily continued with them. Thus, in this area of the Periodic Table simple extrapolations or semi-empirical models turned out to be unreliable. Relativistic calculations at the highest level of theory are, therefore, mandatory to achieve convincing results.

Relativistic theory will undergo further methodical developments like, e.g., creation of special techniques in atomic CC methods to treat open shell systems (needed for predictions of electronic configurations of elements with $Z > 122$), or inclusion of QED effects on the SCF basis in molecular calculations. Also, dispersive potentials for SHE should be worked out in DFT methods in order to treat more accurately weak interactions.

For elements heavier than $Z = 118$, theoretical investigations will be very exciting, since resemblance of properties with their lighter homologs in the groups is expected to be even less pronounced.

The author thanks her former and present collaborators J. Anton, T. Bastug, E. Eliav, U. Kaldor, A. Borschevsky, and M. Ilias for the fruitful joint work. She also appreciates valuable discussions of experimental results with A. Yakushev, J.V. Kratz, Ch. E. Düllmann, A. Türler, and R. Eichler.

References

[1] *The Chemistry of the Superheavy Elements*, 2nd edition, eds. M. Schädel and D. Shaughnessy (Springer, Heidelberg, 2014).
[2] A. Türler and V. Pershina, Chem. Rev. **113**, 1237 (2013).
[3] B. Fricke, Struct. Bond. **21**, 89 (1975).

[4] H.W. Gäggeler and A. Türler, *Gas-Phase Chemistry of Superheavy Elements*. In reference [1], p. 415.
[5] J.V. Kratz and Y. Nagame, *Liquid-Phase Chemistry of Superheavy Elements*. In reference [1], p. 309.
[6] V. Pershina, *Theoretical Chemistry of the Heaviest Elements*. In reference [1], p. 135.
[7] V. Pershina, Nucl. Phys. **A 944**, 578 (2015).
[8] P. Schwerdtfeger and M. Seth, In: *Encyclopedia on Calculational Chemistry* (Wiley, New York, 1998) Vol. **4**, p. 2480.
[9] E. Eliav, S. Fritzsche, U. Kaldor, Nucl. Phys. **A 944**, 518 (2015).
[10] P. Pyykkö, Chem. Rev. **88**, 563 (1988).
[11] P. Schwerdtfeger *et al.*, Nucl. Phys. **A 944**, 551 (2015).
[12] *Relativistic Electronic Structure Theory, Parts I and 2*, ed. P. Schwerdtfeger (Elsevier, Amsterdam, 2002).
[13] BAND2016, SCM, Theoretical Chemistry, Vrije Universiteit, Amsterdam, The Netherlands, `http://www.scm.com`.
[14] P. Pyykkö, Phys. Chem. Chem. Phys. **14**, 14734 (2012).
[15] P. Pyykkö, Chem. Rev. **112**, 371 (2012).
[16] C.S. Nash and B.E. Bursten, J. Am. Chem. Soc. **121**, 10830 (1999).
[17] V. Pershina and J. Anton, J. Chem. Phys. **138**, 174301 (2013).
[18] J. Even *et al.*, Science **345**, 1491 (2014).
[19] R. Eichler *et al.*, Nature, Letters **447**, 72 (2007).
[20] R. Eichler *et al.*, Radiochim. Acta **98**, 133 (2010).
[21] A. Yakushev *et al.*, Inorg. Chem. **53**, 1624 (2014).
[22] N. Gaston *et al.*, Angew. Chem. Int. Ed. **46**, 1663 (2007).
[23] V. Pershina, J. Anton, T. Jacob, J. Chem. Phys. **131**, 084713 (2009).
[24] V. Pershina, in progress.
[25] V. Pershina, Phys. Chem. Chem. Phys. **18**, 17750 (2016).

EPJ Web of Conferences **131**, 07003 (2016) DOI: 10.1051/epjconf/201613107003
Nobel Symposium NS160 – Chemistry and Physics of Heavy and Superheavy Elements

Gas-phase chemistry of element 114, flerovium

Alexander Yakushev[1,2,a] and *Robert* Eichler[3,4]

[1] GSI Helmholtzzentrum für Schwerionenforschung GmbH, 64291 Darmstadt, Germany
[2] Helmholtz-Institut Mainz, 55099 Mainz, Germany
[3] Paul Scherrer Institut, 5232 Villigen PSI, Switzerland
[4] University of Bern, Department of Chemistry and Biochemistry, 3012 Bern, Switzerland

Abstract. Element 114 was discovered in 2000 by the Dubna-Livermore collaboration, and in 2012 it was named flerovium. It belongs to the group 14 of the periodic table of elements. A strong relativistic stabilisation of the valence shell $7s^2 7p^2_{1/2}$ is expected due to the orbital splitting and the contraction not only of the $7s^2$ but also of the spherical $7p^2_{1/2}$ closed subshell, resulting in the enhanced volatility and inertness. Flerovium was studied chemically by gas-solid chromatography upon its adsorption on a gold surface. Two experimental results on Fl chemistry have been published so far. Based on observation of three atoms, a weak interaction of flerovium with gold was suggested in the first study. Authors of the second study concluded on the metallic character after the observation of two Fl atoms deposited on gold at room temperature.

1. Introduction

Man-made superheavy elements (SHE, atomic number $Z \geq 104$) are unique in two aspects. Their nuclei exist only due to nuclear *shell effects*, and their electron structure is influenced by increasingly important *relativistic effects* [1–3]. Recently the synthesis of elements 113, 115, 117 and 118 has been approved, thus, the discovery of all elements of the 7th row in the periodic table of elements with proton number Z up to 118 has been acknowledged [4]. The light transactinides, elements with $Z = 104$–112, are members of the *6d* transition metal series. Element 112, copernicium, Cn, has the closed-shell configuration $[Rn]5f^{14}6d^{10}7s^2$. Subsequently, the filling of the *7p* valence shell is expected starting with element 113 with the electron configuration $[Rn]5f^{14}6d^{10}7s^2 7p^1$ and ending with element 118 with noble gas electron configuration $[Rn]5f^{14}6d^{10}7s^2 7p^6$. Due to the increasing nuclear charge in SHE, the velocity of electrons close to the nuclei approaches the speed of light. This causes a relativistic increase in the electron mass. Hence, the spherical electron shells, having a non-zero electron density at the nucleus, contract in space and become stabilized in energy. This is the so-called direct relativistic effect. As a consequence, the non-spherical atomic orbitals are more efficiently screened from the nucleus, thus undergoing destabilization in energy and expansion in space: the indirect relativistic effect. Finally, the third effect is the spin-orbit splitting of atomic orbitals with $l > 0$. All three relativistic effects scale approximately with Z^2 for valence electron shells, and are thus most pronounced in SHE. The very large spin-orbit splitting in *7p* orbitals results in a quasi-closed-shell $7p^2_{1/2}$ in element 114, flerovium, Fl.

[a] Corresponding author: a.yakushev@gsi.de

This closed sub-shell, being spherical, is strongly stabilized by the direct relativistic effect, similar to the $7s^2$ atomic orbital. Thus, both Cn and Fl are expected to be more inert than their lighter homologues in the groups.

Calculations of the promotion energies for superheavy elements with $Z = 112$, 114, 118 were performed by Pitzer already in 1975 [5]. He concluded that the energy needed to promote electrons from the closed shells $s_{1/2}^2$ and $p_{1/2}^2$ to the valence state electron configurations sp and p^2 in Cn and Fl, respectively, is too high to be compensated by the energy gain from the chemical bond formation. Thus, Cn and Fl should be very inert, like *noble gases* [5]. More recent molecular, cluster and solid-state relativistic calculations on these elements suggest that Cn and Fl are more inert than their lighter homologues in the groups, but still reveal a metallic character [6–9], e.g., upon $M - M'$ interactions ($M =$ Cn or Fl, $M' =$ metal, e.g., Au). In contrast to Pitzer's conclusion, for Cn d-orbitals are expected to be involved in the metallic bonding, whereas Fl is expected to provide both $7p_{1/2}$ and $7p_{3/2}$ atomic orbitals for metal-metal interactions. Therefore, a metal-metal bonding is expected besides pure dispersion forces for either element upon adsorption on metal surfaces [8, 9]. An overview of theoretical predictions for the interaction of Cn and Fl with gold can be found in, for example, Ref. [10]. The absolute values depend on the chosen calculation method and differ from 0.30 to 0.52 eV for Cn and from 0.47 to 0.77 eV for Fl.

2. Experimental methods

For short-lived nuclei produced at the one-atom-at-a-time level, on-line gas-solid chromatography is the most rapid and most effective technique, which is usually applied in chemistry experiments with volatile compounds of superheavy elements in two modes: thermochromatography (TC) and isothermal chromatography (IC) [11, 12]. By TC, the deposition temperature of a volatile species in a column is a measure for the strength of the adsorption enthalpy. By IC, the retention time of a volatile species in a column kept at a varying temperature correlates with the adsorption enthalpy value. The most promising chemical separation method seems to be studies of the adsorption of Fl in the elemental state on a gold surface. Fl, as a volatile and a relative inert metal, can be transported to a chemical device by a carrier gas flow. Taking into account the large uncertainty of the predicted adsorption enthalpy values, a wide range of reactivity should be covered by an experimental setup: from the adsorption on gold at room temperature (metallic behaviour) to the deposition at a low temperature (noble gas behaviour).

This challenge calls for the application of the thermochromatograpy method, where the position of each atom in the column contributes chemical information about the investigated species. If a chromatography channel can be formed by position sensitive nuclear decay detectors, the chemical separation and the detection of the characteristic alpha particle energies are fulfilled in one device. The temperature in the column is usually ranging between room temperature and the temperature reachable by a cooling device, such as a liquid nitrogen cryostat. As an example of such thermochromatography detector channel, the open COMPACT detector is shown in Fig. 1. The silicon PIN diodes are covered with a thin layer of gold and form an array of 32 single detectors 1 cm wide and 32 cm long in each channel.

3. Results and discussion

Several studies aimed at the adsorption of Fl on gold were conducted at the FLNR Dubna, Russia by a collaboration led by the PSI group (Villigen, Switzerland) and at the GSI Darmstadt, Germany by the TASCA collaboration.

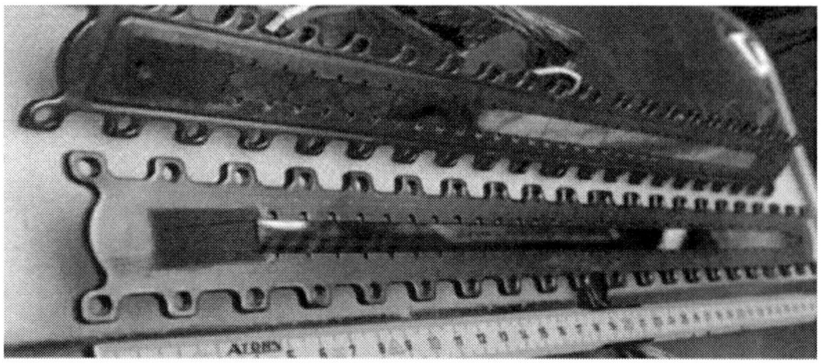

Figure 1. COMPACT detector array – a gas chromatography channel consisting of silicon detectors for α-particle- and SF-fragment spectroscopy.

The first observation of Fl in a chemistry experiment was made in 2007 in Dubna, where Fl isotopes were produced by the irradiation of ^{242}Pu and ^{244}Pu targets with a ^{48}Ca ion beam [13]. Only the long-lived Fl isotopes, ^{287}Fl ($T_{1/2} = 0.48^{+0.16}_{-0.09}$ s), ^{288}Fl ($T_{1/2} = 0.69^{+0.17}_{-0.11}$ s) and ^{289}Fl ($T_{1/2} = 1.9^{+0.7}_{-0.4}$ s) [14] are suitable for the chemical studies. Beam doses of 3.8×10^{18} and 4.5×10^{18} were accumulated during irradiations of ^{242}Pu and ^{244}Pu targets, respectively. The reaction products recoiling from the target were stopped in a recoil chamber made of quartz, which was continuously flushed with a gas flow (He:Ar = 1:1, 1.5 l/min). The species volatile at room temperature were transported through a PFA-Teflon capillary to the detector setup COLD within 2.2 s [13]. Unlike to the detection setup shown in Fig. 1, only one side of the detector channel was covered with gold. The background in the α-particle spectra was caused by other volatile by-products, like Rn and At isotopes, which could be also transported to the detection setup.

Three decay chains were observed (two of them at the temperature of about $-90\,°$C) and assigned to Fl isotopes: one to ^{287}Fl and two to ^{288}Fl. The more long-lived ^{289}Fl was also produced with the ^{244}Pu targets, but no unambiguous identification of the decay chains originating from ^{289}Fl was possible [13]. The observed position of Fl decay chains are presented in Fig. 2, on the panels 3 and 4. The distributions of ^{185}Hg, ^{219}Rn and ^{283}Cn decays are also shown on the panels 1 and 2 for comparison. The solid lines represent the deposition pattern obtained by the Monte Carlo simulations. The simulated deposition patterns for Fl isotopes are obtained for the adsorption enthalpy value $-\Delta H^{Au}_{ads}(\mathrm{Fl}) = 34\,\mathrm{kJ \cdot mol^{-1}}$. This value suggested that the deposition of Fl on gold occurs due to dispersion forces. However, large limits for the 95% confidence interval were determined: $-\Delta H^{Au}_{ads}(\mathrm{Fl}) = 34^{+54}_{-11}\,\mathrm{kJ \cdot mol^{-1}}$ (95% c.l.) [13].

The reaction ^{244}Pu(^{48}Ca;3–4n)^{288,289}Fl was used in the following experiment conducted by the PSI-FLNR group. For background reduction from volatile by-products, a pre-separation with the Dubna Gas-Filled Recoil Separator (DGFRS) was applied prior to the chemical separation.

One decay chain starting from ^{285}Cn was observed during 35 days at a temperature of $-93\,°$C [15]. The deposition temperature of the ^{285}Cn event is in a very good agreement with two Fl events from the first experiment (cf. Fig. 2). This chain can be an incomplete decay chain originating from ^{289}Fl, where the first member was missed. Two additional coincident SF events were registered and tentatively assigned to Fl isotopes or their progenies. Considering that both, Cn and Fl, are volatile elements, no unambiguous chemical conclusion on the observed decay chain could be made [15].

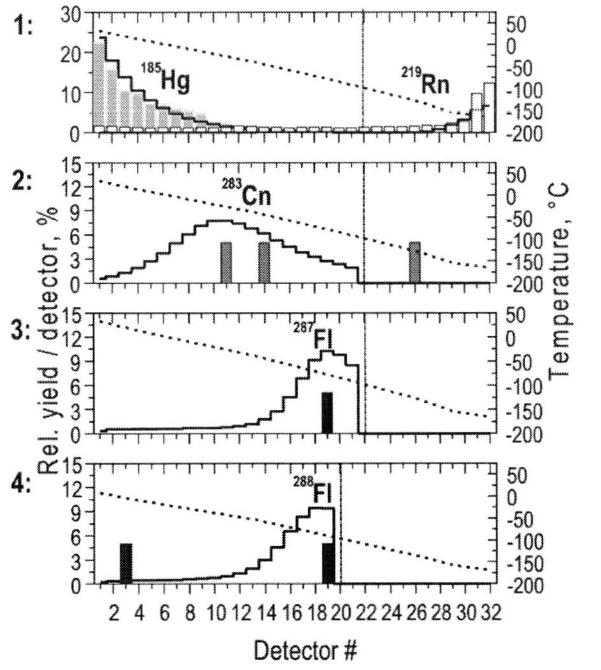

Figure 2. Simulated (solid lines) and measured (bars) deposition patterns of Hg, Rn, Cn, and Fl in COLD (see text for further details). The dotted line depicts the temperature gradient. Adopted from [13].

A different experiment on Fl chemistry behind a pre-separator was conducted at the GSI Darmstadt. The gas filled separator TASCA was used for the pre-separation. Two COMPACT detectors (Fig. 1) in series were mounted directly to the exit of the Recoil Transfer Chamber (RTC). The silicon detectors of both COMPACT arrays were covered with a thin gold layer [16]. The transport time was determined to be 0.81 s [16]. A short PTFE capillary (2 mm in diameter, 30 cm long) connected the detector arrays. To minimize the transport time, the first detector was connected via a 2-cm long PTFE tube to the RTC with a volume of 29 cm^3. The first detector array was kept at room temperature as an isothermal chromatography channel. The negative temperature gradient, which went down to $-162\,°C$, was applied on the second detector array. Again, the nuclear reaction ^{244}Pu(^{48}Ca;3–4n)^{288,289}Fl was used. A beam dose of 4×10^{18} particles was accumulated during 29 days. The chemical yield to the detection setup was measured with short-lived Hg and Pb isotopes. The 100% production rates of Hg and Pb isotopes were determined with a (58×58) mm^2 double-sided silicon strip detector installed in the RTC position. 27% of ^{182,183}Hg atoms and 20% of ^{186}Pb atoms were transported to the first COMPACT detector [16].

Two correlated chains, α–SF and α–α–SF originating from ^{288}Fl and ^{289}Fl, respectively, were observed. Both Fl decays occurred in the first isothermal detector array at room temperature. The positions of these two observed decay chains are presented Fig. 3 in comparison with deposition patterns measured for Pb, Hg, and Rn. Considering the low experimental statistics, a lower limit for the adsorption enthalpy of Fl on gold was determined $-\Delta H_{ads}^{Au} > 48\,\text{kJ} \cdot \text{mol}^{-1}$ (95% c.l.) Unlike the first experiment, a stronger interaction of Fl atoms with gold on the detector surface suggested the formation of weak metal-metal bonds. It was concluded that Fl has demonstrated a behaviour typical for a volatile metal in this experiment [16].

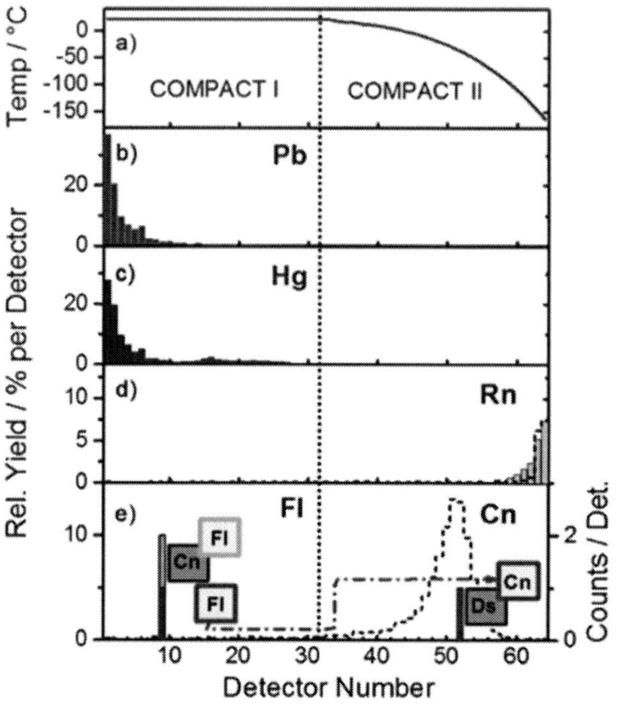

Figure 3. The deposition pattern of Fl decay chains in COMPACT detector arrays in comparison with that of Pb, Hg and Rn. Adopted from [16].

4. Conclusion on chemical properties of Fl and outlook

Two contradicting conclusions on the chemical behaviour follow from the first gas chromatography studies on the flerovium-gold interaction: one suggests weak physisorption of Fl upon the deposition on gold, another suggests a metallic character. The question "is flerovium a metal or a noble gas?" is still waiting for an unambiguous answer. Due to the limited number of observed Fl events so far, the error bars of the determined values of the adsorption enthalpy, representing the interaction strength between flerovium and gold, are very large [13, 16]. Additional experiments with higher statistics are required to obtain an accurate value. Despite the lower overall efficiency, combining the chemistry experiment with a pre-separation stage seems advantageous. Much lower background conditions allow the reliable identification of the decay chains, and thus, increase the sensitivity. Furthermore, the transport time can be reduced by the direct coupling of a chemistry setup to a pre-separator. Most recently, two additional runs were conducted at TASCA resulting in observation of several events from Fl and Cn isotopes. The analysis is being finalized by the TASCA collaboration.

References

[1] B. Fricke, Struct. Bond. **21**, 89 (1975).
[2] P. Pyykkö, Chem. Rev. **88**, 563 (1988).
[3] P. Schwerdtfeger and M. Seth, In: *Encyclopedia on Calculational Chemistry* (Wiley, New York, Vol. **4**, 1998), p. 2480.
[4] P.J. Karol, R.C. Barber, B.M. Sherrill, E. Vardaci, T. Yamazaki, Pure Appl. Chem. **88**, 139 (2016).

[5] K.S. Pitzer, J. Chem. Phys. **63**, 1032 (1975).
[6] V. Pershina, Radiochim. Acta **99**, 459 (2011).
[7] A. Hermann, J. Furthmülller, H.W. Gäggeler, P. Schwerdtfeger, Phys. Rev. B **82**, 155116 (2010).
[8] V. Pershina, J. Anton, T. Jacob, J. Chem. Phys. **131**, 084713 (2009).
[9] A. Zaitsevkii, C. van Wüllen, A. Titov, Russ. Chem. Rev. **78**, 1173 (2009).
[10] V. Pershina, Nucl. Phys. **A944**, 578 (2015).
[11] A. Türler, R. Eichler, A. Yakushev, Nucl. Phys. **A944**, 640 (2015).
[12] I. Zvara, *The Inorganic Radiochemistry of Heavy Elements – Methods for Studying Gaseous Compounds* (Springer, Dordrecht, 2008).
[13] R. Eichler *et al.*, Radiochim. Acta **98**, 133 (2010).
[14] Yu.Ts. Oganessian and V.K. Utyonkov, Rep. Prog. Phys. **78** 036301 (2015).
[15] D. Wittwer *et al.*, Nucl. Instrum. Methods B **268,** 28 (2010).
[16] A. Yakushev *et al.*, Inorg. Chem. **53**, 1624 (2014).

EPJ Web of Conferences **131**, 07004 (2016) DOI: 10.1051/epjconf/201613107004
Nobel Symposium NS160 – Chemistry and Physics of Heavy and Superheavy Elements

Toward an accurate description of solid-state properties of superheavy elements

A case study for the element Og ($Z = 118$)

Peter Schwerdtfeger[a]

Centre for Theoretical Chemistry and Physics, The New Zealand Institute for Advanced Study, Massey University (Albany Campus), Auckland, New Zealand

Abstract. In the last two decades cold and hot fusion experiments lead to the production of new elements for the Periodic Table up to nuclear charge 118. Recent developments in relativistic quantum theory have made it possible to obtain accurate electronic properties for the trans-actinide elements with the aim to predict their potential chemical and physical behaviour. Here we report on first results of solid-state calculations for Og (element 118) to support future atom-at-a-time gas-phase adsorption experiments on surfaces such as gold or quartz.

1. Introduction

Mendeleev arranged the known elements (60 in 1869) into a Periodic Table according to their atomic weights and recurring chemical properties. A proper foundation for this ordering scheme came through the electronic shell model of Bohr, and later through Schrödinger's quantum theory providing proper assignments for the ground state electron configurations of the elements. The Periodic Table now contains 118 known and named elements finishing with the $7p$ electronic shell closure at the recently discovered rare gas element 118 (Oganesson, Og) [1]. Predictions on the placements of even heavier elements have been made recently by Pyykkö [2, 3] for nuclear charges up to $Z = 172$[1] (see also earlier work by Fricke [4]), although attempts to synthesize elements beyond nuclear charge 118 have been unsuccessful so far.

The newest elements in the Periodic Table are rather short-lived undergoing predominantly α-decay within the milli-second to second range [5, 6]. Lifetimes in the second range or above open up the possibility for atom-at-a-time experiments to study the chemical and physical behaviour of such exotic elements [7–9]. For this one requires more neutron-rich superheavy element isotopes near the neutron shell closure at $N = 184$. Future experiments with neutron-rich projectiles or targets might yield longer lived isotopes to study the chemistry of the heaviest elements such as Og. Physisorption studies on gold surfaces already gave estimates for adsorption enthalpies for the superheavy elements 112 (Cn) and 114 (Fl). According to an early study of Eichler *et al.*, these adsorption enthalpies correlate well with the cohesive energies of the bulk elemental substances [10]. It is therefore of current interest to estimate the cohesive energies for such superheavy elements.

---

[a] e-mail: `p.a.schwerdtfeger@massey.ac.nz`

[1] At such high nuclear charges the mean field picture of the Dirac equation starts to breaks down [12].

It is currently a challenge for computational chemistry to obtain accurate lattice parameters and cohesive energies for bulk systems [11], especially for the heavier elements where relativistic and electron correlation effects compete with each other [12–15]. For the rare gas elements, the atomic interaction is of dispersive nature. The total energy of the crystal can therefore be decomposed into the individual many-body interaction terms [16, 17]. At long range such a many-body decomposition converges fast and one already obtains correct trends in cohesive energies along the rare gas series of elements from a simple sum over diatomic contributions [19]. Using such an incremental scheme, Stoll and co-workers were able to obtain accurate cohesive energies for the rare gas fcc crystals of neon, argon, krypton and xenon [20, 21]. Here we present preliminary results using a simple Lennard-Jones (LJ) potential for the two-body term and an Axilrod-Teller (AT) three-body ansatz [18] for the three-body term to obtain first estimates for cohesive energies of the rare gas fcc crystals Rn and Og.

2. Theory

The cohesive energy of a rare gas crystals can be written as a series of n-body interaction terms (called the incremental method) [16, 17],

$$E_{\text{coh}}(r_s) \cong \Delta E^{(2)}_{\text{coh}}(r_s) + \Delta E^{(2)}_{\text{HE-ZPE}}(r_s) + \Delta E^{(2)}_{\text{AE-ZPE}}(r_s) + \Delta E^{(3)}_{\text{coh}}(r_s) + \dots \tag{1}$$

Here r_s is the nearest neighbour distance in the solid, and $E^{(n)}$ is the n-body contribution to the cohesive energy [26],

$$\Delta E^{(2)}_{\text{coh}}(r_s) = \frac{1}{2}\epsilon L_{12}\left(\frac{r_e}{r_s}\right)^{12} - \epsilon L_6\left(\frac{r_e}{r_s}\right)^{6} \tag{2}$$

and

$$\Delta E^{(3)}_{\text{coh}}(r_s) = \frac{1}{3}C_{\text{AT}}\sum_{j<k}\left(1 + 3\cos\theta_0\cos\theta_j\cos\theta_k\right)\left(r_{0j}r_{0k}r_{jk}\right)^{-3} \tag{3}$$

where r_{ij} is the distance between atoms i and j in the crystal, θ_i is the angle between the vectors $\vec{r}_{ij}$ and $\vec{r}_{ik}$, and C_{AT} is the Axilrod-Teller coefficient approximated by known or calculated atomic static dipole polarizabilities α and ionization potentials E_{IP} [27, 28], i.e. we have,

$$C_{\text{AT}} = f_{\text{AT}}\alpha^3 E_{\text{IP}}. \tag{4}$$

The prefactors f_{AT} for Rn and Og are obtained from an extrapolation of known values for the lighter rare gases. We sum from a chosen atom placed at the origin to all other atoms in the lattice at distance r_{0j} with the nearest neighbour distance defined as $r_s = \min_j\{r_{0j}\}$. L_6 and L_{12} are the Lennard-Jones-Ingham coefficients for the fcc crystals given in Ref. [30]. For the zero-point vibrational energies (ZPE) to the two-body cohesive energy we used recently derived analytical expressions for the harmonic (HE) and anharmonic (AE) Einstein contributions (the latter derived from first-order perturbation theory) [29]. The optimization of the lattice geometry has been performed with our program system SAMBA [30].

3. Results

The results are shown in Tables 1 and 2, and summarized for the cohesive energies in Fig. 1. For the lighter rare gas solids vibrational effects are important whilst for the heavier systems the three-body force cannot be neglected anymore. For bulk Ne to Xe our results are in reasonably good agreement with the available experimental data. Deviations are mainly due

Table 1. Optimized bond distances for the rare gas fcc lattices from Ne to Og ($Z = 118$) using a Lennard-Jones (6–12) potential and a three-body Axilrod-Teller term. Equilibrium distances r_e of the diatomic taken from references [22–25], and nearest neighbour distance of the solid r_s^{min} at various levels of theory (see text). The total ZPE (TZPE) includes both harmonic (HE) and anharmonic (AE) terms. All values in [Å], atomic masses M in [u]. Experimental solid-state distances are from Refs. [31–35].

Element	M	r_e	r_s^{LJ}	$r_s^{LJ+HE-ZPE}$	$r_s^{LJ+TZPE}$	$r_s^{LJ+TZPE+AT}$	r_s^{exp}
^{20}Ne	19.992	3.089	3.001	3.125	3.120	3.134	3.157
^{40}Ar	39.962	3.762	3.654	3.697	3.697	3.731	3.756
^{84}Kr	83.912	4.016	3.900	3.925	3.925	3.968	4.009
^{132}Xe	131.904	4.363	4.237	4.254	4.254	4.310	4.336
^{222}Rn	222.018	4.427	4.300	4.310	4.310	4.367	–
^{294}Og	294.000	4.329	4.204	4.211	4.211	4.282	–

Table 2. Cohesive energies at optimized bond distances (see Table 1) for the rare gas fcc lattices from Ne to Og ($Z = 118$) using a Lennard-Jones (6–12) potential and a three-body Axilrod-Teller term. Lennard-Jones binding energies $-\epsilon$ (taken from references [22–25]) and cohesive energies E^{coh} at various levels of theory (see text). The total ZPE (TZPE) includes both harmonic (HE) and anharmonic terms (AE). All values are in [meV]. Experimental cohesive energies are from references [35–37].

Element	$-\epsilon$	E_{coh}^{LJ}	$E_{coh}^{LJ+HE-ZPE}$	$E_{coh}^{LJ+TZPE}$	$E_{coh}^{LJ+TZPE+AT}$	E_{coh}^{exp}
^{20}Ne	−3.63	−31.3	−23.6	−22.9	−22.2	−20.5
^{40}Ar	−12.3	−106.2	−96.9	−96.7	−90.1	−80.0
^{84}Kr	−17.3	−149.0	−141.7	−141.6	−130.3	−116.0
^{132}Xe	−24.3	−209.4	−203.0	−203.0	−183.3	−164.7
^{222}Rn	−34.9	−300.4	−294.5	−294.5	−265.8	(−202)
^{294}Og	−77.4	−666.4	−658.6	−658.5	−577.3	–

to the fact that we chose a simple Lennard-Jones two-body potential, the neglect of phonon dispersion and higher order terms in our many-body decomposition. This suggests that our predictions for the properties of the heaviest rare gas solids, Rn and Og, should be reasonably accurate as well.

The nearest neighbour distances r_s and cohesive energies $E_{coh}(r_s)$ are dependent on our chosen Lennard-Jones parameter in Eq. (6), that is the bond distance r_e and the binding energy ϵ. This can be seen from simple relationships known for Lennard-Jones interactions [30],

$$r_s^{LJ} = \left(\frac{L_{12}}{L_6}\right)^{1/6} r_e \tag{5}$$

and

$$E_{coh}(r_s^{LJ}) = \frac{1}{2}\left(\frac{L_6^2}{L_{12}}\right)\epsilon. \tag{6}$$

We see that the short bond distance for the Og dimer (Table 1) [24] is causing an anomalous behaviour in the trend down the rare gas dimers. This translates directly into an anomaly in the nearest neighbour distances of the rare gas solids down that group. The bond distances r_e and binding energies ϵ for Rn_2 and Og_2 were taken from relativistic coupled cluster calculations of Saue and co-workers, which include corrections for basis set superposition errors and are extrapolated to the basis set limit [24]. Their perturbative triples contributions to the dissociation energy of the rare gas dimers in the coupled cluster procedure are however rather large indicating that one has to include quadruple corrections as well within the coupled cluster procedure, which becomes computationally demanding. However, their CCSD(T)

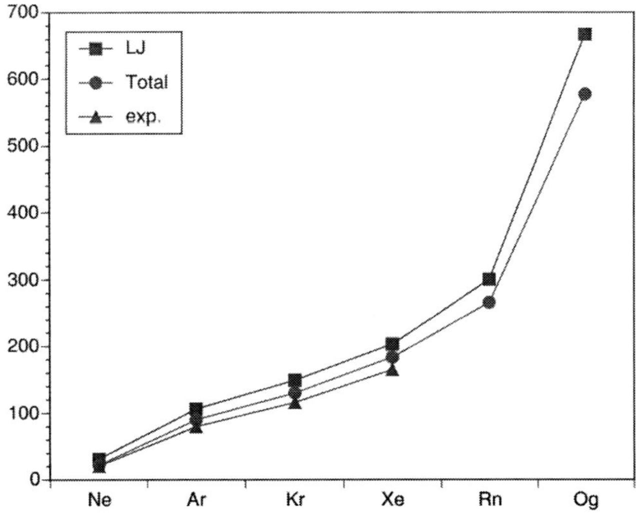

Figure 1. Cohesive energies $-E_{\mathrm{coh}}$ (total) in [meV] for the rare gas crystals in comparison to the available experimental values. The pure two-body Lennard-Jones (LJ) contribution is shown as well for comparison.

results for the bond distance and the dissociation energy of Xe_2 are in good agreement with experiment [35]. We note that our cohesive energies for the bulk systems are overestimated by 8.0% for Ne, to 12.6% for Ar, 11.2% for Kr, and 11.1% for Xe due to the various approximations made in this preliminary study. Taking this into account for the heavier rare gas solids our best estimates for the cohesive energies are 230 and 500 meV for Rn and Og, respectively. The cohesive energy for Og is unusually large for a rare gas element and comparable to the estimated cohesive energies of Cn ($390(^{+120}_{-100})$ meV) [38–40] and Fl (500 meV from theory [41] and $220(^{+230}_{-20})$ meV from experiment [42, 43]). Eichler and Eichler predicted thermochemical data for the superheavy elements empirically [44]. Their estimated cohesive energy is half our predicted value ($\sim$ 250 meV), but interestingly is larger than their empirical data for Cn and Fl.

4. Conclusions

We presented preliminary calculations for the rare gas solids up to the heaviest element Og. The cohesive energy of $\sim$ 500 meV for Og was calculated to be unusually large when compared to the other rare gas solids, which is most likely caused by the strong spin-orbit splitting in the $7p$ shell making the $7p_{3/2}$ orbital rather diffuse. This leads to an increased dipole polarizability for Og (59.0 a.u.) [47] compared to Rn (33.2 a.u.) [48], for example, thus increasing dispersive type of interactions and causing an anomaly in the trend of nearest neighbour distances for the bulk rare gas systems. Hence in this respect Og is not so inert anymore (NB: it is the first rare gas element with a positive electron affinity [45, 46]). We expect that bulk Og melts at much higher temperatures compared to the lighter rare gas elements. More accurate quantum chemical calculations for the potential energy curves and the Axilrod-Teller term of the heaviest rare gas dimers are however required to produce better cohesive energies. Relativistic effects (including and spin-orbit coupling) as a cause for this steep increase from Rn to Og in the cohesive energy need to be analyzed as well. Work in this direction is currently in progress.

This work was supported by a Marsden grant administered by the Royal Society of New Zealand. We thank Dr. Robert Eichler and Prof. Heinz Gäggeler for valuable discussions.

References

[1] Here we adopt the suggested IUPAC name for element 118, Oganesson. See http://iupac.org/.
[2] P. Pyykkö, Phys. Chem. Chem. Phys. **13**, 161 (2011).
[3] P. Pyykkö, Chem. Rev. **112**, 371 (2011).
[4] B. Fricke, in *Structure and Bonding* (Springer, Berlin, Vol. 21, 1975), p. 89.
[5] Yu.Ts. Oganessian *et al.*, Phys. Rev. C **74**, 044602 (2006).
[6] Yu.Ts. Oganessian *et al.*, Phys. Rev. Lett. **104**, 142502 (2010).
[7] M. Schädel (ed.), *The Chemistry of Superheavy Elements* (Kluwer Academic, Dordrecht, 2003).
[8] V. Pershina, Chem. Rev. **96**, 1977 (1997).
[9] A. Türler and V. Pershina, Chem. Rev. **113**, 237 (2013).
[10] R. Eichler, Radiochim. Acta **93**, 245 (2005).
[11] P. Fulde, *Electron Correlations in Molecules and Solids* (Springer Series in Solid-State Sciences, Springer, Heidelberg, 1995).
[12] P. Schwerdtfeger, L.F. Pašteka, A. Punnett, P. Bowman, Nucl. Phys. **A944**, 551 (2015).
[13] P. Pyykkö, Chem. Rev. **88**, 563 (1988).
[14] E. Eliav, U. Kaldor, P.Schwerdtfeger, B.A. Hess, Y.Ishikawa, Phys. Rev. Lett. **73**, 3203 (1994).
[15] P. Schwerdtfeger and M. Seth, in *The Encyclopedia of Computational Chemistry*, P. von R. Schleyer, N.L. Allinger, T. Clark, J. Gasteiger, P. Kollman, H.F. Schaefer III (eds.), Wiley, New York, **1998**, p. 2480.
[16] H. Stoll, Phys. Rev. B **46**, 6700 (1992).
[17] B. Paulus, Phys. Rep. **428**, 1 (2006).
[18] B.M. Axilrod and E. Teller, J. Chem. Phys. **11**, 299 (1943).
[19] A. Hermann, R.P. Krawczyk, M. Lein, P. Schwerdtfeger, I.P. Hamilton, J.J.P. Stewart, Phys. Rev. A **76**, 013202 (2007).
[20] K. Rosćiszewski, B. Paulus, P. Fulde, H. Stoll, Phys. Rev. B **60**, 7905 (1999).
[21] K. Rosćiszewski, B. Paulus, P. Fulde, H. Stoll, Phys. Rev. B **62**, 5482 (2000).
[22] E. Bich, R. Hellmann, E. Vogel, Mol. Phys. **106**, 1107 (2008).
[23] B. Jäger, R. Hellmann, E. Bich, E. Vogel, Mol. Phys. **107**, 2181 (2009).
[24] A. Shee, S. Knecht, T. Saue, Phys. Chem. Chem. Phys. **17**, 10978 (2015).
[25] P. Slavíček, R. Kalus, P. Paška, I. Odvárková, P. Hobza, A. Malijevský, J. Chem. Phys. **119**, 2102 (2003).
[26] J.E. Lennard-Jones and A.E. Ingham, Proc. Roy. Soc. A **107**, 636 (1925).
[27] C.E. Moore, *Atomic Energy Levels*; Natl. Bur. Stand. (U.S.) Circ. No. 467; U.S. GPO: Washington D.C., 1958.
[28] P. Schwerdtfeger, *Table of experimental and calculated static dipole polarizabilities for the electronic ground states of the neutral elements* (CTCP Auckland, 2016); available at http://ctcp.massey.ac.nz/index.php?group=&page=dipole&menu=dipole.
[29] P. Schwerdtfeger, A. Burrows, E. Pahl, unpublished results.
[30] P. Schwerdtfeger, N. Gaston, R.P. Krawczyk, R. Tonner, G.E. Moyano, Phys. Rev. B **73**, 064112 (2006).
[31] C.S. Barrett and L. Meyer, J. Chem. Phys. **41**, 1078 (1964).

[32] O.G. Peterson, D.N. Batchelder, R.O. Simmons, Phys. Rev. **150**, 703 1966.
[33] D.N. Batchelder, D.L. Losee, R.O. Simmons, Phys. Rev. **162**, 767 (1967).
[34] D.L. Losee and R.O. Simons, Phys. Rev. **172**, 944 (1968).
[35] J. Ogilvie and F.Y. Wang, J. Mol. Struct. **273**, 277 (1992).
[36] G.J. McConville, J. Chem. Phys. **60**, 4093 (1974).
[37] L.A. Schwalbe, R.K. Crawford, H.H. Chen, R.A. Aziz, J. Chem. Phys. **66**, 4493 (1977).
[38] R. Eichler *et al.*, Nature **447**, 72 (2007).
[39] N. Gaston, I. Opahle, H.W. Gäggeler, P. Schwerdtfeger, Angew. Chem. Int. Ed. **46**, 1663 (2007).
[40] R. Eichler *et al.*, Angew. Chem. Int. Ed. **47**, 3262 (2008).
[41] A. Hermann, H. Gäggeler, P. Schwerdtfeger, Phys. Rev. B **82**, 155116 (2010).
[42] R. Eichler *et al.*, Radiochim. Acta **98**, 133 (2010).
[43] A. Yakushev *et al.*, Inorg. Chem. **53**, 1624 (2014).
[44] R. Eichler and B. Eichler, in *The Chemistry of Superheavy Elements*, M. Schädel and D. Shaughnessy (eds.), (Springer, Heidelberg, 2014), p. 375.
[45] E. Eliav, U. Kaldor, Y. Ishikawa, P. Pyykkö, Phys. Rev. Lett. **77**, 5350 (1996).
[46] I. Goidenko, L. Labzowsky, E. Eliav, U. Kaldor, P. Pyykkö, Phys. Rev. A **67**, 020102(R) (2003).
[47] V. Pershina, A. Borschevsky, E. Eliav, U. Kaldor, J. Chem. Phys. **129**, 144106 (2008).
[48] T. Nakajima and K. Hirao, Chem. Lett. (Jpn.) **8**, 766 (2001).

EPJ Web of Conferences **131**, 07005 (2016) DOI: 10.1051/epjconf/201613107005
Nobel Symposium NS160 – Chemistry and Physics of Heavy and Superheavy Elements

Complex chemistry with complex compounds

Robert Eichler[1,2,a], *M.* Asai[3], *H.* Brand[4], *N.M.* Chiera[1,2], *A.* Di Nitto[4,5], *R.* Dressler[1], *Ch.E.* Düllmann[4,5,6], *J.* Even[4,6], *F.* Fangli[7], *M.* Goetz[4,5,6], *H.* Haba[8], *W.* Hartmann[4], *E.* Jäger[4], *D.* Kaji[8], *J.* Kanaya[8], *Y.* Kaneya[3], *J.* Khuyagbaatar[4,6], *B.* Kindler[4], *Y.* Komori[8], *B.* Kraus[1,2], *J.V.* Kratz[5], *J.* Krier[4], *Y.* Kudou[8], *N.* Kurz[4], *S.* Miyashita[3,9], *K.* Morimoto[8], *K.* Morita[8,10], *M.* Murakami[8,11], *Y.* Nagame[3], *K.* Ooe[11], *D.* Piguet[1], *N.* Sato[8], *T.K.* Sato[3], *J.* Steiner[4], *P.* Steinegger[1,2], *T.* Sumita[8], *M.* Takeyama[8], *K.* Tanaka[8], *T.* Tomitsuka[11], *A.* Toyoshima[3], *K.* Tsukada[3], *A.* Türler[1,2], *I.* Usoltsev[1,2], *Y.* Wakabayashi[8], *Y.* Wang[7], *N.* Wiehl[5,6], *Y.* Wittwer[1,2], *A.* Yakushev[4,6], *S.* Yamaki[8], *S.* Yano[8], *S.* Yamaki[8], and *Z.* Qin[7]

[1] Paul Scherrer Institute, 5232 Villigen, Switzerland
[2] University of Bern, 3012 Bern, Switzerland
[3] Advanced Science Research Center, Japan Atomic Energy Agency, Tokai, Ibaraki 319-1195, Japan
[4] GSI Helmholtzzentrum für Schwerionenforschung GmbH, 64291 Darmstadt, Germany
[5] Johannes Gutenberg-Universität Mainz, 55099 Mainz, Germany
[6] Helmholtz-Institut Mainz, 55099 Mainz, Germany
[7] Institute of Modern Physics Lanzhou, Chinese Academy of Sciences, 730000 Lanzhou, China
[8] Nishina Center for Accelerator-Based Science, RIKEN, Wako, Saitama 351-0198, Japan
[9] Hiroshima University, Kagamiyama, Higashi-Hiroshima 739-8526, Japan
[10] Kyushu University, Higashi-Ku, Fukuoka, 812-8581, Japan
[11] Niigata University, Niigata, Niigata 950-2181, Japan

Abstract. In recent years gas-phase chemical studies assisted by physical pre-separation allowed for the investigation of fragile single molecular species by gas-phase chromatography. The latest success with the heaviest group 6 transactinide seaborgium is highlighted. The formation of a very volatile hexacarbonyl compound $Sg(CO)_6$ was observed similarly to its lighter homologues molybdenum and tungsten. The interactions of these gaseous carbonyl complex compounds with quartz surfaces were investigated by thermochromatography. Second-generation experiments are under way to investigate the intramolecular bond between the central metal atom of the complexes and the ligands addressing the influence of relativistic effects in the heaviest compounds. Our contribution comprises some aspects of the ongoing challenging experiments as well as an outlook towards other interesting compounds related to volatile complex compounds in the gas phase.

1. Introduction

About two thirds of the transactinide elements known today with $104 \leq Z \leq 111$ are expected to be typical members of the so-called transition metal series. These elements are known as typical metals and reveal an exceptionally low volatility in the elemental state – the lowest known amongst all elements of the periodic table. Hence, their gas-phase chemical investigation relies on the formation of volatile compounds. So far, only very stable

[a] Corresponding author: robert.eichler@psi.ch

and simple inorganic compounds like oxides, oxohydroxides, halides or oxohalides of the transactinides Rf-Hs have been investigated proving their typical transition metal behaviour (see references [1, 2] for recent reviews). Pioneering experiments performed at the Berkeley Gas-filled Separator (BGS) at the Lawrence Berkeley National Laboratory (LBNL), USA between 2003 and 2005 investigated the possibility of in-situ formation of gas-phase complex compounds of group 4 elements of the periodic table Zr and Hf with the goal to investigate the corresponding compound formation also for element 104, rutherfordium (Rf) [3]. It was shown that pre-separated products of complete fusion reactions can be stopped in a carrier gas stream containing a reactive species suited as ligand for gas-phase complex formation. An in-situ formation of volatile species was observed serving as proof-of-principle and paving the way towards the investigation of new fragile compound classes of transactinides. Later in 2012–2013 the in-situ formation of volatile species of transition metal elements Mo, Tc, Ru, Rh were observed produced as short-lived fission products recoiling out from targets of ^{249}Cf and ^{238}U at the TRIGA reactor of the Institute for Nuclear Chemistry at the University of Mainz, Germany, into a carrier gas stream containing carbon monoxide, CO [4, 5]. Experiments at the TASCA separator at GSI Darmstadt, Germany [4–6] as well as at the Institute of Modern Physics, Lanzhou, China, confirmed these observations [7, 8] and demonstrated the formation of volatile species as well for recoiling nuclear fusion reaction products of transition metal elements of the sixth row of the periodic table: W, Re, Os, Ir. Chemical properties established for transition metals in CO containing chemical systems allowed identifying those volatile products likely as gas-phase carbonyl complexes. The metal-ligand bonding in carbonyl complexes is somewhat special, as the central atom stays in its zeroth oxidation state. The bonding emerges from electron pair donation from the CO molecules surrounding the central atom into its empty *d* orbitals and subsequent π-back bonding due to overlapping of the thus filled *d* orbitals with antibonding π^*–orbitals of CO, the latter loosening the C–O bond, but strengthening the M-C bond (see reference [9] for a review). The number of CO ligands is driven by the number of empty *d* orbitals of the central atom. Thus the 18-electron rule applies, suggesting complexes with 18 electrons to be most stable configurations, e.g. $M(CO)_6$ for group 6 elements and $M(CO)_5$ for group 8 elements. Typically, the elements from odd group numbers form dimer or trimer complexes with CO bridges or additional intermetallic cluster bonds to stabilize themselves according to the 18-electron rule. This is obviously impossible for single atomic amounts as observed in these experiments. Thus the volatile species observed for the even-numbered group members could be attributed to predominant formation of the most stable configurations. However, the speciation of the volatile products of odd-numbered elements appears to be tentative.

2. Group 6 carbonyl complex compounds

Interestingly, already in 1999 predictions about the stability of hexacarbonyl complexes of group 6 elements and uranium were published [10] (see Table 1). Relativistic density functional theory was used to calculate the geometry and thermodynamic stability of these carbonyl species for Mo, W, and Sg indicating the hexacarbonyl species as the most stable ones obeying the 18 electron rule and suggesting the stability increase towards the heavier elements. This was attributed mainly to the increasing secondary relativistic effects destabilizing and expanding *d*orbitals. The results for Mo and W were in good agreement with experimentally measured thermodynamic stabilities [11] and calculations from reference [9]. More recently, the physisorption interaction of group 6 hexacarbonyl species on quartz surfaces were quantified using relativistic density functional theory in conjunction with a physisorption model [12] (see Table 1). On-line gas-phase adsorption chromatography

Table 1. Relevant calculated and experimentally determined thermodynamic data available for group 6 hexacarbonyl species: the standard adsorption enthalpy on quartz surfaces at zero surface coverage, ΔH_{ads}(Quartz), and first bond dissociation enthalpy, FBDE.

Element	ΔH_{ads}(Quartz), kJ/mol [exp.] [15]	ΔH_{ads}(Quartz), kJ/mol [theor.] [12]	FBDE, kJ/mol [exp] [11]	FBDE, kJ/mol [theor.] [10]
Mo	49 ± 2	48.5 ± 2	169	169
W	50 ± 2	46.5 ± 2	192	192
Sg	50 ± 2	46.2 ± 2	–	208

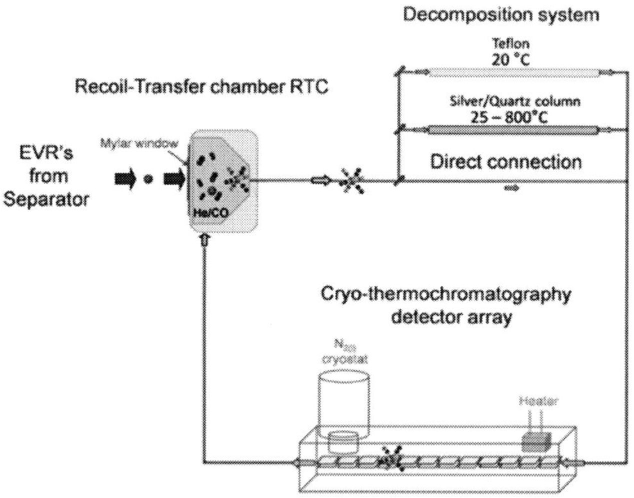

Figure 1. Scheme of the gas loop setup installed at the focal plane of a recoil separator for the online investigation of chemical properties of group 6 hexacarbonyl species regarding the formation, the adsorption, and thermal decomposition (for details see text).

chemistry experiments performed using radioactive isotopes of group 6 elements allowed for the determination of the adsorption interaction of these carbonyl species on quartz surfaces, pointing to a weak physisorption interaction, indeed, as expected by theory and from experiments with macroscopic amounts [4–8]. Thus, indirectly the speciation of the observed compounds was bolstered.

2.1 The experimental observation of $Sg(CO)_6$

The typical scheme of experiments with single atoms behind gas-filled separators is shown in Fig. 1 for the case of the experiments using the GARIS separator at RIKEN Wako, Japan.

The general principle following this pre-separation scheme is described here: Recoiling nuclear fusion reaction products recoiling out of the target are separated according to their momentum-over-charge ratio and are guided by magnetic lens systems to the focal plane of the separator. There they pass through a thin Mylar window separating the (0.5 torr) He-gas filling of the separator from the gas-phase chemistry device starting with the Recoil Transfer Chamber (RTC). RTC is a closed chamber with gas inlets and an outlet mounted behind the Mylar window covering the entire area of this vacuum window. The RTC is flushed with gas mixtures of He and CO of variable content held at room temperature and at pressures typically between 0.1 and 1 bar. After passing this Mylar foil the products are thermalized within the volume of the RTC. Here, the in-situ formation reaction occurs. The formed volatile species are transported with the gas stream through thin PFA® Teflon capillaries with

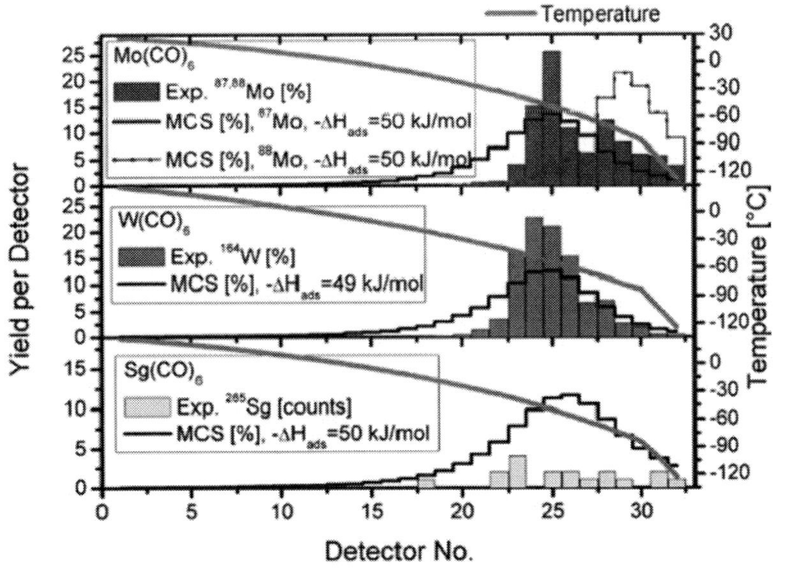

Figure 2. Thermochromatographic deposition patterns (bars) as measured for the group 6 hexacarbonyl compounds including Sg and revealing a very similar weak adsorption interaction of these species with quartz surfaces (adapted from Ref. [15]).

2 mm inner diameter over the direct connection (see Fig. 1) to the detection system. Cryo-thermochromatography setups as used for the chemical investigations of group 8 oxides MO_4 and for the investigation of volatile elements like Hg, Cn, and Fl (see reference [14] for a review) are best suited to study the in-situ chromatographic deposition of the formed volatile species. Using detector surfaces as chromatographic surfaces represents nowadays the most sensitive approach to detect single atomic adsorption. The detectors themselves are sensitive to the characteristic time correlated α and spontaneous fission decays. The following nuclear reactions were applied to produce single atomic isotopes of elements Mo, W, and Sg:

$$^{nat}\mathrm{Zn}(^{24}\mathrm{Mg}, xn)^{87}\mathrm{Mo} \rightarrow (^{87}\mathrm{Mo}\colon T_{1/2} = 14\,\mathrm{s}, E_{\beta+} = 5.3 - 5.5\,\mathrm{MeV})$$

$$^{144}\mathrm{Sm}(^{24}\mathrm{Mg}, 4n)^{164}\mathrm{W} \rightarrow (^{164}\mathrm{W}\colon T_{1/2} = 6.0\,s, E_{\alpha} = 5.15\,\mathrm{MeV})$$

$$^{248}\mathrm{Cm}(^{22}\mathrm{Ne}, 5n)^{265}\mathrm{Sg} \rightarrow (^{265a}\mathrm{Sg}\colon T_{1/2} = 8.5\,s, E_{\alpha} = 8.84\,\mathrm{MeV};$$

$$^{265b}\mathrm{Sg}\colon T_{1/2} = 14.4\,s, E_{\alpha} = 8.69\,\mathrm{MeV}).$$

The formation of volatile carbonyl species was observed and their deposition in the cryo-thermochromatography detector COMPACT [14, 15] was investigated. Example results are shown in Fig. 2. Based on the comparison with measurements performed using detector arrays, where all products passing the separator were implanted at the focal plane of the separator it was possible to estimate a formation and transport yield as 40% for ^{87}Mo, and 10–20% for ^{164}W. Comparison of the number of events attributed to decays of ^{265}Sg observed in COMPACT with results from measurements studying this nuclear fusion reaction in detail [16] allowed for estimation of an overall chemical yield for Sg of about 5% including the detection efficiency. The reasons for these low yields could so far not be understood. However, for the first time a transactinide complex compound featuring a metal-carbon bond as is characteristic of the organometallic compounds was investigated.

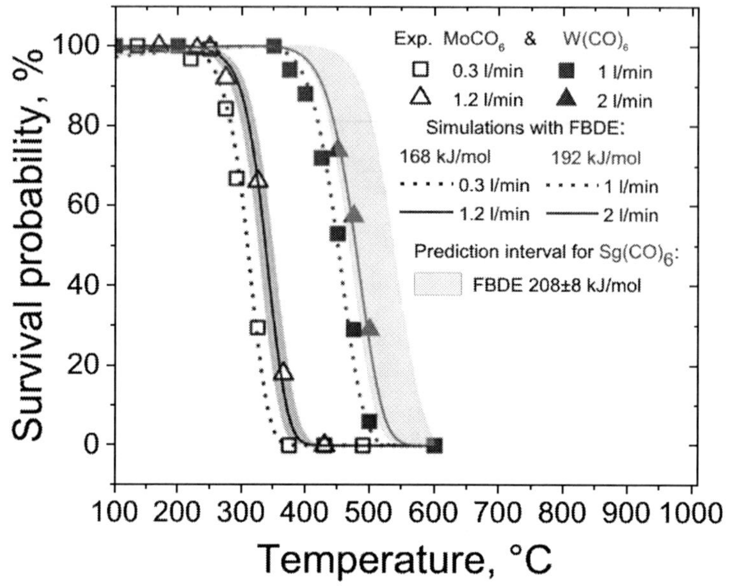

Figure 3. Experimental decomposition curves of $Mo(CO)_6$ (blue symbols) and $W(CO)_6$ (green symbols) at different gas flow rates. Simulated decomposition curves (lines) using for $Mo(CO)_6$: $-\Delta H_{ads} = 48.8$ kJ/mol, FBDE = 169 kJ/mol; and for W(CO)6: $-\Delta H_{ads} = 47.5$ kJ/mol, FBDE = 192 kJ/mol. The blue shaded area shows the response of the model to ΔH_{ads} within uncertainties of 4 kJ/mol. The prediction interval within the uncertainty of 8 kJ/mol for the FBDE of $Sg(CO)_6$ given in Ref. [10] is shown as red shaded area.

2.2 The stability of group 6 hexacarbonyl compounds $M(CO)_6$

The observations described in the previous chapter paved the way towards the recently developed second-generation experiments with group 6 hexacarbonyls aiming for the detailed investigation of the stability of the metal carbon bond in those molecules. The relevant thermodynamic measure for this property is typically represented as the first carbonyl bond dissociation enthalpy (see Table 1, FBDE). Model experiments with isotopes of Mo emerging form a ^{252}Cf spontaneous fission source "Miss Piggy" installed at Bern University [17] were performed. It was shown in Ref. [18] that the simplest approach addressing the stability on a molecular level would be thermal decomposition studies on a silver surface. A typical decomposition setup used for this purpose comprised a tubular flow reactor of 1 m length and 4 mm inner diameter entirely covered inside with thin silver foils. The reactor could be heated to defined temperatures between room temperature and 600 °C. This reactor was placed between the carbonyl production side (for example, the RTC) and the detector for hexacarbonyl species (for example, COMPACT) (see the decomposition setup in Fig. 1). The amount of hexacarbonyl species able to pass this reactor in dependence of the selected isothermal temperature is measured in comparison to the total amount of hexacarbonyl species passing through a PFA® Teflon tube of the same geometry mounted as by-pass of the decomposition reactor as depicted in Fig. 1. The results obtained for $Mo(CO)_6$ and $W(CO)_6$ are shown in Fig. 3 (symbols). The observed shift of the decomposition curves of $W(CO)_6$ of about 100 °C relative to the curves measured for $Mo(CO)_6$ confirm a higher stability of the tungsten complex as expected from Refs. [10, 11]. A kinetic decomposition model was developed superimposing gas chromatographic transport of species through the open tubular reactor with heterogeneous decomposition of the species upon adsorption on the inner surfaces of the reactor [19]. This model was successfully applied to relate the decomposition observations to the FBDE given for the hexacarbonyl complexes [10, 11]. The obtained model describes the decomposition observations for Mo and W at varied experimental parameters very well (see Fig. 3, lines). Therefore, it was also applied in reference [19] to predict the hypothetical decomposition interval for $Sg(CO)_6$ at the same experimental conditions (at 1 l/min gas flow rate) based on the theoretically predicted FBDE [10]. This prediction revealed the suggested method to be sensitive enough for a comparative study. Very recently, first decomposition experiments with $^{265}Sg(CO)_6$ have been performed successfully. The final data analysis is ongoing and will be presented elsewhere.

3. Conclusions

The availability of products of heavy ion induced nuclear fusion reactions pre-separated from the intense heavy ion beams paved the way towards the investigation of highly interesting fragile gas-phase complex compounds. The observation of volatile carbonyl complex formation for group 7–9 elements asks for the development of speciation techniques, for example based on mass spectrometry, to identify those volatile species.

Other organic ligands are known to form volatile gas-phase complexes with transition metals. Here the cyclopentadien forming ferrocene and ruthenocene shall be mentioned. The compound class of acetylacetonates with which this type of experiments have been started are still to be further explored. An interesting feature of the lighter 16 elements is the formation of gaseous "monocarbonyls" MCO M = S [20], Se [21, 22], Te [23], though the bonding situation is different from that in the the complex compounds presented in this paper. However, in this chemical system the formation of volatile PoCO remains to be studied in preparation of future experiments with element 116, livermorium.

References

[1] A. Türler and V. Pershina, Chem. Rev. **113**, 1237 (2013).
[2] *The Chemistry of Superheavy Elements*, 2nd edn. (eds. M. Schädel and D. Shaughnessy, Springer, Berlin, 2014).
[3] Ch.E. Düllmann *et al.*, Radiochim. Acta **97**, 403 (2009).
[4] J. Even *et al.*, Inorg. Chem. **51,** 6431 (2012).
[5] J. Even *et al.*, Radiochim. Acta **102**, 1093 (2014).
[6] J. Even *et al.*, J. Radioanal. Nucl. Chem. **303**, 2457 (2015).
[7] Y. Wang *et al.*, Radiochim. Acta **102**, 69 (2014).
[8] Y. Wang, Z. Qin, F.-L. Fan, H. Haba, Y. Komori, S.-W Cao, X.-L. Wua, C.-M. Tana, Phys. Chem. Chem. Phys. **17**, 13228 (2015).
[9] G. Frenking and N. Fröhlich, Chem. Rev. **100**, 717 (2000).
[10] C.S. Nash and B.E. Bursten, J. Am. Chem. Soc. **121,** 10830 (1999).
[11] K.E. Lewis, D.M. Golden, G.P. Smith, J. Am. Chem. Soc. **106**, 3905 (1984).
[12] V. Pershina and J. Anton, J. Chem. Phys. **138**, 174301 (2013).
[13] H. Haba *et al.*, J. Nucl. Radiochem. Sci. **8**(2), 55 (2007).
[14] A. Türler, R. Eichler, A. Yakushev, Nucl. Phys. **A944**, 640 (2015).
[15] J. Even *et al.*, Science **345**, 1491 (2014).
[16] H. Haba *et al.*, Phys. Rev. C **85**, 024611 (2012).
[17] Ch.E. Düllmann *et al.*, Nucl. Instr. and Meth. **A512**, 595 (2003).
[18] I. Usoltsev *et al.*, Radiochim. Acta **104**, 141 (2016).

[19] I. Usoltsev, R. Eichler, A. Türler, Radiochim. Acta, in press, DOI: 10.1515/ract-2015-2447 (2016).
[20] E.K.C. Lee, Y.N. Tang, F.S., Rowland, J. Phys. Chem., **68**(2), 318 (1964).
[21] R.S. Baldwin, B.O. Peoto, L.B. Church, Radiochim. Acta **16**(2), 94 96 (1971).
[22] M. Wachsmuth, B. Eichler, L. Tobler, D.T. Jost, H.W. Gäggeler, M. Ammann, Radiochim. Acta **88**, 873 (2000).
[23] R.G. Strickert, S. Amiel, A.C. Wahl, Inorg. Nucl. Chem. Lett. **10**, 129 (1974).

EPJ Web of Conferences **131**, 07006 (2016) DOI: 10.1051/epjconf/201613107006
Nobel Symposium NS160 – Chemistry and Physics of Heavy and Superheavy Elements

Superheavy element chemistry at GARIS

Hiromitsu Haba[a]

Nishina Center for Accelerator-Based Science, RIKEN, Wako, Saitama 351-0198, Japan

Abstract. A gas-jet transport system has been installed to the RIKEN GAs-filled Recoil Ion Separator, GARIS to start up SuperHeavy Element (SHE) chemistry. This system is a promising approach for exploring new frontiers in SHE chemistry: background radioactivities from unwanted by-products are suppressed, a high gas-jet transport yield is achieved, and new chemical reactions can be investigated. Useful radioisotopes of $^{261}Rf^{a,b}$, ^{262}Db, and $^{265}Sg^{a,b}$ for chemical studies were produced in the reactions of $^{248}Cm(^{18}O,5n)^{261}Rf^{a,b}$, $^{248}Cm(^{19}F,5n)^{262}Db$, and $^{248}Cm(^{22}Ne,5n)^{265}Sg^{a,b}$, respectively. They were successfully extracted to a chemistry laboratory by the gas-jet method. Production and decay properties of $^{261}Rf^{a,b}$, ^{262}Db, and $^{265}Sg^{a,b}$ were investigated in detail with the rotating wheel apparatus for α- and spontaneous fission spectrometry. Present status and perspectives of the SHE chemistry at GARIS are also briefly presented.

1. Introduction

Studies on chemical properties of SuperHeavy Elements (SHEs, atomic number $Z \geq 104$) are extremely interesting and challenging research subjects in modern nuclear and radiochemistry [1, 2]. SHEs are produced at accelerators in heavy-ion induced nuclear fusion reactions. Due to short half-lives and low production rates of SHEs, any chemistry experiment must be done on a one-atom-at-a-time basis. Thus, rapid and efficient radiochemical procedures must be devised. So far, by coupling a gas- or liquid-chromatography apparatus to a target/recoil-catcher chamber through a gas-jet transport line, chemical properties of SHEs have been investigated for elements 104 (Rf) to 108 (Hs), and recently elements 112 (Cn) to 114 (Fl) [1–3]. At the same time, these successful experiments have definitely demonstrated the limitations of the conventional gas-jet technique just behind the target. Large amounts of background radioactivities from unwanted reaction products become unavoidable for identifications of SHEs with larger Z, especially $Z \geq 106$. High-intensity beams from advanced heavy-ion accelerators give rise to a problem in that the plasma formed by the beam in the target/recoil-catcher chamber significantly reduces gas-jet extraction yields. To overcome these limitations, a coupling of SHE chemistry to a recoil separator for nuclear physics studies has been considered [4].

In 2002, the pioneering SHE experiment with the Recoil Transfer Chamber (RTC) coupled to Berkeley Gas-filled Separator (BGS) at Lawrence Berkeley National Laboratory was first noteworthy [5]. The isotope of ^{257}Rf separated from the large amount of α-decaying products was identified with a liquid scintillator after a liquid-liquid solvent extraction with SISAK. The BGS/RTC system was then applied to a development of the Cryo-Thermochromatographic Separator (CTS) with $^{169-173}OsO_4$ as a homologue of HsO_4 [6]. Further, the production of Zr and Hf isotopes with the BGS/RTC system was investigated

[a] e-mail: haba@riken.jp

using the ^{nat}Ge(^{18}O,xn)^{85}Zr, ^{74}Se(^{18}O,$\alpha 3n$)^{85}Zr, ^{112}Sn(^{50}Ti,$4n$)^{158}Hf, ^{116}Sn(^{50}Ti,$4n$)^{162}Hf, ^{120}Sn(^{50}Ti,$5n$)^{165}Hf, and ^{124}Sn(^{50}Ti,$5n$)^{169}Hf reactions [7]. The solvent extraction of ^{85}Zr and ^{169}Hf with crown ethers [8] and the gas-chromatographic separation of volatile ^{85}Zr and ^{162,165,169}Hf complexes with hexafluoroacetylaceton [9] were conducted as model experiments of Rf.

At GSI Helmholtzzentrum für Schwerionenforschung, a gas-filled recoil separator, the TransActinide Separator and Chemistry Apparatus (TASCA) was built in 2009. In the commissioning of the separator, different RTCs were tested with Pb and Os isotopes produced in the ^{152}Gd(^{40}Ar,xn)$^{192-x}$Pb and ^{nat}Ce(^{40}Ar,xn)$^{175-177}$Os reactions, respectively [10]. The electrodeposition behaviour of $^{175-177}$Os on Pd and Ni electrodes was investigated with an electrolytic cell as a model experiment of Hs. In the final phase of the TASCA commissioning, the fluoride complexation of Rf in a diluted hydrofluoric acid was studied by anion-exchange chromatography with the Automated Rapid Chemistry Apparatus (ARCA) using ^{261}Rf^{a} produced in the ^{244}Pu(^{22}Ne,$5n$)^{261}Rf^{a} reaction. Further, volatile carbonyl complexes of $W(CO)_6$ and $Os(CO)_5$ were synthesized at TASCA using W and Os isotopes produced in the ^{144}Sm(^{24}Mg,4;$5n$)^{163,164}W and ^{152}Gd(^{24}Mg,5;$6n$)^{170,171}Os reactions, respectively [11]. Recently, Fl isotopes were produced in the ^{244}Pu(^{48}Ca,4;$3n$)^{288,289}Fl reactions, and the adsorption behaviour of Fl and their α-decay product ^{284,285}Cn on Au or SiO_2 surface were studied by isothermal gas chromatography and thermochromatography together with Pb, Hg, and Rn [12]. The Dubna Gas-Filled Recoil Separator (DGFRS) was also used for gas-phase chemistry of Fl [13].

At the RIKEN RI Beam Factory, a gas-jet transport system was installed to the GAs-filled Recoil Ion Separator (GARIS) to start up the SHE chemistry at RIKEN. The performance of the system was systematically investigated using ^{88}Nb, ^{170}Ta, ^{206}Fr, ^{245}Fm, and ^{255}No produced in the ^{nat}Ge(^{19}F,xn)^{88}Nb, ^{nat}Gd(^{19}F, xn)^{170}Ta, ^{169}Tm(^{40}Ar,$3n$)^{206}Fr, ^{208}Pb(^{40}Ar,$3n$)^{245}Fm, and ^{238}U(^{22}Ne,$5n$)^{255}No reactions, respectively [14–16]. An accurate and precise excitation function is necessary for effective production of SHE nuclides with extremely low production yields and for planning of long-term experiments with a costly heavy-ion accelerator. The decay properties such as α-particle energy and half-life are also essential to unambiguously assign SHE nuclides and to derive their chemical information. Especially, the half-life value is a direct measure for chemical constants such as equilibrium constant and adsorption enthalpy. Thus, using the GARIS gas-jet system, useful radioisotopes of ^{261}Rf^{a,b}, ^{262}Db, and ^{265}Sg^{a,b} for chemical studies were produced in the ^{248}Cm(^{18}O,$5n$)^{261}Rf^{a,b}, ^{248}Cm(^{19}F,$5n$)^{262}Db, and ^{248}Cm(^{22}Ne,$5n$)^{265}Sg^{a,b} reactions, respectively. Production and decay properties of those isotopes were investigated in detail using a rotating wheel apparatus for α and spontaneous fission (SF) spectrometry [17–20]. In this paper, the recent studies on the production and decay properties of ^{261}Rf^{a,b}, ^{262}Db, and ^{265}Sg^{a,b} are summarized together with the SHE chemistry programs at GARIS.

2. Production of ^{261}Rf, ^{262}Db, and ^{265}Sg

2.1 GARIS gas-jet system

A schematic of the GARIS gas-jet setup is shown in Fig. 1. The ^{18}O, ^{19}F, and ^{22}Ne beams were extracted from the RIKEN linear accelerator for the production of ^{261}Rf^{a,b}, ^{262}Db, and ^{265}Sg^{a,b}, respectively. The $^{248}Cm_2O_3$ targets of about 300 μg cm^{-2} thickness were prepared by a molecular-plating method on a 2-μm Ti backing foil. The eight arc-shaped targets were mounted on a rotating wheel of 100-mm diameter, and the wheel was rotated during the irradiation at 1000 rpm. The EVaporation Residues (EVRs) of interest were separated in-flight from the beam particles and the majority of the nuclear transfer products through

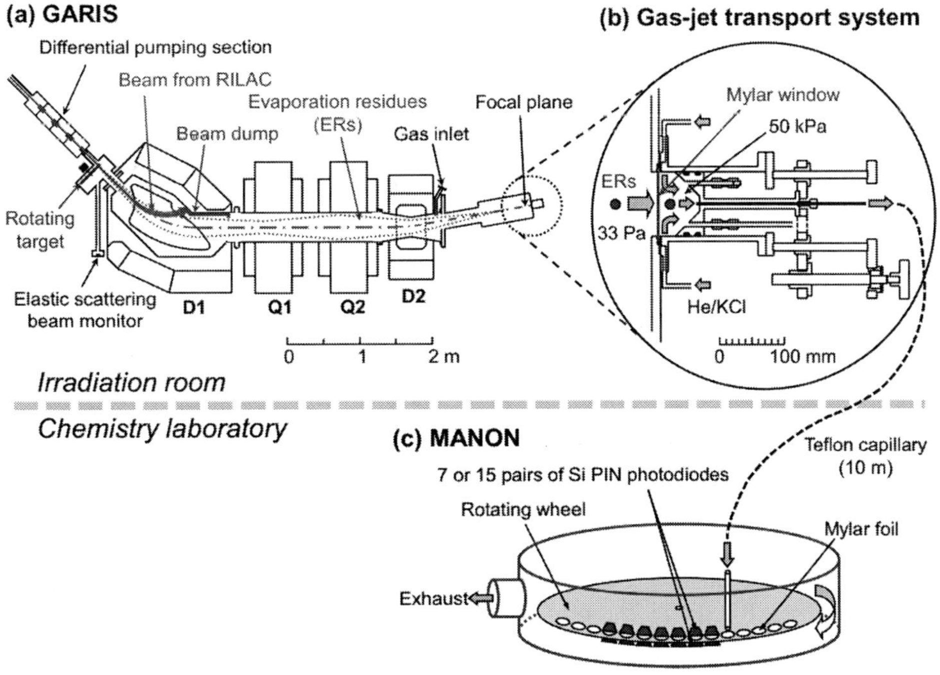

Figure 1. (a) RIKEN GAs-filled Recoil Ion Separator (GARIS). (b) Gas-jet transport system coupled to GARIS. (c) The rotating wheel apparatus MANON for α and SF (Spontaneous Fission) spectrometry at the chemistry laboratory.

D1 (D: dipole), Q1 (Q: quadrupole), Q2, and D2 electromagnets, and were guided to a focal plane of GARIS. GARIS was filled with He gas at a pressure of 33 Pa. The transmissions of GARIS were $(7.8 \pm 1.7)\%$ [17, 18], $(8.1 \pm 2.2)\%$ [20], and ~13% [19] for ^{261}Rf^{a,b}, ^{262}Db, and ^{265}Sg^{a,b}, respectively. At the focal plane, the EVRs passed through a Mylar vacuum window foil of 0.5-μm (^{261}Rf^{a,b} and ^{262}Db) or 0.7-μm (^{265}Sg^{a,b}) thickness which was supported by a honeycomb-hole grid with 84% transparency. In the gas-jet chamber (100-mm inner diameter $\times$ 20-mm depth), the EVRs were thermalized in He gas, attached to aerosol particles of potassium chloride (KCl), and were continuously transported through a Teflon capillary (2.0-mm inner diameter $\times$ 10-m length) to a chemistry laboratory, where the Measurement system for Alpha-particle and spontaneous fissioN events ON line (MANON) was equipped. The flow rate of the He gas was 2.0 l min^{-1}, and the inner pressure of the gas-jet chamber was about 50 kPa. The gas-jet transport efficiency and average transport time were 50% and 3 s, respectively [17, 18].

In MANON, the aerosol particles were deposited on a Mylar foil of 0.5-μm thickness, 40 of which were set on the periphery of a rotating wheel. The wheel was stepped to position the foils between seven or fifteen pairs of Si PIN photodiodes. The step interval of MANON was set to 30.5, 2.0, 15.5, and 20.5 s to effectively detect ^{261}Rf^{a}, ^{261}Rf^{b}, ^{262}Db, and ^{265}Sg^{a,b}, respectively, with their different half-lives. Each Si PIN photodiode had an active area of 18×18 mm^2 and a 38% counting efficiency. The α-energy resolution for the samples deposited on the Mylar foil was about 50-keV full width at half maximum in the top detectors and about 100 keV in the bottom detectors; the latter was larger because of energy degradation in the Mylar foil. All events were recorded in an event-by-event mode together with time information.

2.2 Production and decay properties of ^{261}Rf, ^{262}Db, and ^{265}Sg

2.2.1 $^{261}Rf^{a,b}$

Rutherfordium-261 with a half-life of $T_{1/2} = 68$ s was produced in the ^{248}Cm(^{18}O,5n)^{261}Rf reaction [17]. The beam energy was 95.5 MeV in the middle of the $^{248}Cm_2O_3$ target, and the average beam intensity was 5 particle μA (pμA). In the α-particle spectrum measured with MANON, α peaks of ^{261}Rf and its daughter ^{257}No were clearly identified under low background conditions. The radioactivities due to decays of Po, At, Rn, Fr, Ra, Ac, and Th isotopes, which were largely produced in the transfer reactions on the lead impurity in the target [21], were fully removed by the GARIS gas-jet system.

Although the 68-s ^{261}Rf (^{261}Rf^a) had been known for many decades in the ^{248}Cm(^{18}O,5n)^{261}Rf reaction since 1970 [21], the 3-s isomeric state ^{261}Rf^b found on the decay chains from ^{277}Cn and ^{269}Hs had not been observed as an evaporation residue. Under a low background condition attained with the GARIS gas-jet system, ^{261}Rf^b was successfully identified based on α-α correlations linking α decays of ^{261}Rf^b and its known daughter ^{257}No [18]. The α-particle energy of ^{261}Rf^b was measured to be 8.52 ± 0.05 MeV. The half-life was determined to be 1.9 ± 0.4 s based on both 8.52-MeV α and SF decays. The α and SF branches are $b_\alpha = 0.27 \pm 0.06$ and $b_{SF} = 0.73 \pm 0.06$, respectively. The cross section for the ^{248}Cm(^{18}O,5n)^{261}Rf^b reaction is $\sigma(^{261}\mathrm{Rf}^b) = 11 \pm 2$ nb at 95.1 MeV, which gives a cross-section ratio of $\sigma(^{261}\mathrm{Rf}^a)/\sigma(^{261}\mathrm{Rf}^b) = 1.1 \pm 0.2$.

2.2.2 ^{262}Db

Dubnium-262 was produced in the ^{248}Cm(^{19}F,5n)^{262}Db reaction at beam energies of 97.4 and 103.1 MeV: at the latter energy, the excitation function calculated by the HIVAP code [22] exhibits the maximum cross section. The beam intensity was 4 pμA. Based on the genetically correlated 73 α-α and two α-SF decay chains of ^{262}Db and its known daughter nuclide ^{258}Lr, two α lines at $E_\alpha = 8.46 \pm 0.04$ and 8.68 ± 0.03 MeV with α intensities $I_\alpha = 70 \pm 5\%$ and $30 \pm 5\%$, respectively, were assigned to ^{262}Db, though three energies of $E_\alpha = 8.45$ ($I_\alpha = 75\%$), 8.53 (16%), and 8.67 MeV (9%) had been adopted for ^{262}Db in [23]. The half-life of $T_{1/2} = 33.8^{+4.4}_{-3.5}$ s obtained for ^{262}Db confirmed the literature data [23]. The SF branch of $b_{SF} = 52 \pm 4\%$ is fairly larger than that adopted in [23] ($b_{SF} = \sim 33\%$). The average α-particle energy of $E_\alpha = 8.61$ MeV and the half-life of $T_{1/2} = 3.54^{+0.46}_{-0.36}$ s were obtained for ^{258}Lr, and they agree with the literature values [23]. The EC decay was observed in ^{258}Lr for the first time with $b_{EC} = 0.026 \pm 0.018$. The decay pattern for the chain $^{262}\mathrm{Db} \rightarrow {}^{258}\mathrm{Lr} \rightarrow {}^{258}\mathrm{No}$ is shown in Fig. 2a. The production cross sections for the ^{248}Cm(^{19}F,5n)^{262}Db reaction were $\sigma(5n) = 2.1 \pm 0.7$ and $0.23^{+0.18}_{-0.11}$ nb at 103.1 and 97.4 MeV, respectively. They were reproduced by the HIVAP code [22].

No correlated α-α or α-SF pair was observed for the four-neutron evaporation product ^{263}Db and its α-decay daughter ^{259}Lr. The upper-limit cross sections of $\sigma(4n) \leq 0.064$ nb at 103.1 MeV and ≤ 0.13 nb at 97.4 MeV derived for the ^{248}Cm(^{19}F,4n)^{263}Db reaction are rather smaller than the HIVAP prediction.

2.2.3 $^{265}Sg^{a,b}$

Two isomeric states of ^{265}Sg, i.e., ^{265}Sg^a and ^{265}Sg^b, proposed in [24] were produced in the ^{248}Cm(^{22}Ne,5n)^{265}Sg reaction at the beam energy of 117.8 MeV [19]. The beam intensity was 3 pμA. Based on the correlated α-α(-α) and α-SF decay chains, 18 and 24 events were assigned to ^{265}Sg^a and ^{265}Sg^b, respectively. The half-life and α-particle energy of ^{265}Sg^a were measured to be $T_{1/2} = 8.5^{+2.6}_{-1.6}$ s and $E_\alpha = 8.84 \pm 0.05$ MeV, respectively, and those of

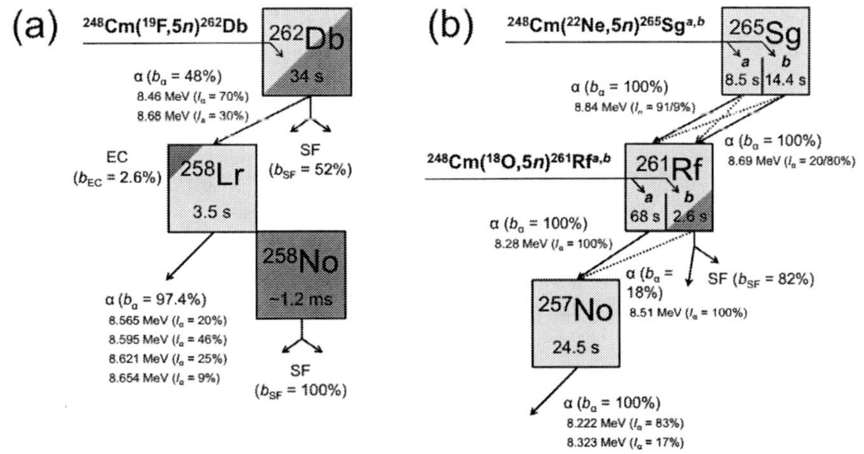

Figure 2. Decay patterns for the chain (a) ^{262}Db $\rightarrow^{258}$ Lr $\rightarrow^{258}$No and (b) ^{265}Sg^{a,b} $\rightarrow^{261}$Rf^{a,b} $\rightarrow^{257}$No.

^{265}Sg^{b} were $T_{1/2} = 14.4^{+3.7}_{-2.5}$ s and $E_{\alpha} = 8.69 \pm 0.05$ MeV. As a daughter product of ^{265}Sg^{a,b}, the decay properties of ^{261}Rf^{b} were also derived: $T_{1/2} = 2.6^{+0.7}_{-0.5}$ s, $E_{\alpha} = 8.51 \pm 0.06$ MeV, and $b_{SF} = 0.82 \pm 0.09$. These results confirmed and refined the complicated decay pattern suggested for the decay chain ^{269}Hs $\rightarrow$ ^{265}Sg^{a,b} $\rightarrow^{261}$ Rf^{a,b} $\rightarrow$ ^{257}No in [24] as shown in Fig. 2b. The production cross sections for ^{265}Sg^{a} and ^{265}Sg^{b} were reliably determined to be $\sigma = 180^{+80}_{-60}$ and 200^{+60}_{-50} pb at 117.8 MeV, respectively.

3. SHE chemistry at GARIS

The GARIS gas-jet system is expected to open new chemical studies that were not accessible before. In the gas-phase chemistry, the beam-free condition in the gas-jet chamber makes it possible to synthesize a large variety of chemical compounds. Even *et al.* [25] succeeded in the synthesis of the first carbonyl complex of SHEs, $Sg(CO)_6$ using the pre-separated ^{265}Sg^{a,b} with GARIS. A next-generation experiment to investigate the stability of the metal carbon bond in $Sg(CO)_6$ is ready with a thermal decomposition setup [26].

So far, aqueous chemistry studies of SHEs have been carried out for Rf, Db, and Sg mainly using batch-wise column chromatography apparatuses such as the Automated Rapid Chemistry Apparatus (ARCA) and the Automated Ion-exchange separation apparatus coupled with the Detection system for Alpha spectroscopy (AIDA) [1, 2]. However, there have been no reports on the aqueous chemistry of Sg and the heavier SHEs after the pioneering studies on fluoride complexation and hydrolysis of Sg in the 1990s [1, 2]. Toward the aqueous chemistry of Sg and element 107, Bh, Komori *et al.* [27] have been developing a rapid solvent extraction apparatus coupled to the GARIS gas-jet system. This new chemistry system consists of a continuous dissolution apparatus (Membrane DeGasser: MDG) [28], the Flow Solvent Extractor (FSE) consisting of a Teflon capillary and a phase separator, and a flow liquid scintillation detector for α/SF-spectrometry. The performance of MDG and FSE was investigated in the extraction system of HNO_3/tri-n-octylamine in toluene using ^{92,94m}Tc and ^{181}Re produced in the ^{nat}Mo(d,xn) and ^{nat}W(d,xn) reactions, respectively, at the RIKEN AVF cyclotron [27]. It was found that the extraction equilibrium of Tc and Re in 1 M HNO_3 can be obtained with a 0.5-mm i.d. × 40-cm capillary at a flow rate of 1 ml min^{-1} for both aqueous and organic solutions. The time required for the solutions to pass through the 40-cm capillary is calculated to be 2.4 s. This is short enough to conduct the solvent extraction of ^{266}Bh ($T_{1/2} = \sim 10$ s) produced in the ^{248}Cm(^{23}Na,5n)^{266}Bh reaction at GARIS. The flow liquid scintillation detector is under development.

4. Summary

The GARIS gas-jet system was developed to open the next generation SHE chemistry. Useful radioisotopes of ^{261}Rf^{a,b}, ^{262}Db, and ^{265}Sg^{a,b} for chemical studies were successfully extracted to the chemistry laboratory, and the production and decay properties of those isotopes were investigated. These results clearly demonstrated that the GARIS gas-jet method is a promising approach for exploring new frontiers in SHE chemistry: the background radioactivities from unwanted reaction products are strongly suppressed, the intense primary heavy-ion beam is absent in the gas-jet chamber, and hence the high gas-jet transport yield is achieved. Further the beam-free conditions make it possible to investigate new chemical systems. The first chemistry behind GARIS, the chemical synthesis and gas-chromatographic separation of $Sg(CO)_6$ were successful. The novel rapid solvent extraction apparatus is under development toward the aqueous chemistry of the heaviest SHEs such as Sg and Bh.

References

[1] M. Schädel and D. Shaughnessy (eds.), *The Chemistry of Superheavy Elements, 2nd edition* (Springer, Heidelberg, New York, Dordrecht, and London, 2013).
[2] A. Türler and V. Pershina, Chem. Rev. **113**, 1237 (2013).
[3] S.N. Dmitriev *et al.*, Mendeleev Commun. **24**, 253 (2014).
[4] Proceedings of "The 1st Workshop on Recoil Separator for Superheavy Element Chemistry", March 20–21, 2002, Darmstadt, Germany.
[5] J.P. Omtvedt *et al.*, J. Nucl. Radiochem. Sci. **3**, 121 (2002).
[6] U.W. Kirbach *et al.*, Nucl. Instr. Meth. **A484**, 587 (2002).
[7] Ch.E. Düllmann *et al.*, Nucl. Instr. Meth. **A551**, 528 (2005).
[8] R. Sudowe *et al.*, Radiochim. Acta **94**, 123 (2006).
[9] Ch.E. Düllmann *et al.*, Radiochim. Acta **97**, 403 (2009).
[10] J. Even *et al.*, Nucl. Instr. Meth. **A638**, 157 (2011).
[11] J. Even *et al.*, Inorg. Chem. **51**, 6431 (2012).
[12] A. Yakushev *et al.*, Inorg. Chem. **53**, 1624 (2014).
[13] D. Wittwer *et al.*, Nucl. Instrum. Meth. B **268**, 28 (2010).
[14] H. Haba *et al.*, J. Nucl. Radiochem. Sci. **8**, 55 (2007).
[15] H. Haba *et al.*, J. Nucl. Radiochem. Sci. **9**, 27 (2008).
[16] M. Huang *et al.*, J. Radioanal. Nucl. Chem. **304**, 845 (2015).
[17] H. Haba *et al.*, Chem. Lett. **38**, 426 (2009).
[18] H. Haba *et al.*, Phys. Rev. C **83**, 034602 (2011).
[19] H. Haba *et al.*, Phys. Rev. C **85**, 024611 (2012).
[20] H. Haba *et al.*, Phys. Rev. C **89**, 024618 (2014).
[21] A. Ghiorso *et al.*, Phys. Lett. **32B**, 95 (1970).
[22] W. Reisdorf and M. Schädel, Z. Phys. A **343**, 47 (1992).
[23] R.B. Firestone and V.S. Shirley, *Table of Isotopes 8th edition* (John Wiley & Sons, New York, 1996).
[24] Ch.E. Düllmann and A. Türler, Phys. Rev. C **77**, 064320 (2008).
[25] J. Even *et al.*, Science **345**, 1491 (2014).
[26] I. Usoltsev *et al.*, Radiochim. Acta **104**, 141 (2015).
[27] Y. Komori *et al.*, RIKEN Accel. Prog. Rep. **49,** in press.
[28] K. Ooe *et al.*, J. Radioanal. Nucl. Chem. **303**, 1317 (2015).

EPJ Web of Conferences **131**, 07007 (2016) DOI: 10.1051/epjconf/201613107007
Nobel Symposium NS160 – Chemistry and Physics of Heavy and Superheavy Elements

Chemical properties of rutherfordium (Rf) and dubnium (Db) in the aqueous phase

Yuichiro Nagame[1,a], *Jens Volker* Kratz[2], and *Matthias* Schädel[3]

[1] Advanced Science Research Center, Japan Atomic Energy Agency, Tokai, Ibaraki 319-1195, Japan
[2] Institut für Kernchemie, Johannes Gutenberg-Universität Mainz, 55128 Mainz, Germany
[3] GSI Helmholtzzentrum für Schwerionenforschung GmbH, 64291 Darmstadt, Germany

Abstract. Recent experimental studies of the chemical characterization of the first two transactinide elements, rutherfordium (Rf) and dubnium (Db), conducted atom-at-a-time in aqueous phases are reviewed. A short description on experimental techniques based on partition methods, specifically automated rapid chemical separation systems, is also given. Perspectives for aqueous-phase chemistry experiments on heavier elements are briefly discussed.

1. Introduction

Chemical studies of the transactinide elements with atomic numbers $Z \geq 104$ provide not only crucial and challenging opportunities to advance our understanding of properties of matter at the limits of existence but they also elucidate the influence of relativistic effects on atomic electrons and the architecture of the Periodic Table of the Elements at the farthest reach [1–6]. These elements are all man-made ones synthesized at accelerators using nuclear reactions of heavy-ion beams with heavy element target materials and they can only be identified by measuring their characteristic nuclear decay, or that of their known daughter nuclei, with sensitive detection techniques. As both half-lives and cross sections of these nuclides are rapidly decreasing with increasing Z, they are usually available in quantities of only a few atoms or often one atom-at-a-time.

The chemical characterization of the transactinide elements in aqueous-phase experiments has been accomplished by partition methods with single atoms, e.g. liquid-liquid extraction, ion-exchange chromatography, and reversed-phase extraction chromatography. The ultimate goal of these atom-at-a-time scale partition experiments is to determine the so-called distribution coefficient, K_d, as a function of ligand concentration. The K_d value is given in its simplest definition as the ratio of the number of atoms – determined by its radioactivity – either in the organic phase (in liquid-liquid extractions) or in the stationary phase (in column chromatography) to that in the aqueous phase. In order to get statistically significant results, it is indispensable to repeat the same experimental procedure several hundred or even several thousand times with cycle times of about the lifetime of the nuclide under investigation. In these processes, the behaviour of the transactinide element is compared with that of its lighter homologs under identical conditions. Investigations have been carried out with automated rapid chemical separation apparatuses to characterize chemical properties of these elements.

[a] e-mail: nagame.yuichiro@jaea.go.jp

Here we summarize recent studies of the chemical separation and characterization of the first two transactinide elements, rutherfordium (Rf) and dubnium (Db), the only ones which so far have been investigated in greater detail in aqueous phases. Some perspectives for further studies are briefly given. Large parts of this paper are based on previous review articles [7–10].

2. Aqueous-phase experiments

Extensive series of detailed investigations were made possible by the development of computer-controlled automated systems that have greatly improved the ability to perform rapidly and reproducibly large numbers of chromatographic separations on miniaturized columns. This made it possible to systematically vary ligand concentrations to determine the stoichiometry of the eluted species. The experiments with automated devices have produced detailed and sometimes surprising results that called for a detailed theoretical modelling of the chemical properties with improved relativistic quantum-chemical calculations. Recent experimental results obtained with automated rapid chemical separation systems are summarized in [10].

The aqueous-phase experiments with the transactinide elements are carried out on the basis of the following steps: (i) synthesis of a specific isotope of a transactinide element, (ii) rapid transport of this nuclide to chemical separation devices by an aerosol loaded gas-jet transport technique, (iii) fast chemical characterization that includes dissolution in an aqueous solution containing inorganic ligands for complex formation, (iv) partition of this complex between the aqueous and an organic phase or an ion-exchange resin, (v) preparation of a sample suitable for nuclear spectroscopy (α-particle spectroscopy), which usually requires the evaporation of an aqueous solution to dryness, and (vi) detection of nuclides through their characteristic nuclear decay properties for an unambiguous identification. For characterization of the transactinide elements, recent partition experiments have been conducted using the automated rapid ion-exchange separation apparatus ARCA II (Automated Rapid Chemistry Apparatus) with a miniaturized computer-controlled liquid chromatography system [11] and AIDA (Automated Ion-exchange separation apparatus coupled with the Detection system for Alpha spectroscopy) [12]; AIDA consists of ARCA II and an automated on-line α-particle detection system. AIDA enables cyclic discontinuous column chromatographic separations of short-lived nuclides in aqueous phases and automated detection of α particles within a typical cycle time of 60 s.

3. Element 104, rutherfordium (Rf)

Pioneering aqueous-phase experiments with 68-s ^{261}Rf produced in the ^{248}Cm(^{18}O, 5n) reaction clearly indicated that the behaviour of Rf is fundamentally different from that of heavy actinides [13, 14]. Although these results were key experiments demonstrating that a new transition element series, the 6d-electron series, begins with $Z = 104$, none of these experiments provided detailed information on Rf chemistry; no K_d values were measured in these first-generation experiments.

Recent systematic investigations of the chromatographic behaviour of Rf performed atom-at-a-time showed that the properties of Rf are similar to those of the group-4 homologs, zirconium (Zr) and hafnium (Hf), although there are sometimes striking differences in the complexation strength between Rf and the lighter homologs. These results confirm that Rf is a member of the group-4 elements in the Periodic Table [7–10]. On the contrary, fluoride complex formation of Rf is significantly different from that of the homologs Zr and

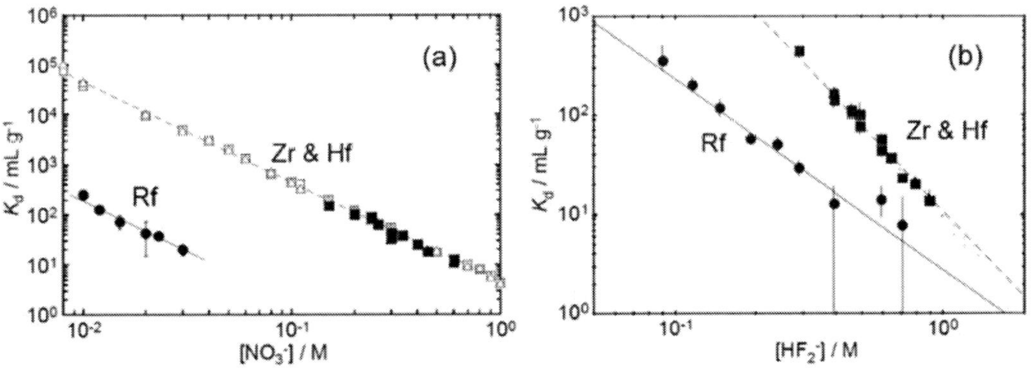

Figure 1. (a) Distribution coefficients (K_d) of Zr and Hf measured under static conditions (open symbols) and those of Zr, Hf, and Rf from column chromatography (solid symbols) as a function of the equilibrated concentration of NO_3^-, $[NO_3^-]$, at fixed $[F^-] = 3.0 \times 10^{-3}$ M. Linear relationships of the log K_d vs. log $[NO_3^-]$ for Rf and the homologs are indicated by the solid and dashed lines, respectively [17]. (b) Distribution coefficients (K_d) for Zr, Hf, and Rf on the anion-exchange resin as a function of the HF_2^- concentration, $[HF_2^-]$. Adapted from [17, 18].

Hf, demonstrating a surprising behaviour of Rf [15–18]. Here, we summarize the fluoride complex formation studies of Rf performed with ARCA II and AIDA.

The successive fluoride complex formation of cationic species of Zr, Hf, and Rf was investigated through cation-exchange chromatography with ARCA II at the Philips Cyclotron, Paul Scherrer Institute [15] and with AIDA at the JAEA tandem accelerator [16–18]. The results demonstrated that the distribution coefficient, K_d, of Rf in mixed solution of HF/0.10 M HNO_3 decreases with increasing concentration of the fluoride ion, $[F^-]$, indicating the consecutive formation of neutral and/or anionic fluoride complexes of Rf as those of homologs. It was also ascertained that at these low F^- concentrations the fluoride complex formation of Rf is significantly weaker than that of Zr and Hf. The sequence of adsorption strength on the cation-exchange resin is Zr $\leq$ Hf $<$ Rf at any given $[F^-]$ [15, 16]. The observed K_d sequence among the group-4 element complexes is in agreement with that predicted theoretically in reference [19]. Here, free energy changes of the complex formation reactions were determined on the basis of fully relativistic density functional theory calculations of various hydrated and hydrolysed fluoride complexes of Zr, Hf, and Rf.

Anionic fluoride complexation has been studied in mixed HF/HNO_3 solutions (dilute HF concentration) [17]. Figure 1a shows the variation of the K_d values for Zr, Hf and Rf as a function of the counter anion NO_3^-, $[NO_3^-]$, at a constant $[F^-]$ of 3.0×10^{-3} M. It is found that the K_d values of Zr and Hf are identical and the logarithmic values of K_d linearly decrease with the logarithmic increase of $[NO_3^-]$ with a slope of -2.0 ± 0.1, as indicated by the dashed line. The K_d values of Rf are indicated by closed circles. As shown in Fig. 1a, at any given $[NO_3^-]$ the K_d values of Rf are much smaller than those of Zr and Hf, and they smoothly decrease with increasing $[NO_3^-]$ with the slope of -2.2 ± 0.2, as indicated by the solid line. The results indicate that, at constant $[F^-]$ of 3.0×10^{-3} M, anionic complexes of Zr, Hf, and Rf are present as $[MF_6]^{2-}$ (M = Zr, Hf, and Rf) [17]. The formation of an anionic fluoride complex of Rf was interpreted in detail by taking into account not only chemical equilibria of anion-exchange reactions and of formation reactions of fluoride complexes but also comparing those with the ones of the homologs Zr and Hf. Upper limits of formation constants for the fluoride complexes of Rf were experimentally evaluated for the first time.

The result demonstrated that the formation of the hexafluoride complex of Rf, $[RfF_6]^{2-}$, is at least one order of magnitude smaller than those of the homologs Zr, $[ZrF_6]^{2-}$, and Hf, $[HfF_6]^{2-}$ [17].

In addition, also anion-exchange experiments in HF solution (higher HF concentration) showed surprising results [18]. Figure 1(b) depicts the K_d values for Zr, Hf, and Rf as a function of HF_2^- concentration, $[HF_2^-]$. The K_d values of the group-4 elements decrease linearly with $[HF_2^-]$ in a $\log K_d$ vs. log $[HF_2^-]$ plot. It should be noted here that the slopes for Zr and Hf are clearly −3 (dashed line), while that for Rf is significantly different, i.e. −2 (solid line). This implies that Rf is likely to be present as the hexafluoride complex, $[RfF_6]^{2-}$, similar to the well-known $[ZrF_6]^{2-}$ and $[HfF_6]^{2-}$ at lower concentration of HF, while Zr and Hf are likely to be present in the form of the heptafluoride complexes, $[ZrF_7]^{3-}$ and $[HfF_7]^{3-}$. This shows a strikingly different fluoride complexing behaviour between Rf and its homologs. As mentioned above, the experimental finding on the fluoride complexation sequence $Zr \geq Hf > Rf$ is in qualitative agreement with relativistic molecular density functional calculations [19].

The fast centrifuge system SISAK (Short-lived Isotopes Studied by the AKUFVE-technique, where AKUFVE is a Swedish acronym for an arrangement of continuous investigations of distribution ratios in liquid extraction) was applied to perform continuous liquid-liquid extraction and detection of short-lived Rf using small-volume centrifuges and a liquid-scintillation detection system. SISAK, employed to extract 4-s ^{257}Rf produced in the ^{208}Pb(^{50}Ti,n) reaction, was coupled to the recoil separator BGS (Berkeley Gas-filled Separator) at the Lawrence Berkeley National Laboratory (LBNL) 88-inch cyclotron [20]. Recoiling ^{257}Rf, pre-separated from the primary beam and nuclear reaction by-products, was thermalized in a stopping-gas cell and was transferred to SISAK through the aerosol loaded gas-jet system. The extraction behaviour of Rf was similar to that of the homologs Zr and Hf [20]. This pilot experiment demonstrated that the SISAK system in combination with liquid scintillation detectors and a physical pre-separator can be used for future investigations of chemical properties of transactinide elements.

4. Element 105, dubnium (Db)

First studies of the aqueous-phase chemistry of Db were conducted manually using 34-s ^{262}Db produced in the irradiation of ^{249}Bk with ^{18}O ions at the LBNL 88-inch cyclotron [21]. Like the homologs niobium (Nb) and tantalum (Ta), Db was adsorbed on a glass surface upon fuming with nitric acid. The other chemical separation attempted in [21] involved the extraction of anionic fluoride species into methyl isobutyl ketone (MIBK). Under these conditions, Ta was extracted into MIBK nearly quantitatively, while Nb was extracted to only a small extent. It was found that Db is not behaving chemically like its lighter homolog Ta. The non Ta-like behaviour of Db might indicate that Db forms polynegative anions like $[DbF_7]^{2-}$ under the chosen conditions.

To investigate this unexpected finding and more facets of the Db chemistry, a large number of automated separations were conducted with ARCA II. Here, we describe typical results from reversed-phase extraction chromatography of Db. In the first experiments, separations with the liquid anion-exchanger triisooctyl amine (TiOA) on an inert support were performed [22]. It was found that TiOA extracts all group-5 elements including the pseudo-homolog protactinium (Pa) from HCl solutions above 10 M. At lower HCl concentration, however, Db showed a striking non Ta-like behaviour and it followed its lighter homolog Nb and Pa. From this similarity, it was concluded that the complex structure of

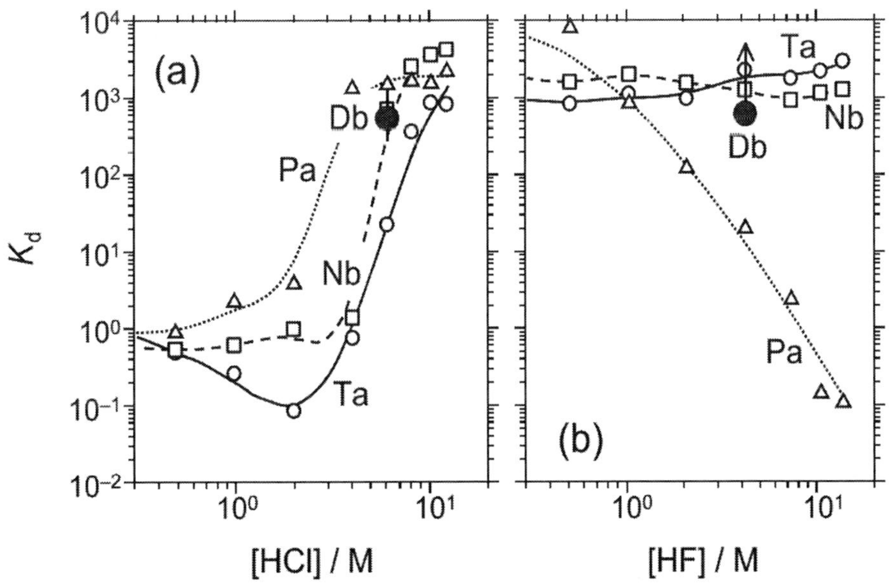

Figure 2. (a) Variation of the K_d values (mL · g^{-1}) of Nb, Ta, and Pa from HCl solutions into Aliquat 336/HCl. The K_d value of Db in 6 M HCl (with error bars encompassing the 68% confidence limit) is indicated by the bold dot. (b) Extraction behaviour of Nb, Ta, and Pa from HF solutions in Aliquat 336/HF. The lower limit for the K_d value of Db in 4 M HF (representing the 68% confidence limit) is indicated by the bold dot with the arrow. Adapted from [24].

Db was $[DbOCl_4]^-$ or $[Db(OH)_2Cl_4]^-$. In addition, the preferential formation of oxyhalide complexes of Db was predicted theoretically [23].

Then, amine extraction experiments with Aliquat 336 (trioctyl-methylammonium chloride) of group-5 elements were systematically performed [24]. Chromatographic column separations with ARCA II were developed to study separately the fluoride and chloride complexation of Db. In the system Aliquat 336/HCl, a K_d value of 438^{+532}_{-166} mL · g^{-1} for Db in 6 M HCl was deduced, which is close to that of Nb and differs from the values for Pa and Ta; see Fig. 2a. Thus, the extraction sequence Pa > Nb ≥ Db > Ta is established exactly as theoretically predicted [25]. In the system Aliquat 336/HF, the result is depicted in Fig. 2b. The interpretation of these results indicates that the amine extraction behaviour of Db halide complexes is similar to that of its lighter homolog Nb. In HF solution, the behaviour of Db differs clearly from that of Pa, while in HCl it differs considerably from Ta.

The anion-exchange behaviour of Db produced in the $^{248}Cm(^{19}F, 5n)^{262}Db$ reaction at the JAEA tandem accelerator was investigated together with Nb, Ta, and Pa in mixed HF/HNO_3 solution [26]. The K_d value of Db in 0.31 M HF/0.10 M HNO_3 solution ($[F^-] = 3 \times 10^{-3}$ M) was measured, where Nb and Ta form $[NbOF_4]^-$ and $[TaF_6]^-$, respectively. It was found that the adsorption of Db on the resin is considerably weaker than that of Ta and it is similar to that of Nb and Pa. This result suggests that Db forms a fluoro-oxo complex $[DbOF_4]^-$, like Nb, but not $[DbF_6]^-$, like Ta. Note that the K_d value of Db is also close to that of Pa that forms $[PaOF_5]^{2-}$ and/or $[PaF_7]^{2-}$. The formation of complexes such as $[DbOF_5]^{2-}$ and $[DbF_7]^{2-}$ could also be suggested for Db. To unequivocally clarify the fluoride complexation of Db, further systematic studies of the Db behaviour as a function of $[F^-]$ and $[NO_3^-]$ are required. Presently available results from studies in various aqueous solutions indicate that Db behaves more like the homolog Nb, sometimes like its pseudo-homolog Pa, but not like Ta.

As discussed in [24], the extraction sequences in Figs. 2a and 2b are in good agreement with the theoretical calculations in [25] and [27], respectively. Although the non-Ta behaviour of Db in [26] is qualitatively consistent with the calculation [27], we need further investigations, such as determination of a chemical species of Db, for a quantitative interpretation.

5. Perspectives

The first chemical separation and characterization of element 106, seaborgium (Sg), in the aqueous phase was performed using ^{265}Sg ($T_{1/2} = 7.4$ s) produced in the reaction of ^{22}Ne with ^{248}Cm at the UNILAC (UNIversal Linear Accelerator) at the GSI (Gesellschaft für Schwerionenforschung) [28, 29]. (Today's knowledge on the decay of ^{265}Sg shows the presence of two states; one with a half-life of about 9 s and one with about 15 s [30]). 3900 identical cation-exchange separations in 0.1 M $HNO_3/5 \times 10^{-4}$ M HF were performed with ARCA II. Three correlated mother-daughter (α-α) decays were observed that were assigned to the decay of ^{261}Rf and ^{257}No, as the descendants of ^{265}Sg. The mother decays were not observed because it presumably decayed in the time interval between the end-of-separation and the start-of-measurement. However, it was suggested that Sg behaves like a typical group-6 element and forms neutral or anionic oxide or oxide halide compounds; in contrast to its pseudo-homolog uranium (U). Further investigation with a newly developed continuously-working system, however, is required to clarify the properties of Sg as mentioned below.

New technical and methodological developments have been carried out and are still ongoing, which provide a basis for future aqueous-phase chemistry experiments. These have the potential to provide new insights into properties of the lighter transactinides and to push the frontier even beyond Sg. Successive complex formation experiments with Sg and beyond are most fascinating. Another option is to perform electrochemical studies of transactinide elements using redox reactions. Advanced theoretical model calculations have already provided quantitative predictions [31]. Preparations have started to measure the redox potential of hexavalent Sg [32]; a complex and continuously working system is envisioned which consists of an electrochemistry apparatus [33, 34] for the potentially controlled Sg reduction connected to a separation device like SISAK for the extraction and a flow-through liquid-scintillation detection system [35]. By means of a newly designed interface [36], this shall be coupled with a recoil separator like GARIS (Gas-filled Recoil Ion Separator) at RIKEN [37]. In this future program, the ^{265}Sg nuclide produced in the ^{248}Cm(^{22}Ne, 5n) reaction is planned to be pre-separated from the primary beam and nuclear reaction by-products by GARIS, followed by thermalization in a stopping-gas cell and transfer to the chemistry set-up in an aerosol loaded gas-jet system. The products are fed into the electrochemistry apparatus to reduce the Sg^{6+} to the presumably most likely tetravalent state. Then, ions in the hexavalent cationic and in a low oxidation states are separated from each other by solvent extraction in an automated apparatus like SISAK. Both streams of liquid will be assayed by α spectrometry. If successful, this will give the first information on redox properties of a transactinide element.

Interesting perspectives arise from the observation of relatively long-lived isotopes of all elements from element 108, hassium (Hs) to element 114, flerovium (Fl), first observed in the pioneering works at the Flerov Laboratory of Nuclear Reactions (FLNR) [38]. Mostly synthesized in ^{48}Ca-induced reactions (^{26}Mg in the case of $^{269-271}$Hs [39]) not only these directly produced isotopes are of interest but also the α-decay grand-grand-daughter nuclei in the Rf to Db region, some having half-lives of the order of one day. On the one hand, this makes chemical investigations of the heaviest element in the Periodic Table all the way up to Fl conceivable. The chemistry of these elements is extremely interesting due to the predicted dramatic influence of relativistic effects [40]. In addition, being able to investigate isotopes of Rf and Db with half-lives as long as a day opens up new perspectives for the application of new techniques which require longer times but provide deeper insights. However, for all these studies at first new technical developments are mandatory to cope with cross section in the range between 10 pb and 1 pb.

References

[1] M. Schädel, Angew. Chem. Int. Ed. **45**, 368 (2006).
[2] J.V. Kratz, *Handbook of Nuclear Chemistry, 2nd ed., vol. 2*, eds. A. Vértes, S. Nagy, Z. Klencsár, R.G. Lovas, F. Rösche (Springer, Dordrecht, 2011), p. 925.
[3] M. Schädel, Radiochim. Acta **100**, 579 (2012).
[4] A. Türler and V. Pershina, Chem. Rev. **113**, 1237 (2013).
[5] M. Schädel and D. Shaughnessy, *The Chemistry of Superheavy Elements, 2nd edition* (Springer, Berlin, 2014).
[6] M. Schädel, Phil. Trans. R. Soc. A **373**, 20140191 (2015).
[7] J.V. Kratz, Radiochim. Acta **99**, 477 (2011).
[8] J.V. Kratz and Y. Nagame, *Liquid-Phase Chemistry of Superheavy Elements.* In *The Chemistry of Superheavy Elements. 2nd edition*, eds. M. Schädel and D. Shaughnessy (Berlin, Springer, 2014), p. 309.
[9] Y. Nagame *et al.*, J. Radioanal. Nucl. Chem. **300**, 77 (2014).
[10] Y. Nagame, J.V. Kratz, M. Schädel, Nucl. Phys. **A 944**, 614 (2015).
[11] M. Schädel *et al.*, Radiochim. Acta **48**, 171 (1989).
[12] Y. Nagame *et al.*, Radiochim. Acta **93**, 519 (2005).
[13] R.J. Silva *et al.*, Inorg. Nucl. Chem. Lett. **6**, 871 (1970).
[14] E.K. Hulet *et al.*, J. Inorg. Nucl. Chem. **42**, 79 (1980).
[15] E. Strub *et al.*, Radiochim. Acta **88**, 265 (2000).
[16] Y. Ishii *et al.*, Bull. Chem. Soc. Jpn **84**, 903 (2011).
[17] A. Toyoshima *et al.*, Radiochim. Acta **96**, 125 (2008).
[18] H. Haba *et al.*, J. Am. Chem. Soc. **126**, 5219 (2004).
[19] V. Pershina *et al.*, Radiochim. Acta **90**, 869 (2002).
[20] J.P. Omtvedt *et al.*, J. Nucl. Radiochem. Sci. **3**, 121 (2002).
[21] K.E. Gregorich *et al.*, Radiochim. Acta **43**, 223 (1988).
[22] J.V. Kratz *et al.*, Radiochim. Acta **48**, 121 (1989).
[23] V. Pershina *et al.*, Radiochim. Acta **64**, 37 (1994).
[24] W. Paulus *et al.*, Radiochim. Acta **84**, 69 (1999).
[25] V. Pershina, Radiochim. Acta **80**, 75 (1998).
[26] Y. Kasamatsu *et al.*, Chem. Lett. **38**, 1084 (2009).
[27] V. Pershina and T. Bastug, Radiochim. Acta **84**, 79 (1999).
[28] M. Schädel *et al.*, Nature **388**, 55 (1997).
[29] M. Schädel *et al.*, Radiochim. Acta **83**, 163 (1998).
[30] H. Haba *et al.*, Phys. Rev. C **85**, 024611 (2012).
[31] V. Pershina, E. Johnson, B. Fricke, J. Phys. Chem. A **103**, 8463 (1999).
[32] A. Toyoshima *et al.*, J. Radioanal. Nucl. Chem. **303**, 1169 (2015).
[33] A. Toyoshima *et al.*, Radiochim. Acta **96**, 323 (2008).
[34] A. Toyoshima *et al.*, Inorg. Chem. **52**, 12311 (2013).

[35] L. Stavsetra *et al.*, Nucl. Instr. Meth. **A 543**, 509 (2005).
[36] K. Ooe *et al.*, J. Radioanal. Nucl. Chem. **303**, 1317 (2015).
[37] H. Haba *et al.*, Chem. Lett. **38**, 426 (2009).
[38] Yu.Ts. Oganessian, V.K. Utyonkov, Nucl. Phys. **A 944**, 62 (2015).
[39] J. Dvorak *et al.*, Phys. Rev. Lett. **97**, 242501 (2006).
[40] V. Pershina, *Theoretical Chemistry of the Heaviest Elements*. In *The Chemistry of Superheavy Elements. 2nd edition*, eds. M. Schädel and D. Shaughnessy (Berlin, Springer, 2014), p. 135.

EPJ Web of Conferences **131**, 07008 (2016) DOI: 10.1051/epjconf/201613107008
Nobel Symposium NS160 – Chemistry and Physics of Heavy and Superheavy Elements

Chemistry aided nuclear physics studies

Julia Even[a]

KVI – CART, University of Groningen, 9747 AA Groningen, The Netherlands

Abstract. Studies of the superheavy elements bring several challenges through low production yields, short half-lives, and high background rates. This paper describes the possibilities of chemical separations as techniques to overcome the background problematic and to investigate the nuclear properties of the heaviest nuclides.

1. Introduction

The transuranium elements up to nobelium (No, $Z = 102$) have been discovered applying radiochemical separation methods (see, for example, reference [1]). The discoveries of the heaviest actinoide element lawrencium (Lr, $Z = 103$) and the first transactinoide rutherfordium (Rf, $Z = 104$) have been achieved through the application of nuclear physics techniques and followed shortly after by radiochemical separations. With increasing proton number, the production cross sections and the half-lives decrease. Thus, from the discovery of dubnium (Db, $Z = 105$) on, physics techniques were the driving forces for the discovery of new elements. In the early eighties, the area of superheavy elements at electromagnetic separators started with the discovery of bohrium (Bh, $Z = 107$), hassium (Hs, $Z = 108$), and meitnerium (Mt, $Z = 109$) at SHIP at GSI, Darmstadt (see contribution of S. Hofmann in this proceedings, and reference [2]). Although physics techniques dominate the research on heavy elements in our days, chemists are still actively involved in this field. Besides studies of the chemical properties, chemical separations play a role in nuclear physics studies. In this paper, examples will be given how the combination of chemistry and physics can overcome recent challenges.

2. Identification of the proton number

The elements flerovium (Fl, $Z = 114$) up to element $Z = 118$ have been so far only synthesized in ^{48}Ca-induced fusion reactions with actinoide targets [3]. However, an unambiguous identification of the proton number of these elements could still not be achieved. All these elements decay into previously unknown spontaneous fission isotopes. Thus, there is no direct link to the well-known part of the nuclear chart. In case of ^{288}115, the α-decay chains end in a long-lived nuclide which decays through spontaneous fission. This long-lived fission nucleus was assigned to ^{268}Db or after decay through electron capture to ^{268}Rf. One option to identify the proton number of ^{288}115 is the chemical isolation and identification of the daughter nuclide ^{268}Db or ^{268}Rf [4, 5]. For this approach, ^{288}115 was produced in the reaction ^{243}Am(^{48}Ca,3n) and implanted in a copper block. After irradiation, the upper copper layer was dissolved, elements of group four and group five were

[a] e-mail: `j.even@rug.nl`

separated through ion-exchange chromatography. Samples of the chromatography fractions were prepared for spontaneous fission and α spectroscopy. Each sample was measured for 960 hours. In the group four and group five fraction, several fission events were observed, which were assigned to the decay of long-lived Rf or Db isotopes [4, 5]. These results were yet not included in the assignment of element 115 by the joint working party of IUPAP and IUPAC [6]. Their report notes that a new chemical separation scheme for Db and Rf was applied. In the past, Db and Rf chemistry had already been studied with the isotopes ^{262,263}Db and ^{257,261}Rf, which were directly produced in fusion-evaporation reactions [7, 8]. Relativistic effects influence the chemical behaviour of Db and Rf. In the future, a chemical system which has been already used to study transactinoides directly produced in fusion evaporation reaction, should be used for the identification of the proton number. Thus, miss-interpretation of data due to unknown or unexpected chemical properties of these elements can be excluded.

3. In-situ gas-phase chemistry

Besides ion-exchange chromatography also gas-phase chemical techniques are playing an important role in superheavy element research. Especially, studies of isotopes in the vicinity of the neutron sub-shell closure around $N = 162$ and the proton sub-shell closure $Z = 108$ can benefit from gas-phase chemical techniques. These nuclides are primarily accessible in reactions with relatively light projectiles (such as ^{26}Mg or ^{22}Ne) and actinoide targets. Studies of reaction products of these reactions are challenging for an electromagnetic separator as the target transfer products have almost the same magnetic rigidity as the fusion-evaporation products of interest. This leads to a high background rate in the focal plane. Thus, alternative or additional separation methods need to be found. One possibility is selective in-situ gas-phase chemistry.

3.1 Volatile group 8 tetroxides

In case of hassium (Hs, $Z = 108$), the separation, and transport as volatile hassium tetroxide (HsO_4) has been established. Hassium tetroxide was first synthesized and investigated in an experiment focusing on its chemical properties [9]. ^{269}Hs was produced in the reaction ^{248}Cm(^{26}Mg,3n), and thermalized directly behind the target in a recoil chamber which contained a helium-oxygen mixture. Hassium reacts with the oxygen and forms like its lighter homologue element osmium a volatile tetroxide. Within a few seconds, HsO_4 can be transported in a gas stream out of the recoil chamber through capillaries to detector devices. HsO_4 can be deposited at low temperatures on the surface of an α-detector.

In contrast to Hs the nuclear by-products do not form volatile oxides and are thus not transported. Only noble gases and volatile elements reach the detector through the gas stream. Chemistry is thus more efficient than any electromagnetic separator. However, the technique is limited by the half-life of the Hs isotopes. Only isotopes which live a few seconds can be investigated.

HsO_4 has successfully applied to separate Hs isotopes for studies near the neutron sub-shell closure at $N = 162$. To optimize spectroscopy of Hs, the thermochromatography detector COMPACT was build [10, 11]. The detector consists of two arrays of 32 PIN diodes each. These two arrays form a 0.6-mm wide channel with a negative temperature gradient. The transport gas enters this channel at the warm side and leaves the detector through the cold end. HsO_4 is deposited in the middle of this channel at around $-45°$C. The detection efficiency for α particles is about 80% [10]. HsO_4 was used to study the excitation function of the reaction ^{248}Cm(^{26}Mg,3-5n) [10–12]. With this approach, the isotopes ^{270}Hs and ^{271}Hs were discovered [10, 11]. Also the reaction ^{238}U(^{36}S,4n) has been investigated [13].

3.2 Metal carbonyl complexes

To study other nuclei with the COMPACT detector, new suitable chemical systems have to be investigated. Recently, the in-situ chemistry of metal carbonyl complexes has been explored [14–18] and it is still under investigation (see contribution R. Eichler *et al.* in this proceedings). Group 6 and group 8 elements were known to form volatile complexes with carbon monoxide. However, these complexes are not stable at temperatures of a few hundred degrees Celsius [18]. Direct synthesis in a recoil chamber behind the target is not possible due to the presence of a plasma induced by the primary beam. The fusion products need to be pre-separated from the primary beam to allow the formation of carbonyl complexes. In a first experiment at the gas-filled separator TASCA, volatile complexes of short-lived isotopes of tungsten, rhenium, osmium, and iridium were synthesized and transported in the gas stream to the chromatography detector COMPACT [14, 16]. These studies demonstrated the potential that carbonyl chemistry bears for superheavy element research. In a follow-up experiment at the GARIS separator at RIKEN, the first carbonyl complex of a superheavy element was synthesized – seaborgium hexacarbonyl [15]. Seaborgium could be isolated from the huge amount of target-like transfer products which also reached the focal plane of GARIS. It was transported in the gas stream to the chromatography detector. Figure 1 shows a sum spectrum of all 64 α detectors in the COMPACT array [17]. As it can be clearly seen, the interesting energy region above 8 MeV is background free. The α event at 8.56 MeV was followed 214 s later by the coincident detection of two fission fragments at 11 MeV and 74 MeV, respectively, in the same detector pair. This experiment was the first experiment which was sensitive for such a long correlation time. The observed decay chain of ^{265}Sg which is shown in Fig. 1 gives a hint for new decay properties of ^{261}Rf [17]. Two isomeric states of ^{261}Rf have been reported in literature [19, 20]. ^{261a}Rf decays through α emission with a half-life of $T_{1/2} = 70$ s. ^{261b}Rf is comparably short lived with $T_{1/2} =$ 1.9 s and decays through α emission or spontaneous fission. Observations in the carbonyl chemistry experiment led to the speculation that ^{261a}Rf might have an unknown spontaneous fission branch, or that there might be a ^{261}Rf isomer. This isomer may decay through electron capture into ^{261}Lr which has a half-life of $T_{1/2} = 39$ min. Further investigations are required.

The studies of carbonyl complexes demonstrated their potential for low-background spectroscopy of seaborgium. Furthermore, the studies of the lighter homologue elements of group 7, 8, and 9 showed that bohrium, hassium, and meitnerium are accessible using carbonyl chemistry.

4. Chemical reactions in gas catchers

Gas-catchers are commonly used in nuclear physics experiments to thermalize recoiling ions in an inert gas atmosphere and extract these through electric fields to prepare samples for precision studies. Generally, the gas in a gas catcher is kept as clean as possible. Gas catchers have been already implemented at the heavy element separators SHIP, BGS, and GARIS to couple these to mass spectrometers or mass analyzers [21–23]. So far chemical reactions in a gas catcher have not been established at these facilities.

Nevertheless, ion chemistry can be used as a powerful tool to prepare samples for nuclear physics studies. At the Low Energy Beam and Ion Trap (LEBIT) facility at the National Superconducting Cyclotron Laboratory (NSCL), chemistry has already applied to separate nuclear isobars. LEBIT is a Penning-trap facility built for high precision mass measurements of rare isotopes produced by fast-beam fragmentation [24]. LEBIT consists of a gas catcher in which the ions delivered from the fragment separator are thermalized. Subsequently, a set

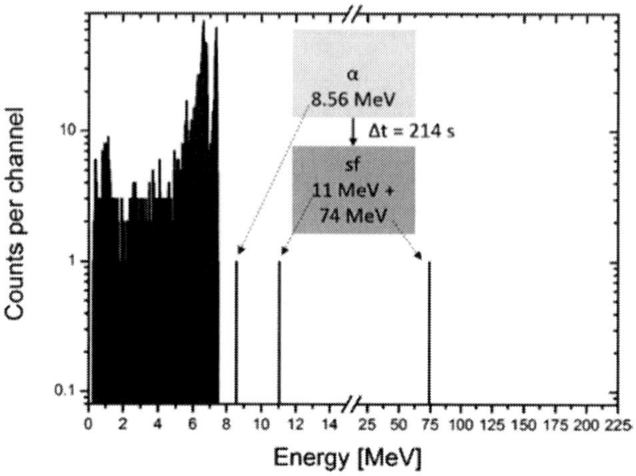

Figure 1. Sum spectrum of all 64 α detectors of the COMPACT detector array. The spectrum was collected while $4.35 \cdot 10^{17}$ ^{22}Ne particles hit the target. The spectrum is free from background activity in the interesting energy range above 7.5 MeV. The α peaks at energies below 7.5 MeV originate from the decay of ^{219}Rn and its daughter isotopes ^{215}Po and ^{211}Bi. A ^{227}Ac-emanation source was implemented into the gas loop for continous calibration of the COMPACT detectors. Graphic according to reference [17].

of radio-frequency quadrupole (RFQ) structures guides the ions through different pressure regimes and the species of interest is selected in an RFQ section which operates as a mass filter. Then the ions pass a beam cooler and buncher before they are injected into the Penning trap for mass measurements.

For mass measurements of ^{32}Si at LEBIT, chemical reactions in a gas catcher were applied to prepare clean samples [25]. ^{32}Si was produced in in-flight fragmentation of ^{40}Ar and extracted along with ^{32}S and $^{16}O_2$ from the gas catcher. The isobaric contamination was several orders of magnitude higher. Thus, element-selective separation was required. Therefore, water vapor was added to the gas in the stopping cell. Si formed $^{32}SiOH(H_2O)^+$ with a molecular mass of 67 u while S formed $^{32}SO_2H^+$ with a molecular mass of 65 u, and O_2 did not react. Thus, these species could be separated in a mass analyzer behind the gas cell. The molecular bonds were subsequently broken in the beam cooler and buncher through collisions with the gas before injection of the ions into the Penning trap [25].

For mass measurements of the heaviest elements, so far an extra separation step was not required. However, this method will become interesting for superheavy chemistry research. Here physics techniques can aid chemistry studies. Until now, the interpretation of the chemistry studies strongly relies on the support from theoretical predictions and studies of the lighter homologue elements. The structure of the chemical compounds cannot unambiguously be determined. Thus, the combination of gas-chemistry with more sophisticated methods such as mass spectrometry would be beneficial. The approach of chemical reactions in a gas catcher combined with mass measurements appears feasible. A good system to start with are the transition metal carbonyl complexes. In mass measurements of Fe and Co at LEBIT, it was observed that Fe and Co formed mono-, bi-, and even tri-carbonyl complexes with small impurities of carbon monoxide in the buffer gas [24]. Controlling the amount of carbon monoxide in the buffer gas will open the door for kinetic studies of the carbonyl complex formation of even the transition metals of the 7th period of the periodic table.

5. Conclusion and outlook

Although the period of chemically driven element discovery passed, nuclear physics experiments can still benefit from chemical techniques. Chemical studies can be used to identify the proton number of an isotope. Furthermore, fast gas-phase transport of volatile species can be applied for background free decay spectroscopy. Group 8 tetroxide and the volatile transition metal carbonyl complexes are powerful chemical separators. Currently, new detectors are under development to improve the detection efficiency and allow α, β, and γ coincidence measurements aided by gas-phase chemistry [26]. This new detector type will have a sandwich structure. The inner chromatography channel consists of an α-detector array followed by layers of detectors sensitive for electrons and photons.

However, also superheavy element chemistry can benefit from developments in experimental nuclear physics. Ion-chemistry combined with single molecule mass spectrometry will open the door to unambiguously identify the structure of chemical compounds of the heaviest elements. It will open the door for studies of chemical reaction kinetics. A wise choice of combination of the different experimental methods will lead to many interesting discoveries in the future.

The author acknowledges support through the Rosalind Franklin fellowship program of the University Groningen co-funded by the EU.

References

[1] Y. Nagame and M. Hirata, Radiochim. Acta **99**, 377 (2011).
[2] G. Münzenberg and K. Morita, Nucl. Phys. **A944**, 3 (2015).
[3] Y.T. Oganessian and V.K. Utyonkov, Nucl. Phys. **A944**, 62 (2015).
[4] D. Schumann *et al.*, Radiochim. Acta **93**, 727 (2005).
[5] N.J. Stoyer *et al.*, Nucl. Phys. **A787**, 388 (2007).
[6] P.J. Karol *et al.*, Pure and Appl. Chem. **88**, 139 (2016).
[7] J.V. Kratz, Radiochim. Acta **99**, 477 (2011).
[8] M. Schädel, Phil. Trans. R. Soc. A **373**, 20140191 (2015).
[9] C.E. Düllmann *et al.*, Nature **418**, 859 (2002).
[10] J. Dvorak *et al.*, Phys. Rev. Lett. **97**, 242501 (2006).
[11] J. Dvorak *et al.*, Phys. Rev. Lett. **100**, 132503 (2008).
[12] J. Dvorak *et al.*, Phys. Rev. C **79**, 037602 (2009).
[13] R. Graeger *et al.*, Phys. Rev. C **81**, 061601 (2010).
[14] J. Even *et al.*, Inorg. Chem. **51**, 6431 (2012).
[15] J. Even *et al.*, Science **345**, 1491 (2014).
[16] J. Even *et al.*, Radiochim. Acta **102**, 1093 (2014).
[17] J. Even *et al.*, J. Radioanal. Nucl. Chem. **303**, 2457 (2015).
[18] I. Usoltsev *et al.*, Radiochim. Acta **104**, 141 (2016).
[19] H. Haba *et al.*, Phys. Rev. C **83**, 034602 (2011).
[20] H. Haba *et al.*, Phys. Rev. C **85**, 0246411 (2012).
[21] M. Block *et al.*, Eur. Phys. J. A **50**, 49 (2005).
[22] P. Schury *et al.*, Nucl. Instr. Meth. **B317**, 537 (2013).
[23] J.M. Gates, these proceedings.
[24] R. Ferrer *et al.*, Phys. Rev. C **81**, 044318 (2010).
[25] A.A. Kwiatkowski *et al.*, Phys. Rev. C **80**, 051302 (2009).
[26] A. Di Nitto *et al.*, in *GSI Scientific Report 2014*, GSI Report 2015-1, p. 184.

EPJ Web of Conferences **131**, 08001 (2016) DOI: 10.1051/epjconf/201613108001
Nobel Symposium NS160 – Chemistry and Physics of Heavy and Superheavy Elements

Status and perspectives of the Dubna superheavy element factory

Sergey Dmitriev[a], *Mikhail* Itkis, and *Yuri* Oganessian

Flerov Laboratory of Nuclear Reactions, Joint Institute for Nuclear Research, 141980 Dubna, Russian Federation

Abstract. In the Flerov Laboratory of Nuclear Reactions, Joint Institute for Nuclear Research (FLNR JINR), construction of a new experimental complex is currently in progress (Superheavy Element Factory), aimed at the synthesis of new superheavy nuclides and the detailed study of those already synthesized. The project includes the construction of a new accelerator of stable and long-lived isotopes in the mass range $A = 10$–100 with an intensity of up to 10 pμA and energy up to 8 MeV/nucleon; construction of a new experimental building and infrastructure for housing the accelerator with five channels for the transportation of beams to a 1200-m^2 experimental hall that is equipped with systems of shielding and control for operations with radioactive materials; development of new separators of reaction products; upgrade of the existing separators and development of the new detection modules for the study of nuclear, atomic, and chemical properties of new elements. The first experiments are planned for 2018.

1. Introduction

The six heaviest chemical elements with atomic numbers 113 to 118 that fill the 7th row of Mendeleev's Periodic Table were synthesized in reactions of ^{48}Ca ions with actinide targets in the experimental studies carried out over the recent years. Over 50 new isotopes of elements 104 to 118 with maximum neutron excess were for the first time produced and their decay properties were determined in these investigations [1, 2]. The new isotopes considerably filled up the Chart of the Nuclides and expanded it up to $Z = 118$ and $N = 177$ superheavy elements. Fundamental conclusions of the modern theory concerning the limits of existence of nuclear matter have for the first time received experimental confirmation.

The discovery of the new domain (island) of stability and the very fact of existence of superheavy elements (SHE) have posed a number of new questions associated with fundamental properties of nuclear matter:

- Can even heavier nuclei exist? Is the "Island of Stability of SHE" the last one on the Chart of the Nuclides?
- Can the superheavy nuclei be formed in the process of nucleosynthesis like those stable and long-lived nuclei in the groups of Pt, Pb, and U-Th found in Nature?
- What is the electron structure of a superheavy atom in strong Coulomb field?
- What is the limit of Mendeleev's Table? How much are the chemical properties of superheavy elements similar to those of their lighter homologues?

---

[a] Corresponding author: `dmitriev@jinr.ru`

In search for answers to these and many related questions we have to considerably widen our knowledge about the subject of investigation. Obviously, to reach this aim we need to greatly broaden the field of work and involve new facilities. However, judging by first attempts, going beyond the area of already synthesized superheavy elements is quite a difficult task.

2. SHE factory

Direct synthesis of elements with $Z > 118$ in fusion reactions means using projectiles heavier than Ca, since the capability of high-flux reactors to produce target material is limited to Cf isotopes. A number of attempts undertaken in various laboratories and aimed at the synthesis of elements 119 and 120 did not yield any result. Limits of production cross sections of nuclei with $Z = 120$ in, for example, the reaction $^{54}Cr+^{248}Cm$ [3] and nuclei with $Z = 119$ via $^{50}Ti+^{249}Bk$ [4] were ten to twenty times lower than those of production of isotopes of elements 114 and 115 in experiments with ^{48}Ca.

Advance in the other direction – towards the desired area of more neutron-rich isotopes of the known SHE closer to the predicted shell closure at $N = 184$ – is possible only with use of radioactive projectiles that are more neutron-rich than ^{48}Ca. Unfortunately, the intensities of beams of radioactive nuclei at the most advanced accelerators and even those in the design stage are extremely low for performing this kind of experiments.

Moving in the opposite direction – towards neutron deficit in superheavy nuclei – leads to strong decrease of production cross section when going off the shell $N = 184$.

Under these circumstances, it seems appropriate to investigate in detail the already synthesized nuclei: to determine their masses, forms, and fission barriers, find their excited states, etc. Yet here we need to considerably improve the statistics. Individual events of formation and decay of superheavy nuclides that are registered in present experiments with a rate of one per day to one per month (this is considered a great achievement of the last years) are far from being sufficient for these purposes. Experimental results should be presented not by single events but by hundreds and thousands. Only then one will see rare decay channels of the superheavy nuclei or study in detail nuclear transitions to various states, which means to make complete measurements in spite of extremely low cross sections of producing superheavy elements.

Note that the present situation in some aspects reproduces the beginning of work on synthesis of superheavy elements twenty years ago when a new approach to the synthesis of SHE (reactions ^{48}Ca+actinides) was found and a significant breakthrough in experimental technique was made.

We came to the conclusion that at present, based on accumulated knowledge in study of

- reactions of synthesis of new elements (fusion reactions with evaporation of neutrons);
- cross sections of production of the nuclei of SHE reaction channels at various beam energies (excitation functions);
- competing reaction channels (background conditions);
- decay characteristics of the isotopes of SHE (decay modes, energies, decay times, etc.), as well as the achievements of the last twenty years in the fields of
- technique of physical experiment;
- physics and technology of charged particle accelerators and plasma physics;
- new highly sensitive precision detectors for nuclear radiation;
- production of new target materials in high-flux nuclear reactors and development of new target technologies, etc.

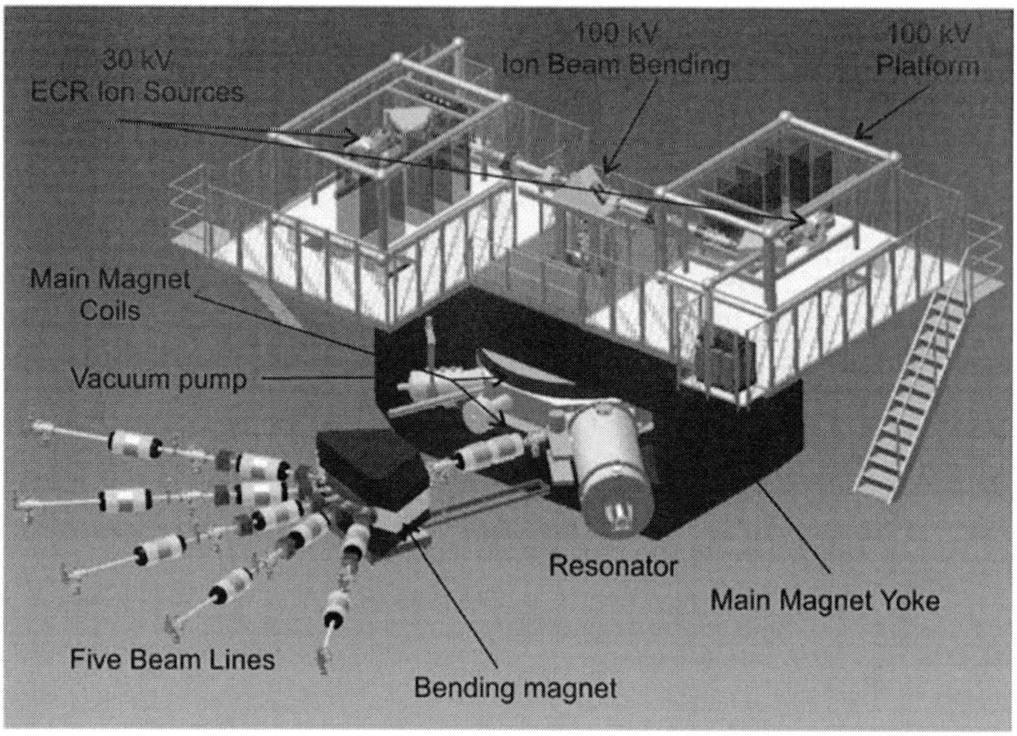

Figure 1. DC-280 cyclotron view.

Table 1. Expected parameters of the DC280 cyclotron.

DC280 (*expected parameters*)		
Ion	Ion energy [MeV/A]	Output Intensity [pps]
^{7}Li	4	1×10^{14}
^{18}O	8	1×10^{14}
^{40}Ar	5	6×10^{13}
^{48}Ca	5	$6 \times 10^{13} - 1.2 \times 10^{14}$
^{54}Cr	5	2×10^{13}
^{58}Fe	5	1×10^{13}
^{136}Xe	5	1×10^{14}
^{238}U	7	5×10^{10}

it is possible to increase the production rate of the known isotopes of SHE by one to two orders of magnitude and carry out experiments aimed at the synthesis of new elements with $Z>118$ by employing a new experimental complex, the "SHE Factory", and thus make a second breakthrough in the world of the heaviest nuclei and elements.

The ideas incorporated in the project of the SHE Factory are associated with developing the experimental basis in several directions. These directions are:

- construction of new powerful accelerator of stable and long-living isotopes in mass range $A = 10$–100 with intensity up to 10 pμA and energy up to 8 MeV/nucleon (see Fig. 1 and Table 1 for more details);
- construction of a new experimental building and infrastructure for placing the accelerator with five channels for transportation of beams to a 1200-m^2 experimental

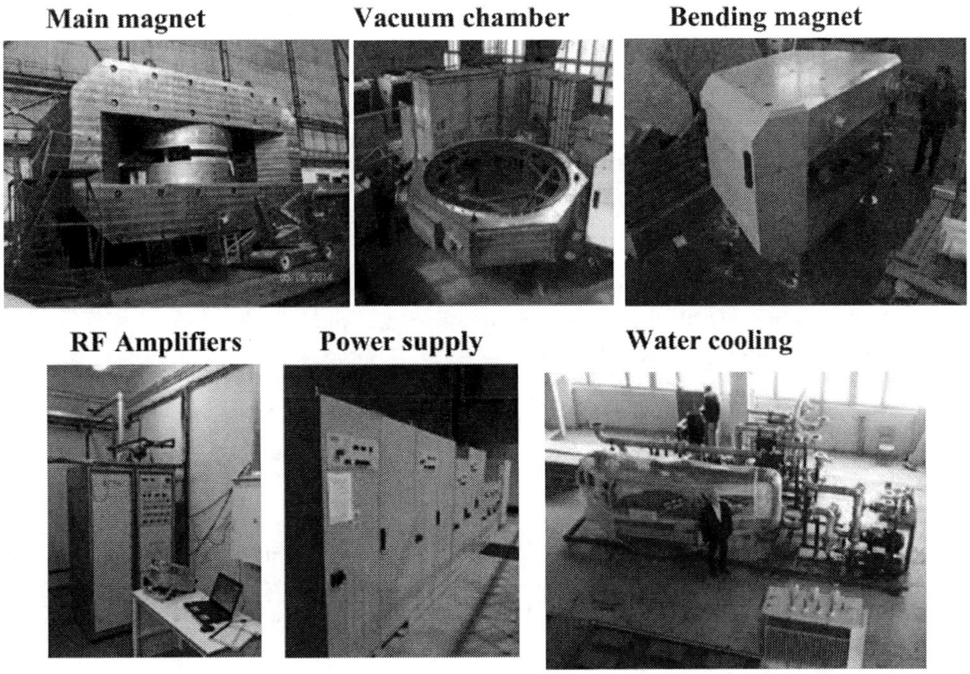

Figure 2. The main parts of cyclotron DC-280.

hall, equipped with systems of shielding and control matching the class two of operations with radioactive materials;

- development of new separating channels and development of new detection modules (including use of the already existing ones) for the study of nuclear, atomic, and chemical properties of new elements;
- production of new target materials and development of techniques of making targets with high thermal and radiation stability (running and projected nuclear reactors and enrichment facilities);
- development of a base for research with intense ion beam in related fields of science and technology.

All major parts of new cyclotron DC-280 are produced and delivered to Dubna (Fig. 2).

In general, studies of heavy and superheavy nuclei are related to atomic physics, nuclear physics, astrophysics, and radiochemistry. They cover a wide range of phenomena, the study of which is far more than experiments that are to be carried out at the SHE Factory. However, when experiments are performed with extremely heavy nuclei at the limit of stability and thus require extremely high sensitivity, advantages of the SHE Factory become highly important. This is the specificity of the new complex. Its program of experimental research will be directly connected with the achievements of world science and is designed for broad international cooperation.

A long-term program of research of SHE should be developed with regard to the development of the new complex, its phased commissioning, production of the primary beam, and start of the first experiment in 2018. Experimental research at the existing complex ECR-4M+U-400+DGFRS and at the functioning setups will continue at the same pace until the first full-scale experiment at SHE Factory (2017–2018). However, priority is given to the studies that will be continued at the SHE Factory.

The program already begun in 2014. It can be nominally divided, with some overlap, in three phases:

- construction of the SHE Factory, design and construction of new experimental setups, research work with beams of ^{48}Ca and ^{50}Ti ions with aim of obtaining additional data for the experimental program at the SHE Factory (2014–2017);
- putting into operation of the new accelerator complex, production of the primary beam at target position, preparation of detector modules and experimental equipment for running the first experiment;
- setting up the first experiment;
- producing results that demonstrate the capabilities of the SHE Factory;
- reaching of design parameters (2018).

During the first phases (2014–2018) the scientific and technical studies are aimed at getting answers to key questions that determine further research at SHE Factory. This research includes the following:

- Study of the nuclei at the border of the "Island of Stability" of SHE. Synthesis of neutron-deficient isotopes of Fl ($Z = 114$) in reactions ^{48}Ca+^{239,240}Pu and of new isotopes of element 118 in reactions ^{48}Ca+$^{249-251}$Cf. The first experiment was carried out in 2015 [5].
- Change from ^{48}Ca to a beam of ^{50}Ti with aim of preparing experiments on the synthesis of elements 119 and 120. Model experiment on the synthesis of the isotope ^{285}Fl in reactions ^{48}Ca+^{240}Pu and ^{50}Ti+^{238}U. The ^{50}Ti ion beam was obtained with the ECR+U-400 complex by using MIVOC method [6].
- Experimental investigation of characteristics of spontaneous fission of the neutron-rich nuclei with $Z = 104$ to 112 obtained in decay chains of superheavy elements.
- Preparation of experiments at SHE Factory aiming on study of chemical behaviour of elements 112 to 118 compared to their light homologues. Running control experiments with a beam of ^{48}Ca. Development of separator and instrumentation.
- Choice of the modes of operation of the cyclotron DC-280 and ECR source for producing intense beam of ^{48}Ca ions and carrying out the first experiment at the SHE Factory.
- Choice of design and calculation of characteristics of the separator of recoil nuclei for the first experiment with high-intensity ^{48}Ca beam at the SHE Factory. New gas-filled recoil separator has been already designed and contracted with the SigmaPhi company in France. Its commissioning is planned for the beginning of 2018.
- Alpha spectroscopy of the deformed nuclei ^{257}Rf and ^{258}Db using high-resolution detector of α particles in cold fusion reactions ^{50}Ti+^{208}Pb,^{209}Bi in view of carrying out experiments on spectroscopy of superheavy nuclei ^{287}Fl and ^{288}115 and their decay products.

3. Summary

Currently, the Superheavy Elements Factory project is being implemented at the FLNR. It is one of the key projects in the programme of development of JINR. The launch of the Factory is scheduled for 2017/2018. The DC-280 cyclotron commissioning is planned for December 2017. First experiments with intense beams of ^{48}Ca with the use of new gas-filled separator should be carried out in the mid of 2018.

The SHE Factory has no analogue in the world practice. It is developed with a perspective of 20 to 25 years (as all the heavy-ion accelerators at JINR) and should be in demand as an experimental complex that provides leadership in study of heavy and superheavy elements.

The reported study was partially supported by the special grant of the Moscow region government (RF).

References

[1] Yu.Ts. Oganessian and V.K. Utyonkov, Rep. Prog. Phys. **78**, 036301 (2015).
[2] Yu.Ts. Oganessian *et al.*, Phys. Rev. C **87**, 014302 (2013).
[3] S. Hofmann *et al.*, Eur. Phys. J. **A52**, 180 (2016).
[4] J. Khuyagbaatar *et al.*, Phys. Rev. Lett. **112**, 172501 (2014).
[5] V.K. Utyonkov *et al.*, Phys. Rev. C**92**, 034609 (2015).
[6] A.V.Yeremin *et al.*, Phys. Part. Nucl. Lett. **12**, 43 (2015).

How to calculate α-decay rates in the future?

B. Gillis Carlsson[a], *Daniel E.* Ward, and *Sven* Åberg

Division of Mathematical Physics, LTH, Lund University, 22100 Lund, Sweden

Abstract. New elements discovered during past decades have been created in fusion reactions where a lighter nucleus is collided with a heavier one. The new elements created often decay by emitting α particles. From the half-lives of the decays and the energies of the emitted particles one may extract some properties of the new elements. In this talk the recent work performed by the Lund group to model α decay starting from nuclear density-functional theory is reviewed and a possible extension is mentioned.

1. Introduction

From theoretical predictions one expects around 7000 [1] different nuclear isotopes to exist. The current experimental tabulations contain around 2350 of these nuclei [2]. Out of all these different isotopes only around 300 are predicted to have half-lives longer than the expected life of the Solar System [1]. The remaining ones decay in various ways until they finally end up as stable isotopes. Many of the nuclei heavier than lead decay by sending out α particles. An emitted α particle travels with around 5% of the speed of light. Due to collisions with the molecules in the air the particle typically only travels a few centimeters before it attracts electrons and combines with them to form an He atom. It is through this process that Helium gas is produced in the bedrock.

The new superheavy elements recently discovered also decay by emitting α particles. Typically the experimental groups can measure the average time before the created nucleus sends one out and its kinetic energy. Since energy is conserved in the process the measured kinetic energy tells us about the difference in mass between the initial nucleus and the resulting decay product. Sometimes the decay is repeated in several steps until one finally reaches a more long-lived isotope. The extracted mass differences obtained in each step allows us to connect the unknown mass of the new element to the mass of a known element. The mass, which is a complex quantity that cannot be calculated directly from the underlying theory of Quantum Chromo Dynamics (QCD), tells us how bound the nuclei are.

The time it takes for the decay to occur is even more complicated to calculate. The initial nucleus is created in a non-stationary state and the wave function will therefore evolve in time and explore the available phase space. After some time there is a chance that the initial nucleus is divided into a lighter nucleus and an α particle.

One way to model this process is to reduce the degrees of freedom of the system. Typically this is done by assuming that the system is composed of protons and neutrons treated as point particles. The interaction between these particles has been carefully deduced from scattering experiments. In this way ab-initio calculations can be performed for lighter elements but for

[a] Corresponding author: gillis.carlsson@matfys.lth.se

heavy nuclei such calculations become intractable due to the complexity associated with the many interacting particles.

A method that enables calculations for the entire chart of nuclei and that can be extended to deal with descriptions of nuclear reactions is density-functional theory (DFT). During recent years the field of nuclear DFT has seen an amazing progress and in this talk we present the results of merging DFT models with a time-dependent description to obtain predictions for the half-lives of the heavy elements.

2. Predictions of Q_α values

The liquid drop model was originally proposed by Niels Bohr in 1936 and was later used by Bohr and Wheeler in 1939 [3] to explain nuclear fission. In this model, the probability for fission becomes large for heavy nuclei due to the repulsive effect of the Coulomb interaction. For nuclei with more than roughly 110 protons the fission barrier vanished and they were thus expected to be unstable and short lived. In more refined models, quantum mechanical effects due to the shell structure are predicted to have a stabilizing effect, enhancing the half-lives and thus opening the possibility for the existence of superheavy nuclei. In recent years the advancement of experimental methods have allowed the first exploration of this part of the nuclear chart.

The created superheavy nuclei often decay by sending out α particles. By summing the kinetic energy of the α particle and the recoil energy of the nucleus one obtains the Q_α value that specifies how much energy is released in the reaction. For nuclei with odd particle numbers the reaction often leaves the daughter nucleus in an excited state that may then further cool down by emitting photons. Summing the Q_α energy and the energy of the emitted photons one obtains the $Q_\alpha(gs)$ defined as the difference in ground-state binding energy before and after the reaction.

The measured $Q_\alpha(gs)$ values for the superheavy region of nuclei are illustrated with filled circles in Fig. 1. They should be compared with the full lines that show calculations based on the Skyrme-Hartree-Fock-Bogoliubov (SHFB) method using the SLy4 parametrization and a pairing functional of mixed type [4, 5]. As seen in this figure the $Q_\alpha(gs)$ values are expected to decrease as one approaches the region around 180 neutrons. As the Q_α values become lower the half-lives increase and can lead to some very long-lived isotopes. Isotopes with both even and odd neutron numbers are shown in Fig. 1. Calculations for the isotopes with odd neutron numbers have also recently been carried out based on a Woods-Saxon potential [6] and a comparison between different models is presented in [7]. Most models give similar results for the $Q_\alpha(gs)$ values but can have important differences in details, especially as one approaches the spherical region around $N = 184$ where collective correlations are expected to play a role.

For the $_{98}$Cf, $_{100}$Fm, $_{102}$No and $_{104}$Rf chains it has been possible to measure α decay to several different final states. The most common scenario is that the nucleus prefers to keep the odd neutron in the same orbital before and after the decay. The emitted α particle is thus formed by nucleons occupying a set of double-degenerate time-reversed orbitals. This type of decay is labeled favoured decay. The corresponding released energy Q_f is shown in Fig. 1 with open symbols. In most of the situations this decay leaves the nucleus in an excited state that further deexcites through internal conversion or emission of γ radiation. It is of interest to know when this will happen since the γ radiation gives information of the level structure and the internal conversion process can be used to identify the elements. Whether or not the nucleus will decay through the favoured channel depends on a balance between the difficulty of forming the α particle and the difficulty for the α particle of escaping the nucleus.

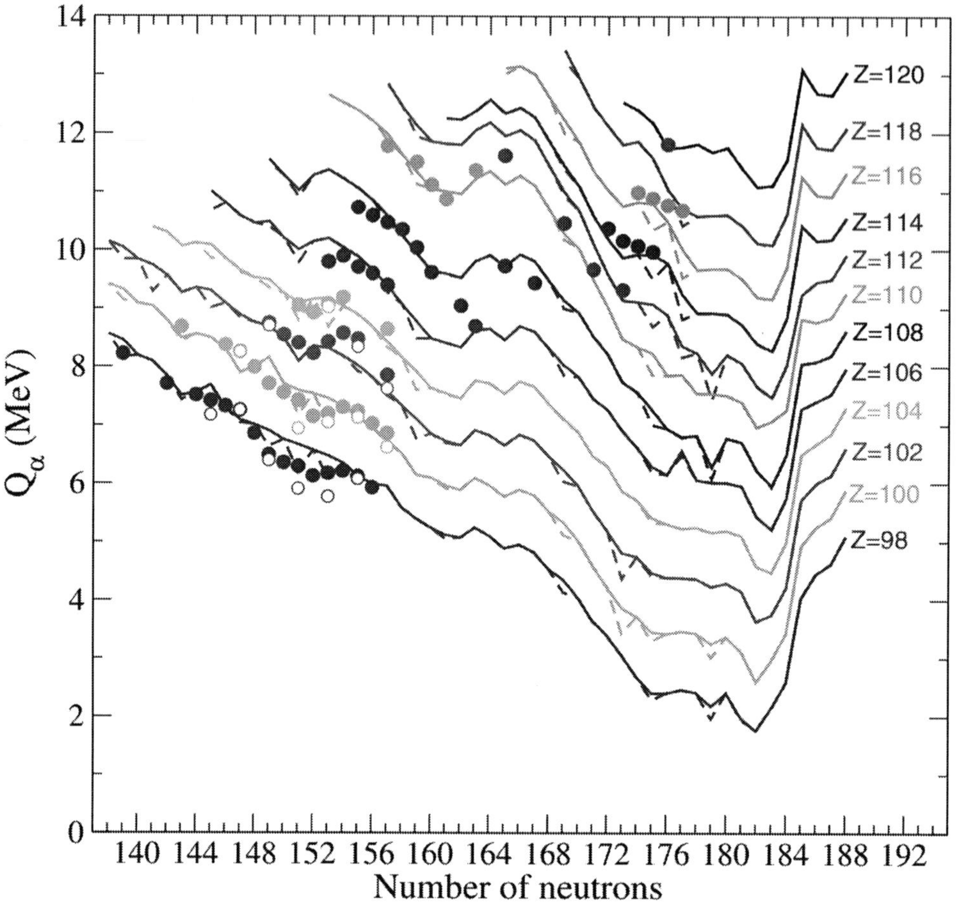

Figure 1. Q_α values of heavy and superheavy nuclei calculated using the SHFB model. Nuclei with odd particle numbers are calculated taking into account polarization effects in the equal-filling approximation [4]. The experimental data is taken from Refs. [6, 8].

The chance of formation increases for favoured decays but the chance of escaping often decreases due to a lowering of the Q_α value.

For illustrative purposes we may assume that if picking the favoured decay channel does not decrease the Q_α value by more than 0.5 MeV then it decays through the favoured channel. The dashed lines in Fig. 1 illustrate the Q_f value in the cases where $Q_\alpha(gs) - Q_f < 0.5$. Along each isotope sequence one finds several cases where the odd nucleus is predicted to decay through the favoured channel enabling the observation of additional radiation. Taking the $_{98}$Cf chain as an example the isotopes with $N = 147$, 149, 151, and 153 have predicted favoured decays that are indeed observed, and the spins and parities of the levels are also predicted correctly. The change in energy from choosing the favoured decay compared to the ground-state decay is also predicted roughly correct. In the case of $N = 145$ and 155 the model predicts the favoured decays to be the ground-state decays while in experiment the favoured decays lead to excited states in the daughter.

In general for the odd nuclei there is a competition between several decay channels populating different excited states in the daughter nuclei. To correctly model the decay scenario requires calculating the decay probabilities that depend on both the energies of the excited states as well as the structure of the mother and daughter wave functions.

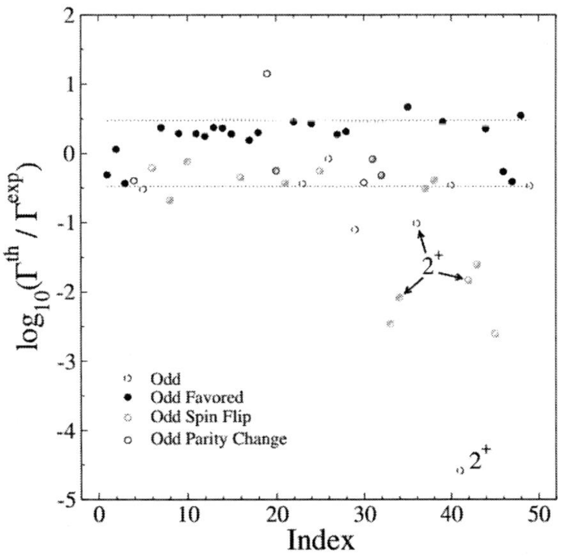

Figure 2. Logarithm of the ratio of the theoretical decay width to the experimental value for α-decaying odd-A nuclei. The Skyrme parametrization SLy4 combined with mixed pairing is used in the calculations. The dotted lines indicate a deviation from experiment by a factor of three.

2.1 Hindrance factors

Determining the α-decay path requires calculating the probability of decay to different excited states in the daughter nucleus. Such probabilities can be calculated by combining the SHFB model with a formalism to treat the α-emission process. A microscopic description of the emission is obtained from the *R*-matrix approach [9]. The combination of these two frameworks was recently explored for α decay of spherical even-even [10] and odd nuclei [5]. The results show that within this formalism the nuclei are predicted to be much too long lived. A suggested explanation comes from missing many-body correlations that are not captured in the nuclear wave functions. Since relative decay rates were well reproduced the method could however be renormalized by a single constant. This implies that the missing correlation effects always increase the probability for the α particle to form with the same percentage. This assumption was tested by comparing calculations for all even-even [10] and odd spherical nuclei with experimental data [5].

Figure 2 shows the ratio of theoretical and experimental partial decay widths ($\Gamma = \hbar \ln(2) / t_{1/2}$). The theoretical widths are calculated by combining the SHFB approach with the *R*-matrix approach [5]. All the data for spherical nuclei with odd particle numbers are included in the picture [5]. As seen from the figure most of the data can be reproduced with high accuracy. In a few cases the calculated widths were much smaller than the experimental values. Some of these points had previously been classified as having an odd particle in the daughter state coupled to a 2^+ vibration of the core. For such states our basic assumption of a pure one-quasiparticle excitation is violated. A description of these states could be obtained from a particle-vibration coupling model but this is outside the scope of our present investigation.

The decays can be classified according to whether the odd particle stays in the same orbital (favoured), changes direction of the intrinsic spin (spin flip), changes parity (parity change), or changes orbital but keeps spin and parity the same (orbital change). From the experimental data we can extract the hindrance factors for the decays belonging to the

Table 1. Experimental hindrance factors for different groups of α decays extracted from spherical or close to spherical nuclei [5]. "No." denotes the number of data points in each group.

Group	No.	Geometric Mean	Geometric Standard Deviation
Even-even	48	1	2.74
Favoured odd	22	2.57	2.97
Spin flip	14	163	3.22
Parity change	6	462	7.19
Orbital change	6	33.2	5.42
All unfavoured odd	23	117	5.44

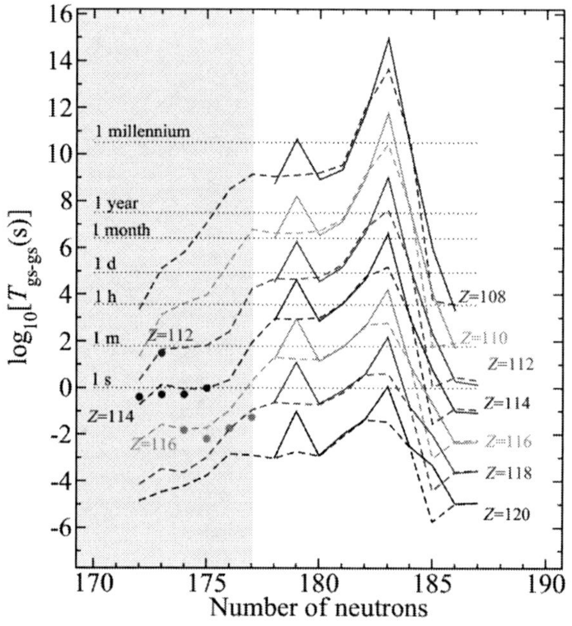

Figure 3. Alpha-decay half-lives of heavy and superheavy nuclei calculated using the SHFB model. The shaded area marks where the mother nuclei ground states are well deformed $|\beta| > 0.1$.

different groups. Table 1 shows the hindrance factors extracted from experiment for the different types of decay. For even-even nuclei the average hindrance factor is taken to be 1. For the favoured decays of odd nuclei they are on average hindered by a factor 2.57 meaning that they are roughly 2.57 times more long lived than an even-even decay having the same Q_α value. The geometric standard deviations for the different groups are quite large. For the favoured decays the half-lives vary with a factor 2.97 which is mainly due to the fluctuations in the pairing obtained when crossing shell gaps. Such fluctuations are not captured with the simple rules of thumb. Taking all the unfavoured decays as a single group the variation is a factor of 5.44 around an average hindrance of 117. This can be compared to the results in Fig. 2 where most of the decays fall within the dotted lines implying a difference between theory and experiment that is less than a factor three.

2.2 Lifetimes of superheavy nuclei

Figure 3 shows the calculated ground-state-to-ground-state α-decay half-lives of the even-even and odd nuclei in the superheavy island. The calculation is based on the SHFB model using the SLy4 parametrization and the predicted $Q_\alpha(gs)$ values shown in Fig. 1.

The dashed lines corresponds to half-lives calculated without taking into account the hindrance due to differing wave functions. The solid lines show calculations where the structure of the mother and daughter nuclei are considered [5]. As seen in the figure one expects increased half-lives for ground-state-to-ground-state decays for $N = 179$ corresponding to a hindered decay where the odd particle changes from a $d_{3/2}$ orbital to a $d_{5/2}$ orbital. The hindrance may cause the decay to instead follow the favoured path providing the difference in Q_α values are not too large. This competition was analyzed for the case of ${}_{116}\mathrm{Lv}_{179}$ [5] where it was predicted that the decay will mainly go to the ground state with the odd particle moving from $d_{3/2}$ to $d_{5/2}$ resulting in an enhanced half-live for this isotope.

In general there are two main scenarios that will be the situation for many of the odd nuclei:

1. The main decay path is of unfavoured type and leaves the daughter in the ground state. In this case we expect enhanced half-lives as illustrated in Fig. 3. These half-lives will show a dependence on the orbitals involved.
2. The decay mainly follows the favoured channel leaving the daughter in an excited state. The measured Q_f values will then tend to be smaller than for the neighbouring even-even nuclei. In this case one further expects the nucleus to deexcite through γ radiation or internal conversion that can give valuable information concerning the structure.

3. Conclusions and outlook

In conclusion, a short summary of our recent progress in the study of α decay of superheavy nuclei has been presented. The microscopic modeling of α decay is a challenge for theory. The approach discussed was able to give an improved description of hindrance factors that are important for predicting decay and half-lives of superheavy elements. Future experiments may utilize reactions such as titanium on berkelium that could lead to the creation of element 119. The α-decay chains will then go through the deformed region and are likely to lead to additional nuclear radiation being emitted. To simulate the decay chains the present formalism can be extended to treat deformed nuclear shapes. A combined theory and experimental study in this direction could lead to deeper insight into the structure of the nuclei around the superheavy island.

References

[1] J. Erler *et al.*, Nature **486**, 509 (2012).
[2] G. Audi *et al.*, Chin. Phys. C **36**, 1287 (2012).
[3] N. Bohr and J.A. Wheeler, Phys. Rev. **56**, 426 (1939).
[4] M.V. Stoitsov *et al.*, Comp. Phys. Comm. **184**, 1592 (2013).
[5] D.E. Ward, B.G. Carlsson, S. Åberg, Phys. Rev. C **92**, 014314 (2015).
[6] P. Jachimowicz, M. Kowal, J. Skalski, Phys. Rev. C **89**, 024304 (2014).
[7] P.-H. Heenen *et al.*, Nucl. Phys. **A944**, 415 (2015).
[8] The ENSDF database, `http://www.nndc.bnl.gov/ensdf/` (2016).
[9] R.G. Lovas *et al.*, Phys. Rep. **294**, 265 (1998).
[10] D.E. Ward, B.G. Carlsson, S. Åberg, Phys. Rev. C **88**, 064316 (2013).

Prospects of *A* and *Z* identification experiments at LBNL

Jacklyn M. Gates[a]

Lawrence Berkeley National Laboratory, Berkeley, CA 94720, USA

Abstract. The identification of six new elements within the last 15 years and with proton numbers, $Z = 113$–118 has transformed the heavy element field. However, one key piece of information on these nuclei remains unmeasured: their proton and mass numbers, A. At Lawrence Berkeley National Laboratory, the heavy element group has undertaken a program to study these new elements to perform experiments aimed at measuring the Z and A.Here, an overview of recent experiments aimed towards identifying the Z of SHE, and the prospects for Z and A identification experiments at LBNL are presented.

1. Introduction

Over the last 15 years, multiple experiments have been performed to produce elements with proton numbers, $Z = 113$–118 at the Flerov Laboratory for Nuclear Reactions using reactions of ^{48}Ca on actinide targets [1]. Many of these experiments have been successfully reproduced at laboratories around the world [2–8]. Several of the new nuclei have been produced in multiple reactions, i.e. ^{283}112, produced directly in the ^{238}U(^{48}Ca,3n) reaction and as a (grand)daughter nucleus in the ^{242}Pu(^{48}Ca,3n),^{244}Pu(^{48}Ca,5n) and ^{245}Cm(^{48}Ca,2n) reactions [9–11]. Taken together, the experiments present a self-consistent set of data regarding the production and decay properties of these new nuclei, hereafter referred to as Superheavy Nuclei (SHE). While no measurement of Z or link to nuclei with known Z has been made, the new nuclei have been assigned to elements with $Z = 113$–118 based on the large amount of data from cross bombardments and excitation function measurements [12, 13].

While there is little reason to doubt the assigned proton numbers, a direct measurement of SHE Z and mass numbers, A, is of great interest. Presently, there are three main ways to determine Z and A of SHE: (i) A and Z determination by linking to known elements, (ii) Z determination through the observation of characteristic Kx-rays, and (iii) A determination through mass measurements. At Lawrence Berkeley National Laboratory (LBNL) we are currently working on methods (ii) and (iii) for further investigating SHE. Here an overview of recent experiments aimed towards identifying the Z of SHE, and the prospects for Z and A identification experiments at LBNL are presented.

[a] e-mail: jmgates@lbl.gov

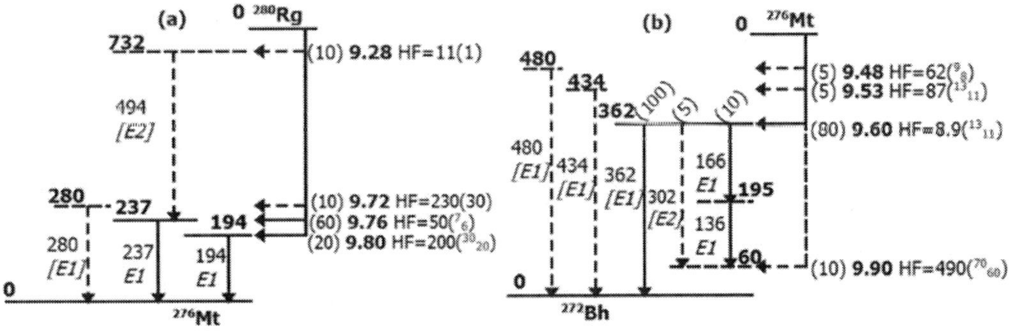

Figure 1. Proposed level schemes for ^{276}Mt ($Z = 109$) and ^{272}Bh ($Z = 107$), adapted from [7].

2. Prospects for *Z* identification experiments at LBNL

At Lawrence Berkeley National Laboratory (LBNL), separation of evaporation residues (EVRs) of interest from unwanted background is achieved using the Berkeley Gas-filled Separator (BGS) [14]. The BGS couples high EVR efficiency with high separation factors, which has allowed for the study of nuclides produced at rates of atom-per-day to atom-per-month [2, 7, 8, 15] in recent years. At the focal plane of the BGS is the corner-cube-clover (C3) detector, which consists of three double-sided silicon-strip detectors (DSSD) in the shape of the corner of a cube, and three high-purity germanium clover-type detectors (HPGe) press up against the face of each DSSD [7, 15]. The combination of DSSDs and HPGe allows for the detection of γ- and x rays emitted with the decay of SHE. The C3 detector has been discussed before and has a measured full energy α-particle detection efficiency of 77(5)% and a maximum photon efficiency of 30(2)% for E_{ph} =120 keV [7].

2.1 Attempt using the ^{243}Am(^{48}Ca, xn)$^{291-x}$ 115 reaction

The ^{243}Am(^{48}Ca,xn) experiment was performed in 2013 and previously published [7]. During the experiment, 46 decay chains were observed that could be attributed to the decay of element 115. Forty-three of the decay chains were “long” chains consisting of EVR-α-α-α-α-α-SF decay chain while the other three chains were short EVR-α-(α)-SF chains. All 46 decay chains were attributed to the decay of ^{288}115 based on the observed decay properties [16, 17].

Over the course of the experiment, multiple α-photon coincidences were observed with the decay of ^{280}111 and ^{276}109 allowing for level schemes for the low-lying levels above ^{276}109 and ^{272}107 to be devised. The level schemes are reproduced here in Fig. 1. Unfortunately, no SHE *Z* identification could be made based on the observed data.

3. Prospects for *A* identification experiments at LBNL

Without a definitive *Z* identification in the ^{243}Am(^{48}Ca,xn) reaction, the focus at LBNL has shifted to determining the *A* of a SHE. While the BGS, has a high efficiency and great suppression of beam and unwanted reaction products, these come at a cost: the BGS has a mass resolution of $A/\Delta A{\sim}20$, a relatively large focal plane image [14], and the detector station is in a high neutron and γ-ray background region. Currently, the BGS is being upgraded to couple to the mass analyzer, FIONA (For the Identification Of Nuclide *A*).

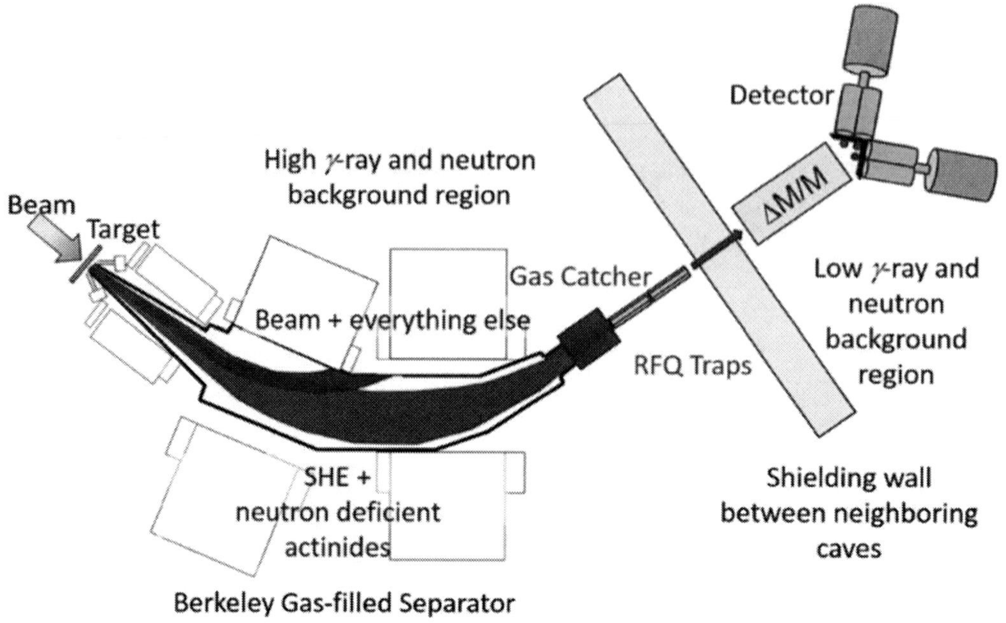

Figure 2. Basic schematic of the new BGS + FIONA apparatus in LBNL. See text for details.

A basic schematic of the setup is shown in Fig. 2. The goal of the upgrade is to provide a low energy, mass separated "beam" of SHE to a low neutron and γ-ray background region for A identification and nuclear structure studies, which includes Z identification through Kx ray observation.

Coupling the mass analyzer to the BGS is done using a technique that has been employed with setups such as CARIBU at Argonne National Laboratory [18], and SHIPTRAP at the GSI [19, 20]. The EVRs are thermalized in a radiofrequency (RF) gas catcher at the BGS focal plane, cooled and bunched in a radiofrequency quadrupole (RFQ) trap and then re-accelerated through a shielding wall and into a neighbouring cave, where they are separated by mass. By having the mass analyzer in a separate cave from the BGS, we can study the transmitted EVRs in a low neutron and γ-ray background region.

The gas catcher, RFQ traps and acceleration were developed and built at Argonne National Laboratory (ANL), while LBNL designed the mass analyzer. There were several requirements to take in account in the design of the mass analyzer:

- $A/\Delta A \geq 300$ – with production rates of atoms-per-day, mass determination should require the observation of only three atoms.
- High dispersion – the pixelation on the BGS focal plane detector is 2×2 mm, therefore, adjacent masses should be separated by about one centimeter.
- Low extraction voltage from RFQ and acceleration across a gap of <10 kV.
- High efficiency – the efficiency of transporting EVRs from the BGS target to the mass analyzer is expected to be 20%, therefore the mass analyzer should be as efficient as possible. The target was set to > 95% efficiency.
- Accept the emittance from the RFQ trap – The longitudinal emittance is expected to be 15 $\pi\cdot$ mm$\cdot$ mrad and the transverse emittance is expected to be 1 $\pi\cdot$ eV$\cdot\mu$s.
- Ability to determine implantation time.
- Fit within existing space.

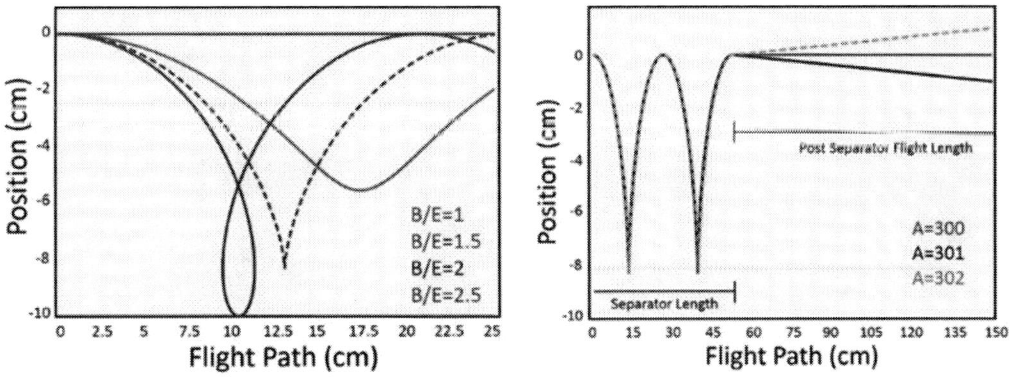

Figure 3. Trajectories of ions of $A = 300$–302 in perpendicular $E$and B fields with B/E ratios of 1.0, 1.5, 2.0, and 2.5 (left) and trajectories of ions of $A = 300$–301, $q = 2+$ in and field with $B/E = 2.0$ (right).

3.1 BGS + FIONA – Concept

Multiple different mass analyzers were simulated, including multi-reflection time-of-flight, Wien filter, and 180–270° bend magnet. However, none of these separators were capable of satisfying the requirements listed above. Therefore, we have designed a mass analyzer where particles are separated based on differing trochoidal trajectories in perpendicular electric and magnet fields.

Equations of motion for particles in perpendicular electric and magnetic fields are well known and can be found in references [21, 22], for example, and can be used for a first approximation of the separation principles. The projection of the path of a charged particle in perpendicular electric and magnetic fields is a trochoid in the plane perpendicular to the magnetic field. The equations of motion for a charged particle of mass m and charge Ze passing through a Wien filter with the electric field E and the magnetic field B, are given by:

$$x = a\omega t + \frac{v\sin(\theta)}{\varpi}(1 - \cos(\omega t)) - \left(a - \frac{v\cos(\theta)}{\omega}\right)\sin(\varpi t) \tag{1}$$

$$y = \frac{v\sin(\theta)}{\varpi}\sin(\varpi t) + \left(a - \frac{v\cos(\theta)}{\omega}\right)(1 - \cos(\omega t)). \tag{2}$$

Where v is the initial velocity in the x–y plane perpendicular to the magnetic field, θ is the angle between v and the x-axis, and ω and a are defined by the relations $\varpi = ZeB/m$ and $a = mE/ZeB^2$. In a traditional Wien filter, the E and B fields are adjusted such that ions of a given velocity pass straight through the fields, which those of differing velocities are bent away. This is shown in Fig. 3 (left) by the line $B/E = 1.0$. However, if the magnetic field is raised relative to the E field, then the ions begin to take trochoidal trajectories through the field region. This is shown in Fig. 3 by the lines $B/E = 1.5$, 2.0, and 2.5. The pitch of each trochoid is almost directly related to the A to charge, q ratio of the traversing ion. As such, it is possible to separate neighbouring isotopes based on their different A/q ratios, as shown in Fig. 3 (right).

To determine whether this separation principle would be suitable to differentiate between neighbouring SHE, simulations were performed with the charged particles simulation software *SIMION*®, version 8.1 [23]. *SIMION*® allows the user to construct three-dimensional representations of electrical elements and then the program calculates the electric

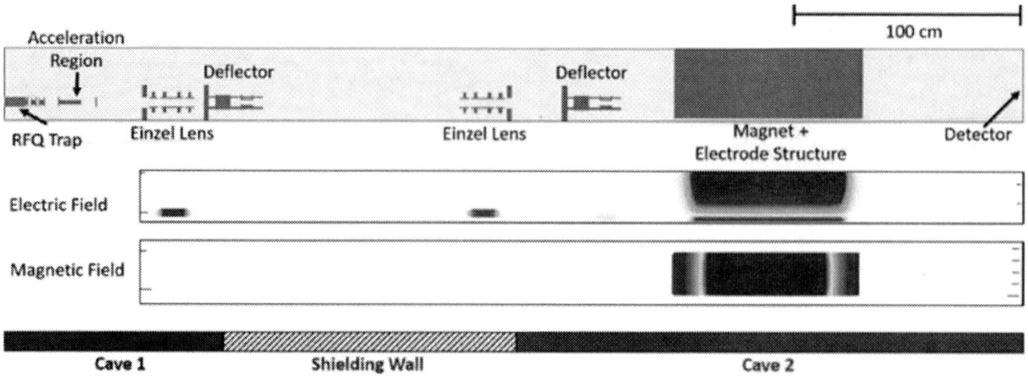

Figure 4. *SIMION*® representation of the FIONA setup (top), and the electric and magnetic fields observed by ions passing through FIONA (middle). Along the bottom is the physical location of the element at LBNL.

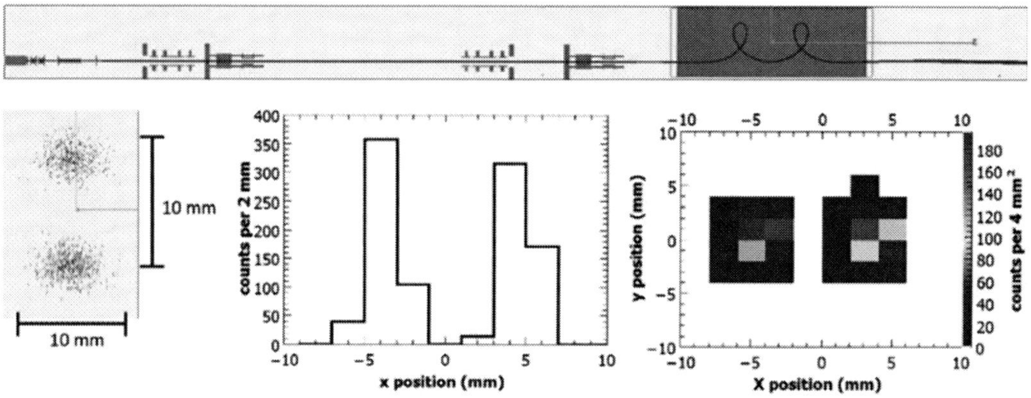

Figure 5. *SIMION*® simulations of ions with $A = 288$ (red) and $A = 289$ (black) and $q = 2+$ through FIONA (top), expected distribution of ions reaching the FIONA focal plane (bottom left), with expected C3 pixelation (bottom right) and a two-dimensional projection of expected ion distribution in the C3 detector (bottom center).

fields around the element and the trajectories of charged particles through those elements. User programs can be added to include the effects of magnetic fields or gas on the charged particle trajectories. Figure 4 has an example of the simulation used to test the FIONA principle. The simulation includes an RFQ trap, acceleration region, two einzel lenses for beam focusing, two horizontal and vertical beam deflectors and an electrode structure inside a magnet to give perpendicular electric and magnetic fields. Figure 4 shows the equal potential lines for the electric and magnetic fields.

To determine the expected performance of FIONA, ions with $A = 288$–289 and $q = 2+$ were cooled and trapped in a *SIMION*® RFQ trap, the trap was dumped, the ions were accelerated to energies of 5 keV and allowed to pass through the rest of the simulation and terminate in the C3 detector at the FIONA focal plane. The simulations were performed using $B/E = 2.0$. Figure 5 shows the trajectories of groups of 1000 ions of each mass, as well as the expected distribution of those ions on the C3 detector. The expected resolution is $A/\Delta A \approx 1000$, sufficient for SHE A determination. The simulated efficiency through FIONA was approximately 100%.

4. Conclusion and future outlook

FIONA is being installed at the focal plane of the BGS. Presently, the gas catcher, RFQ trap, acceleration region and magnet with an internal electrode structure have been installed. A control system for the beam line and the electronics has been developed and is being tested. Testing of the full apparatus is expected to begin in fall 2016. During experimental campaigns, the C3 detector will be placed at the focal plane of FIONA, allowing for detection of mass separated isotope and identification of photons emitted with those isotopes. This will open up a wide range of experimental studies at BGS + FIONA that will continue for the next decade.

References

[1] Yu.Ts. Oganessian and V.K. Utyonkov, Rep. Prog. Phys. **78**, 036301 (2015).
[2] L. Stavsetra *et al.*, Phys. Rev. Lett. **103**, 132502 (2009).
[3] Ch.E. Düllmann *et al.*, Phys. Rev. Lett. **104**, 252701 (2010).
[4] J. Khuyagbaatar *et al.*, Phys. Rev. Lett. **112**, 172501 (2014).
[5] S. Hofmann *et al.*, Eur. Phys. J. **A48**, 62 (2012).
[6] D. Rudolph *et al.*, Phys. Rev. Lett. **111**, 112502 (2013).
[7] J.M. Gates *et al.*, Phys. Rev. C **92**, 021301(R) (2015).
[8] P.A. Ellison *et al.*, Phys. Rev. Lett. **105**, 182701 (2010).
[9] Y.T. Oganessian *et al.*, Phys. Rev. C **70**, 064609 (2004).
[10] Y.T. Oganessian *et al.*, Phys. Rev. C **69**, 054607 (2004).
[11] Y.T. Oganessian *et al.*, Phys. Rev. C **74**, 044602 (2006).
[12] P.J. Karol *et al.*, Pure Appl. Chem. **88**, 139 (2016).
[13] P.J. Karol *et al.*, Pure Appl. Chem. **88**, 155 (2016).
[14] K.E. Gregorich, Nucl. Instr. Meth. **A711**, 47 (2013).
[15] J. Rissanen *et al.*, Phys. Rev. C **88**, 044313 (2013).
[16] U. Forsberg *et al.*, Nucl. Phys. **A953**, 117 (2016).
[17] Y.T. Oganessian *et al.*, Phys. Rev. C **87**, 014302 (2013).
[18] G. Savard *et al.*, Hyperfine Int. **199**, 301 (2011).
[19] J. Dilling *et al.*, Hyperfine Int. **127**, 491 (2000).
[20] M. Block *et al.*, Eur. Phys. J. **A25**, 49 (2005).
[21] W. Bleakney and J.A. Hipple, Phys. Rev. **53**, 521 (1938).
[22] K. Yano and S.H. Be, Jpn. J. Appl. Phys. **19**, 1019 (1980).
[23] D.A. Dahl, Int. J. Mass Spectrom. **200**, 3 (2000).

EPJ Web of Conferences **131**, 08004 (2016) DOI: 10.1051/epjconf/201613108004
Nobel Symposium NS160 – Chemistry and Physics of Heavy and Superheavy Elements

On the search for elements beyond *Z* =118. An outlook based on lessons from the heaviest known elements

Christoph E. Düllmann[1,2,3,a]

[1] Institute of Nuclear Chemistry, Johannes Gutenberg University, 55128 Mainz, Germany
[2] GSI Helmholtzzentrum für Schwerionenforschung, 64291 Darmstadt, Germany
[3] Helmholtz Institute Mainz, 55099 Mainz, Germany

Abstract. Recently, IUPAC approved all elements up to $Z = 118$ as discovered. Search experiments for the heavier elements with $Z = 119$ and 120 have been performed in recent years, but have so far not led to their discovery. I will review some aspects associated with the study and identification of the heaviest known elements that are relevant for future search experiments for elements beyond $Z = 118$ and highlight pressing issues that should be addressed, both on the experimental as well as on the theory side, to allow for performing these future experiments under improved and better informed conditions.

1. Introduction

At the time of the Nobel Symposium NS160, held in beautiful Bäckaskog Castle, Sweden, in May/June 2016, all elements up to $Z = 118$ have been officially accepted by IUPAC as discovered [1, 2]. One week later, the recommended names for the four most recent elements were announced. Figure 1 shows the corresponding cut-out of the chart of nuclei (with decay properties updated according to [3]), where all known or claimed nuclei are indicated in coloured boxes, overlaying the calculated shell-correction energy landscape according to a macroscopic-microscopic model [4]. Also indicated (by white boxes for yet unknown isotopes) are the hypothetical decay chains originating from the $3n$ and $4n$ compound nucleus (CN) evaporation channels following complete fusion of ^{50}Ti $+^{249}$Bk leading to $^{299}119^*$ (red bordered boxes), ^{50}Ti $+$ ^{249}Cf leading to ^{299} 120^* (light blue bordered boxes), and ^{54}Cr $+^{248}$Cm, ^{58}Fe $+^{244}$Pu, and ^{64}Ni $+$ ^{238}U, which all lead to $^{302}120^*$ (dark blue bordered boxes). Decay modes of unknown isotopes were assumed based on [5–7].

On the first day of this Nobel Symposium, the introductory speaker for the physics of the superheavy elements, Matti Leino, listed among his conclusions that it is "difficult to increase Z, and N. How about deepening our understanding of already synthesized species?" I follow this recipe and devote Sect. 2 to lessons from known elements that bear relevance to the search for new elements. Section 3 will be devoted to an outlook, going beyond $Z = 118$, picking up on the reactions mentioned above.

[a] e-mail: duellmann@uni-mainz.de

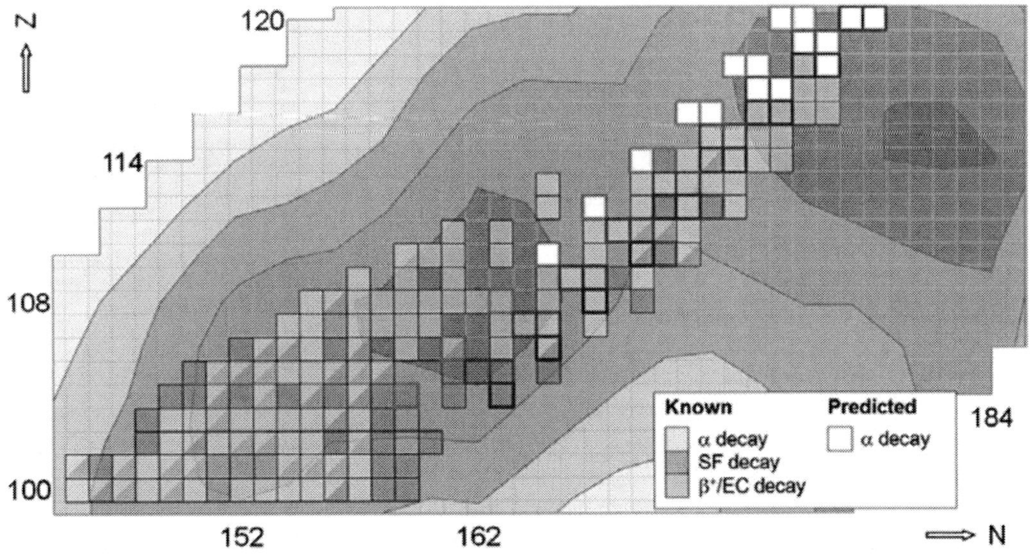

Figure 1. Chart of nuclei of the heaviest elements. The underlying contour plot shows calculated shell-correction energies [4]. Known nuclei are indicated by coloured boxes, new nuclei occurring in the hypothetical $3n$ and $4n$ decay chains produced in complete fusion of different target and projectile combinations leading to $Z = 119$ and $Z = 120$ are indicated by white boxes. See text for more details.

2. Lessons from known elements

For this, I will look at the case of the element 115. In experiments performed at the Flerov Laboratory for Nuclear Reactions (FLNR) in Dubna, Russia [8, 9], at the GSI Helmholtzzentrum in Darmstadt, Germany [3, 10], and at the Lawrence Berkeley National Laboratory (LBNL) in Berkeley, CA, USA [11], over 100 decay chains have been observed following the irradiation of ^{243}Am targets with ^{48}Ca beams and subsequent separation of evaporation residues in gas-filled recoil separators. All were assigned to originate from element 115, produced in complete fusion, neutron-evaporation reactions. Indirect arguments on the reaction mechanism and nuclear structure of the produced nuclei were invoked to support this assignment [8, 9], along with the IUPAC-preferred tool of "cross-bombardments", in which a certain nucleus is produced in multiple pathways; for example, in the case of element 115, the claimed production of ^{289}115 via ^{243}Am(^{48}Ca,2n) and via ^{249}Bk(^{48}Ca,4n)^{293}117 $\rightarrow$ α decay [1]. None of the experiments performed to date were able to directly identify the Z or the A of any nucleus ascribed to belong to any element 115 decay chain (or, in a broader perspective, of any nucleus produced in a ^{48}Ca + actinide target reaction). While the current assignment appears correct based on indirect arguments, K.E. Gregorich argues in his contribution to these proceedings that the assignment appears "good, but not good enough" [12]. Still, IUPAC approved these data as originating from element 115 [1], and especially highlighted the cross-bombardment aspect in its assessment, which formed the basis for their approval. The IUPAC-approved cross bombardment case of elements 115 and 117, originally invoked by the FLNR-led collaborations has since been challenged [13], and the identification of scientifically sound connections of element 117 and element 115 chains is ongoing [14]; even if found, they will be unable to prove the Z of the originating element, though. The safe, proven assignment of the α decay of the first member of any decay chain to element 115 is only possible if the nuclear reaction can be shown to lead to an isotope of element 115 (i.e., if complete fusion is achieved, followed by

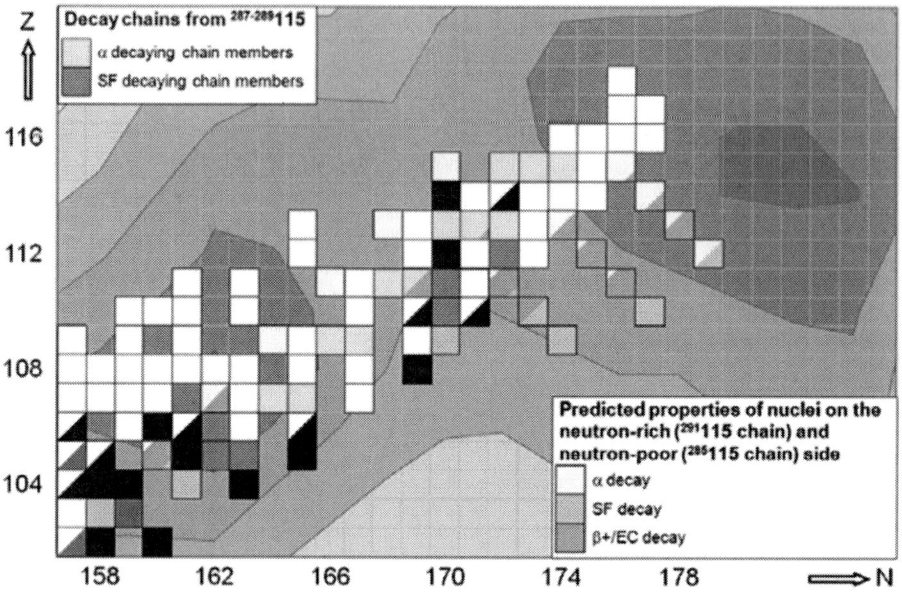

Figure 2. Black-outlined coloured boxes: decay chains populated in ^{243}Am(^{48}Ca,2-4n) reactions, as given in "scenario 2" in [3]. White (black) boxes: other known α-decaying (SF-decaying) nuclei. Red-bordered coloured boxes: predicted decay properties [16] of element 115 decay chains located on the neutron-rich (^{291}115) and neutron-deficient (^{285}115) side of those from the ^{243}Am(^{48}Ca,2-4n) reactions. The endpoints of the ^{285}115 chain (^{261}Lr and ^{265}Rf) are known and decay by SF.

the evaporation exclusively of neutrons), and if its decay is uniquely detected. I will focus here on the second aspect. For this, it is illustrative to review predictions on partial half-lives of different decay modes in the relevant region of the nuclear chart. The growing importance of β^+/electron capture (EC) decay, especially towards the neutron-rich region in the isotopic chains of element 113 and 115, has been noted, e.g., in [15, 16], see Fig. 2.

In none of the presently reported decay chains assigned to proceed through element 115 are there members that were originally (neither by IUPAC) assigned to possess a β^+/EC branching. To fully appreciate the relevance of this, one should bear in mind that the detection techniques used for the study of the ^{48}Ca $+^{243}$Am reaction at the recoil separators in Dubna, Darmstadt, and Berkeley, are sensitive to α as well as to spontaneous fission (SF) decay; however, the direct detection of EC decay is not possible. The presence of an EC-decaying member in the chain could only be inferred indirectly by comparing decay properties of later chain members with known data – which is only possible if decay properties of (*i*) nuclei populated if EC decays are present, as well as of (*ii*) those populated in the absence of EC decay, are known. None of the nuclei populated by the decay chains obtained in the ^{48}Ca $+^{243}$Am reaction have been known before, though.

Chemical approaches able to separate individual elements have the power to achieve Z identification. For this, chemical properties of the element need to be established and exploited. The heaviest element, which had its chemical properties reproducibly studied, is Cn ($Z = 112$) using the isotope ^{283}Cn produced as α-decay daughter in the reaction ^{242}Pu(^{48}Ca,3n)^{287}Fl, see review in [17]. Z identification is still not fully direct, but relies on chemical similarities with the lighter members in group 12 of the Periodic Table [18]. For element 113, first attempts for its chemical study have been performed, and for Fl, conflicting results on its chemical behaviour are reported [19, 20], see discussion in [17]. The power of chemical separators can only be fully employed, if the chemical properties of the element to be studied are already established. Once this is achieved, chemical studies of isotopes with unknown Z will allow verifying its atomic number.

Coming back to the decay chains allegedly originating from element 115, more specifically to the 96 five-α long chains from $^{288}115$, we note that there are arguments supporting that no chain member is present which decays (exclusively or in large fractions) by an intermediate β^+/EC decay, which would render the chain to continue through even-even isotopes: first, statistical analyses of lifetime distributions of each individual generation within these chains showed these distributions to be consistent with what would be expected if every specific decay step occurred from the same nuclear state in all chains. Significant abundance of branched decay in any member is thus unlikely and would only go undetected by this method if a partially β^+/EC-decaying member with a lifetime that is very long or very short compared to that of its daughter were present. Only in these two extreme cases would the distribution of apparent lifetimes (which would then sometimes be the sum of lifetimes of β^+/EC decay and subsequent α decay) at this position in the decay chain fit with what is expected for a decay of a single nuclear state. Secondly, β^+/EC decay in this chain would populate isotopes known or expected to be fast fissioning even-even nuclei. Therefore, experimental evidence suggests that no decay branches other than α (and terminating SF, potentially preceded by β^+/EC decay) are present in these long chains.

For $^{289}115$, data is scarcer, precluding as definite statements. Interestingly, isotopes of elements 115, 113, and 111 that are only slightly more neutron-rich than those produced in the ^{48}Ca+^{243}Am reaction are predicted to possess sizable β^+/EC branches. As follows from Fig. 4 of [15], this decay mode is expected also in lighter isotopes of these elements, also in those populated in the decay chains of $^{288,289}115$. On the neutron-deficient side, taking the $^{285}115$ chain as an example, β^+/EC branches are furthermore predicted for the element 107 and 105 members. The α and β^+/EC daughters of the element 105 member, i.e., ^{261}Lr and ^{265}Rf, are known to decay by SF, in agreement with the calculations. The question thus arises how safe a Z assignments can be if β^+/EC decay as a prominent possibility is neglected (or, more accurately: not detectable). The problem is illustrated in Fig. 3. Due to the multiple decay modes with significant branches as they are predicted to occur in several of the populated nuclei (assuming decay properties as predicted in [16]) each decay chain from $^{291}115$ will pass through one out of in total 19 different pathways. In Fig. 3, just five out of these pathways are highlighted. All include two α decays and different numbers of β^+/EC-decaying members at different positions in the chains, and are terminated by SF. Table 1 presents these five pathways in more detail.

As can be seen, all five pathways lead to the detection of α-α-SF decay chains in the absence of sensitivity to β^+/EC decays. None of the three detected nuclei can be fixed in A or Z. Focusing on the Z of the first detected decaying nucleus, we see that this will be 115 in three of the five pathways, 114 in another one, and 113 in the last discussed one!

A safe assignment of the first α particle to element 115 is thus impossible. It thus appears crucial to expand current techniques to be sensitive also to β-decay processes. Detection of this decay mode and its detailed study are routinely performed in other scientific communities. As just one example, neutrino mass determination via a "missing-mass" type of approach, using EC decaying isotopes – e.g., ^{163}Ho, which is used due to its very low Q_{EC} value [21] – is mentioned, with the "Electron Capture in Holmium-163" (ECHo) collaboration [22, 23] being one of the groups working in this field. Metallic magnetic microcalorimeters [24] are used. In an exploratory study, similar detectors have been applied for studies of fusion-evaporation reaction products behind the SHIP separator at GSI [25], but the technique has not yet reached maturity to be directly applicable in superheavy element research. Approaches based on more traditional techniques may appear better suited to

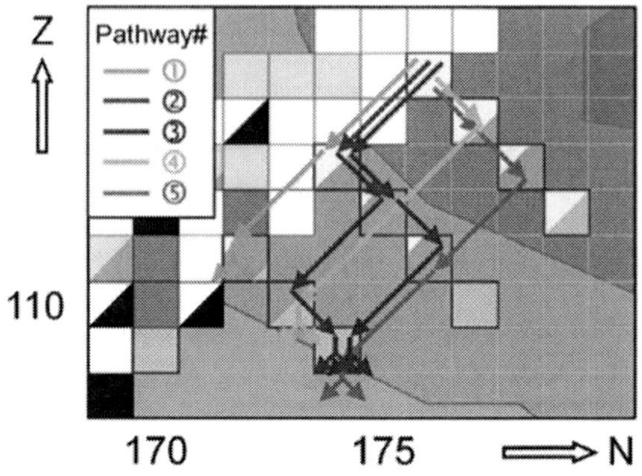

Figure 3. Cut-out from Fig. 2, indicating predicted decay modes and their relative importance for $^{291}115$ and its daughters. Among all possible 19 decay pathways, five are selected; see text for more details.

Table 1. Five decay pathways given in Fig. 3, originating from $^{2}91115$.

#	Decay pathway	Apparent chain[1]	α_1	α_2	SF
1	α–α–SF	α–α–SF	^{291}**115**	^{287}**113**	^{283}**111**
2	α–β^+/EC–α–β^+/EC–SF	α–α–SF	^{291}**115**	^{287}**112**	^{283}**109**
3	α–β^+/EC–β^+/EC–α–SF	α–α–SF	^{291}**115**	^{287}**111**	^{283}**109**
4	β^+/EC–α–α–SF	α–α–SF	^{291}**114**	^{287}**112**	^{283}**110**
5	β^+/EC–β^+/EC–α–α–SF	α–α–SF	^{291}**113**	^{287}**111**	^{283}**109**

[1]In case of non-detection of β^+/EC decay.

achieve direct EC observation [26], including delayed X-ray coincidence observation, which is possible if the detection system is operated in a low-background environment as present, e.g., at the SHIP [27] or the planned BGS-FIONA setup [28].

Having discussed Z assignment and the problem of detecting β^+/EC-decaying isotopes, it is worthwhile to discuss the assignment of A to observed decay chains. This is discussed in [3, 13, 14] for chains produced in ^{48}Ca $+^{243}$Am reactions. Important aspects include that nuclei having significant branches for different decay modes are prevalent also among the heaviest known elements. Therefore, the length of a decay chain (i.e., the number of chain members) cannot be used to judge which set of chains originates from the same nucleus. Not even elaborate statistical analyses as presented in [3] are sufficient to judge whether the A of one type of chain is identical or different to that of a distinctly different set of chains, not least because the occurrence of isomeric states, which decay, e.g., by α decay to analog states in the daughter rather than to their own ground state, is expected according to theoretical calculations (see Sect. 6 in [3]).

To conclude, despite all progress made in recent years, culminating in the IUPAC-approved discovery and naming of all elements up to $Z = 118$, the direct identification of A or $Z$of any nucleus produced in ^{48}Ca + actinide reactions has not been achieved yet. Detection to date relies exclusively on the observation of α and SF decay. β^+/EC decay, expected to be present also in the region that is already being explored, cannot be detected experimentally.

Table 2. Fusion reactions employed to search for elements beyond $Z = 118$.

	Compound nucleus	Beamtime invested	Cross section limit[1]	Ref.
$^{50}Ti + ^{249}Bk$	$^{299}119^*$	4.5 months	[2]	[30]
$^{64}Ni + ^{238}U$	$^{302}120^*$	4 months	≥ 90 fb	[31]
$^{58}Fe + ^{244}Pu$	$^{302}120^*$	2 months	≥ 400 fb	[32]
$^{54}Cr + ^{248}Cm$	$^{302}120^*$	$>$ 1 month	≥ 580 fb	[33]
$^{50}Ti + ^{249}Cf$	$^{299}120^*$	$>$ 1 month	[2]	[30]

[1] 63.2% confidence level ("one-event limit").
[2] Under final analysis.

3. The way to elements beyond *Z* =118

Recent work suggests fusion reactions with actinide targets to provide the highest cross sections for the synthesis of elements beyond $Z = 118$. To proceed on the successfully established ^{48}Ca beam route, targets beyond $_{98}Cf$ would be necessary. Current technology for the production of transuranium isotopes, e.g., at the ORNL's High Flux Isotope Reactor (HFIR) is capable of producing μg and pg amounts of long-lived $_{99}Es$ and $_{100}Fm$ isotopes, respectively [29]. This is orders of magnitude less than the experiments discovering and studying the heaviest known elements require. Thus, reactions induced by beams heavier than ^{48}Ca will be needed. The most commonly suggested reactions giving access to $Z \geq 119$ include $^{48}Ca + ^{249}Bk$ (leading to $Z = 119$) and $^{64}Ni + ^{238}U$, $^{58}Fe + ^{244}Pu$, $^{54}Cr + ^{248}Cm$, and $^{50}Ti + ^{249}Cf$ (all four leading to $Z = 120$). All five have been employed in search experiments, but none has led to the conclusive discovery of a new element. 2 summarizes the efforts invested to date to find elements beyond $Z = 118$.

As can be seen, more than one year of beamtime has been devoted to search experiments, which were carried out at GSI Darmstadt (^{50}Ti+^{249}Bk, ^{64}Ni+^{238}U, ^{54}Cr+^{248}Cm, and ^{50}Ti+^{249}Cf) and FLNR Dubna (^{58}Fe+^{244}Pu). Most likely, non-observation of new elements was due to insufficient sensitivity. It is noteworthy that some of the predictions of the nuclear half-lives of the produced isotopes, especially in the case of $Z = 120$, suggest these to be in the order of microseconds, i.e., comparable to the flight-time through the separator. Hence, losses due to decay-in-flight might have negatively affected these experiments. Preparations for new, improved experiments are under way at several laboratories. Thus, it appears appropriate to reflect on which aspects should get most attention to perform the best possible experiments, thus maximizing the likelihood of success. Obvious, purely technical aspects include higher primary beam intensities, targets that can withstand such intensities over extended periods [34, 35], more efficient separators, more sensitive detection systems that allow safe identification of single decay chains from new elements, and longer irradiation times. I will focus here on a different aspect: numerous theoretical calculations of maximum cross sections of the reactions presented in Table 2 have been published in recent years (see overview in [30] for the reactions leading to $Z = 120$), which span ranges of several orders for each reaction. This translates to unknown minimum irradiation times needed to guarantee the discovery of a new element with high confidence. In general, maximum cross sections scale with reaction asymmetry, but theoretical uncertainties are comparable to differences between different reactions in many cases. ^{54}Cr+^{248}Cm is favoured among the reactions leading to $^{302}120^*$. The more asymmetric $^{50}Ti + ^{249}Cf$ leads to $^{299}120^*$. The influence of the three extra neutrons in the former is often assumed to be favourable, as the reaction product is situated closer to the next spherical shell closure expected at $N = 184$. The

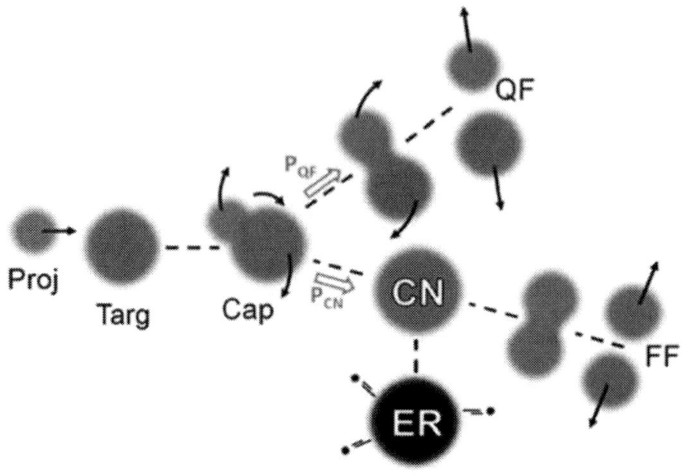

Figure 4. Schematic of the relevant processes in nucleus-nucleus collisions aimed at synthesizing the heaviest elements. Proj: projectile; Targ: target; Cap: captured system; CN: compound nucleus; ER: evaporation residue; QF: quasifission; FF: fusion-fission; P_{CN}: probability for the captured system to form a compound nucleus; P_{QF}: probability for the captured system to fission. Obviously, P_{QF} =1-P_{CN}.

opposite conclusion can also be reached, e.g., considering the fission barrier landscape as depicted in Fig. 6 of [36], which shows fission barrier heights calculated in the macroscopic-microscopic approach, which is very successful in describing experimental data in the region of the heaviest known elements. Attempts to extract fission barrier heights from experimental observables have been performed [37] but are inherently model-dependent. Thus, no safe guidance on the optimum reaction, based purely on experimental data, exists.

Another aspect that is at least as critical as the proper choice of the nuclear reaction concerns the proper choice of the irradiation energy, as excitation functions are only a few MeV wide. Reliable experimental systematics exists for the ^{48}Ca + actinide reactions (e.g., Fig. 4 in [38]). However, the situation is dramatically different as soon as the reactions given in Table 2 are concerned. As detailed in [30], where the system ^{50}Ti +^{249}Cf is employed to illustrate the current situation, optimum beam energies suggested by different theoretical frameworks differ by more than 20 MeV! Recalling that typical energy ranges covered in a single experiment are roughly 5 MeV highlights the need for a well-informed choice of the irradiation energy. Performing at multiple beam energies searches that are significantly more sensitive than the results shown in Table 2 (which were all obtained at one single energy) would consume years of beamtime. In this situation, performing complementary experiments studying these fusion reactions, which yield information that guides future new element search efforts, appears well justified. For this, we recall that the formation of a superheavy nucleus is parametrized as the product of three individual steps, each characterized with its own probability: $\sigma_{\text{ER}} = \sigma_{\text{cap}} \cdot \text{P}_{\text{CN}} \cdot \text{W}_{\text{sur}}$ with σ_{ER} the evaporation residue (ER) cross section, σ_{cap} the capture cross section, P_{CN} the compound nucleus formation probability, and W_{sur} the exit channel survival probability. The formation of superheavy elements is significantly suppressed by the quasifission (QF) process, in which two colliding nuclei reseparate on a fast time-scale before coalescing into a compound nucleus. See [39] for a detailed account, and Fig. 4 for a schematic of the relevant processes.

QF is the complementary process to compound nucleus formation, thus $P_{\text{QF}} = 1 - P_{\text{CN}}$, where P_{QF} can reach large values. In-depth studies of superheavy element formation reactions relying on ER observation are hampered by the tiny σ_{ER}, which renders such studies very (beam)time consuming. The competing QF process profits from much higher rates for systems where P_{CN} is small. The understanding of QF processes has matured to a point where its detailed study, e.g., by recording mass angle distributions, provides significant insight. Thus, it appears timely to perform such studies of the reactions given in Table 2, and to gain an experimental data set of P_{CN} values, and of the heights of the Coulomb barrier in these reactions. Currently, theoretical frameworks only rarely predict observables of the QF process, but mostly focus on ER cross sections. Ideally, these frameworks will be extended to deliver predictions on observables of this "most likely outcome" of nucleus-nucleus collisions, i.e., QF (at least in cases where the underlying physics is contained in the framework). Such observables include mass angle distributions [40] or mass energy distributions [41]. Combined with the corresponding experimental data, which can be recorded in short experiments, this will allow testing different models, thus providing guidance concerning which ones deliver the most trustworthy predictions. Better informed selection of, e.g., the optimum irradiation energy will be invaluable. First steps in this direction have been made recently [39, 42]. A combined effort encompassing theoretical as well as experimental work will provide an improved basis for the certainly coming future efforts to discover new elements with atomic numbers $Z > 118$ and to thus open a new row in the Periodic Table of the Elements.

References

[1] P.J. Karol *et al.*, Pure Appl. Chem. **88**, 139 (2016).
[2] P.J. Karol *et al.*, Pure Appl. Chem. **88**, 155 (2016).
[3] U. Forsberg *et al.*, Nucl. Phys. **A953**, 117 (2016).
[4] Z. Patyk, R. Smolanczuk, A. Sobiczewski, Nucl. Phys. **A626**, 337c (1997).
[5] I. Muntian, Z. Patyk, A. Sobiczewski, Phys. At. Nucl. **66**, 1015 (2003).
[6] I. Muntian *et al.*, Acta Phys. Pol. **B34**, 2073 (2003).
[7] R. Smolańczuk, J. Skalski, A. Sobiczewski, Phys. Rev. C **52**, 1871 (1995).
[8] Yu.Ts. Oganessian *et al.*, Phys. Rev. C **69**, 021601 (2004).
[9] Yu.Ts. Oganessian *et al.*, Phys. Rev. C **87**, 014302 (2013).
[10] D. Rudolph *et al.*, Phys. Rev. Lett. **111**, 112502 (2013).
[11] J.M. Gates *et al.*, Phys. Rev. C **92**, 021301(R) (2015).
[12] K.E. Gregorich, EPJ Web Conf., these proceedings (2016).
[13] U. Forsberg *et al.*, Phys. Lett. **B760**, 293 (2016).
[14] D. Rudolph *et al.*, (in preparation) (2016).
[15] A.V. Karpov *et al.*, Int. J. Mod. Phys. E **21**, 1250013 (2012).
[16] V.I. Zagrebaev, A.V. Karpov, W. Greiner, Phys. Rev. C **85**, 014608 (2012).
[17] A. Türler, R. Eichler, A. Yakushev, Nucl. Phys. **A944**, 640 (2015).
[18] R. Eichler *et al.*, Angew. Chem. Intl. Ed. **47**, 3262 (2008).
[19] R. Eichler *et al.*, Radiochim. Acta **98**, 133 (2010).
[20] A. Yakushev *et al.*, Inorg. Chem. **53**, 1624 (2014).
[21] S. Eliseev *et al.*, Phys. Rev. Lett. **115**, 062501 (2015).
[22] L. Gastaldo *et al.*, J. Low Temp. Phys. **176**, 876 (2014).
[23] L. Gastaldo *et al.*, Eur. Phys. J. Special Topics, (in preparation) (2016).
[24] L. Gastaldo *et al.*, Nucl. Instrum. Methods A **711**, 150 (2013).
[25] S. Kraft-Bermuth, PhD Thesis, Fachbereich Physik, JGU University Mainz, 2004.
[26] F.P. Heßberger *et al.*, Eur. Phys. J. A **52**, 192 (2016).
[27] F.P. Heßberger *et al.*, Eur. Phys. J. A, (submitted) (2016).

[28] J.M. Gates, EPJ Web Conf., these proceedings (2016).
[29] J.B. Roberto *et al.*, Nucl. Phys. **A944**, 99 (2015).
[30] Ch.E. Düllmann *et al.*, (in preparation) (2016).
[31] S. Hofmann, J. Phys. G **42**, 114001 (2015).
[32] Yu.Ts. Oganessian *et al.*, Phys. Rev. C **79**, 024603 (2009).
[33] S. Hofmann *et al.*, Eur. Phys. J. A **52**, 180 (2016).
[34] S.N. Dmitriev and A.G. Popeko, J. Radioanal. Nucl. Chem. **305**, 927 (2015).
[35] E. Jäger *et al.*, J. Radioanal. Nucl. Chem. **299**, 1073 (2014).
[36] M. Kowal, P. Jachimowicz, A. Sobiczewski, Phys. Rev. C **82**, 014303 (2010).
[37] S. Hofmann *et al.*, Eur. Phys. J. A **52**, 116 (2016).
[38] Yu.Ts. Oganessian, and V.K. Utyonkov, Nucl. Phys. **A944**, 62 (2015).
[39] D.J. Hinde *et al.*, EPJ Web Conf., these proceedings (2016).
[40] R. du Rietz *et al.*, Phys. Rev. C **88**, 054618 (2013).
[41] M.G. Itkis *et al.*, Nucl. Phys. **A944**, 204 (2015).
[42] H.M. David *et al.*, (in preparation) (2016).

AUTHOR INDEX

AUTHOR INDEX

EDP Sciences – Web of Conferences
17, Avenue du Hoggar
Parc d'Activité de Courtabouf
BP 112
F-91944 Les Ulis Cedex A
France

ISBN 978-1-5108-3368-5

CURRAN ASSOCIATES INC.
proceedings.com